I0041829

Advanced IoT Technologies and Applications in the Industry 4.0 Digital Economy

The application of internet of things (IoT) technologies and artificial intelligence (AI)-enabled IoT solutions has gradually become accepted by business and production organizations as an effective tool for automating several activities effectively and efficiently and developing and distributing products to the global market. Within this book, the reader will learn how to implement IoT devices, IoT-equipped machines, and AI-equipped IoT applications using models and methodologies along with an array of case studies.

Advanced IoT Technologies and Applications in the Industry 4.0 Digital Economy covers the basics of IoT-equipped machines in developing and managing various activities in many industries. It discusses all of the key points of an AI-enabled IoT solution, which includes predictive analytics, robotic process automation, predictive maintenance, automated processes, IoT technologies and IoT-equipped sensors related to machines and processes, production testing systems, and product assessment processes in the production environment. The book presents the concepts and interactive methods using datasets, processing workflow charts, and architectural diagrams along with additional real-time systems for easy and fast understanding of the application of IoT-equipped machines and AI-enabled solutions in organizations and includes many case studies throughout the book to enforce reader comprehension.

This book is an ideal read for industry specialists, practitioners, researchers, scientists, and engineers working or involved in the fields of Robotics, IT, Computer Science, Soft Computing, IoT, AL/ML/DL, Data Science, the Semantic Web, Knowledge Engineering, and other related fields.

Advanced IoT Technologies and Applications in the Industry 4.0 Digital Economy

Edited by
Alex Khang, Vugar Abdullayev, Vladimir Hahanov
and Vrushank Shah

CRC Press
Taylor & Francis Group
Boca Raton London New York

CRC Press is an imprint of the
Taylor & Francis Group, an **informa** business

Font cover image: Photon/Shutterstock

First edition published 2024
by CRC Press
2385 NW Executive Center Drive, Suite 320, Boca Raton FL 33431

and by CRC Press
4 Park Square, Milton Park, Abingdon, Oxon, OX14 4RN

© 2024 selection and editorial matter, Alex Khang, Vugar Abdullayev, Vladimir Hahanov and Vrushank Shah; individual chapters, the contributors

CRC Press is an imprint of Informa UK Limited

Reasonable efforts have been made to publish reliable data and information, but the author and publisher cannot assume responsibility for the validity of all materials or the consequences of their use. The authors and publishers have attempted to trace the copyright holders of all material reproduced in this publication and apologize to copyright holders if permission to publish in this form has not been obtained. If any copyright material has not been acknowledged please write and let us know so we may rectify in any future reprint.

Except as permitted under U.S. Copyright Law, no part of this book may be reprinted, reproduced, transmitted, or utilized in any form by any electronic, mechanical, or other means, now known or hereafter invented, including photocopying, microfilming, and recording, or in any information storage or retrieval system, without written permission from the publishers.

For permission to photocopy or use material electronically from this work, access www.copyright.com or contact the Copyright Clearance Center, Inc. (CCC), 222 Rosewood Drive, Danvers, MA 01923, 978-750-8400. For works that are not available on CCC please contact mpkbookspermissions@tandf.co.uk

Trademark notice: Product or corporate names may be trademarks or registered trademarks and are used only for identification and explanation without intent to infringe.

ISBN: 978-1-032-55204-0 (hbk)
ISBN: 978-1-032-56182-0 (pbk)
ISBN: 978-1-003-43426-9 (ebk)

DOI: 10.1201/9781003434269

Typeset in Times
by Deanta Global Publishing Services, Chennai, India

Contents

Contents

Preface

The application of IoT technology and its applications is gradually accepted by business and production organizations as an effective tool for automating a number of works effectively and efficiently in developing and distributing the products in the global market. Nowadays, most IoT technologies and AI-equipped IoT applications can support most of the tasks for offices, schools, homes, buildings, hospitals, healthcare services, cities, factories, airports, transportation systems, grid systems, agriculture, etc.

In industry 4.0 economy, an IoT-equipped machine and its applications can replace humans in most of the key activities in daily life or work at a company or factory. For instance, these applications can generate basic decisions or actions from a sequence of commands transferred from the IoT devices or IoT-equipped machines used to detect vehicle violations in traffic or defective products in assembling or broken packages in distribution systems.

With this book, you'll be able to implement IoT devices, IoT-equipped machines, IoT-integrated drones, and AI-equipped IoT applications using models and methodologies along with covering smart IoT technologies to support automating actions, monitoring activities, manage tasks, predict outcomes, forecast risks, and make informed decisions that facilitate improving in the various services and production activities in the fields of healthcare, manufacturing, higher education, smart home, smart building, smart city, power and grid system, transportation, airport, population management, finance services, banking services, urban management, workforce management, agriculture, aquaculture, petroleum exploration, forest management, and cultural heritage management.

This book would be a great collection of advanced IoT technologies, IoT-equipped machines, and AI-equipped IoT applications for overall concepts, models, designs, innovations, developments, and implementation in a high-tech industry and digital business ecosystem. This book targets a mixed audience of students, engineers, scholars, researchers, academics, and professionals who are learning, researching, working, and applying AI and IoT for different industries.

Happy reading!

Alex Khang
Vugar Abdullayev
Vladimir Hahanov
Vrushank Shah

About the Editors

Alex Khang is Professor in Information Technology, AI expert, data scientist, software industry expert, and the chief technology officer (AI and Data Science Research Center) at the Global Research Institute of Technology and Engineering, North Carolina, United States. He has over 28 years of teaching and research experience in information technology at universities and institutions in Vietnam, India, and the United States. He has 52 authored books (2000–2010), and 2 authored books (2020). He has published 65 documents indexed in Scopus, 50 book chapters, 15 edited books and 11 edited books (calling for book chapters) for AI ecosystem. He has over 28 years of working experience as a software product manager, data engineer, AI engineer, cloud computing architect, solution architect, software architect, and database expert in the foreign corporations of Germany, Sweden, the United States, Singapore, and multinationals (former CEO, former CTO, former Engineering Director, Product Manager, and Senior Software Production Consultant).

Vugar Abdullayev, Doctor of Technical Sciences, is Associate Professor at Information Technologies and Systems, Azerbaijan University of Architecture and Construction in Baku, Azerbaijan. He has completed his PhD in computer science in the year 2005. He is currently Associate Professor of the "Computer Engineering" department at the Azerbaijan State Oil and Industry University, Baku, Azerbaijan. He is the author of 61 scientific papers. His researchers are related to the study of cyber-physical systems, IoT, big data, smart city, and information technologies. He has published four book chapters and two edited books (calling for book chapters— Taylor and Francis) in the healthcare ecosystem.

Vladimir Hahanov, Doctor of Science, is Professor of Computer Engineering Faculty, Design Automation Department, at Kharkiv National University of Radio Electronics, Ukraine. His previous positions were Acting Science Vice-Rector (2016) and Dean of Computer Engineering Faculty (2003–2016). He is a supervisor for 4 Doctor of Science, 35 PhDs, and 150 more engineers from 27 countries. He is also General Chair of the IEEE East-West Design & Test Symposium for 19 years since 2003; IEEE Senior Member since 2010; IEEE Computer Society Golden Core Member since 2010; IEEE member (No 41407206) since 2001; and SAE Member since 2016.

Vrushank Shah is Assistant Professor and Head of the Electronics and Communication Engineering Department at the Indus Institute of Technology and Engineering, Indus University, Rancharda, Via Shilaj Ahmedabad, Gujarat, India. He has organized over 50 workshops, STTPs, FDPs, and webinars for the benefit of students. He has worked as a Deputy Director, Indus Center for Startups, Incubation and Innovation, Indus University and played a crucial role to develop 25+ Startups/ PoCs/IPRs. He was involved in guiding the startups toward sustainability and upbringing of incubation center.

Contributors

Sriram A.
Department of Electronics and
  Communication Engineering, SRM
  TRP Engineering College
Tiruchirappalli, Tamil Nadu, India

Nikita Agarwal
Rashtriya Raksha University,
  Ministry of Home Affairs,
  Lavad-Gandhinagar
Gandhinagar District, Gujarat 382305,
  India

Shahzada Asif
Ashoka Institute of Technology and
  Management
Varanasi, Uttar Pradesh, India

Libin Baby
School of Sciences, CHRIST (Deemed
  to be University)
Bangalore, Karnataka, India

Mina Bahadori
Department of Industrial Engineering,
  Clemson University
South Carolina, USA

Chellaswamy C.
Department of Electronics and
  Communication Engineering, SRM
  TRP Engineering College
Tiruchirappalli, Tamil Nadu, India

Neil Manoj C.
School of Sciences, CHRIST (Deemed
  to be University)
Bangalore, Karnataka, India

Jaydeep Chakravorty
Electrical Engineering Department
  Indus University
Ahmedabad, India

Sachin Chaudhary
Assistant Professor, School of Computer
  Science and Application, IIMT
  University Meerut
Uttar Pradesh, India

Puspraj Singh Chauhan
ECE Department, Pranveer Singh
  Institute of Technology
Kanpur, India

Nilesh G. Chothani
Electrical Engineering Department,
  Pandit Deendayal Energy University
Ahmedabad, India

Mehdi Davari
Department of Management, Isfahan
  Branch, Islamic Azad University
Isfahan, Iran

Shruti Dixit
SAGE University Bhopal, Bhopal
Madhya Pradesh 462041, India

Vandana Dubey
Ashoka Institute of Technology and
  Management
Varanasi, India

Ushaa Eswaran
Principal and Professor, Indira Institute
  of Technology and Sciences,
  Markapur
Andhra Pradesh 523320, India

Niveditha G.
School of Sciences, CHRIST
(Deemed to be University)
Bangalore, Karnataka, India

Lokpriya M. Gaikwad
Assistant Professor, SIES Graduate
School of Technology
Navi Mumbai, India

Chima Jude Iheaturu
Institute of Geography, University of
Bern
Hochschulstrasse 4, 3012 Bern,
Switzerland

Javeed S. Imran
Vel Tech Rangarajan Dr. Sagunthala
R&D Institute of Science and
Technology
Chennai, Tamil Nadu, India

Ankit Jain
ECE Department, Pranveer Singh
Institute of Technology
Kanpur, India

Pooja Jain
Amity Business School, Amity
University, Gwalior
Madhya Pradesh, India

Rachit Jain
Madhav Institute of Technology &
Science, Gwalior
Madhya Pradesh, India

Sankar Sam Jose
School of Sciences, CHRIST
(Deemed to be University)
Bangalore, Karnataka, India

Kassim Kalinaki
Department of Computer Science,
Islamic University, In Uganda
(IUIU)
Mbale, Uganda

Sandip Kanase
Assistant Professor, Bharati Vidyapeeth
College of Engineering
Navi Mumbai, India

Krishna Kanodia
Department of Computer Science and
Engineering, National Institute of
Technology
Rourkela 769008, Odisha, India

Navneet Kaur
SAGAR Institute of Research and
Technology, Bhopal
Madhya Pradesh 462041, India

Imran Ullah Khan
ECE Department, Integral University
Lucknow, India

Alex Khang
Professor, AI Expert, Data Scientist,
Software Industry Expert, Global
Research Institute of Technology and
Engineering
North Carolina 27612, United States

Priyanka Kujur
Department of Computer Science and
Engineering, National Institute of
Technology
Rourkela 769008, Odisha, India

Bhupendra Kumar
Associate Professor, School of
Computer Science and Application,
IIMT University Meerut
Uttar Pradesh, India

Priti Kumari
Ashoka Institute of Technology and
    Management
Varanasi, Uttar Pradesh, India

Lalitha Mary
School of Sciences, CHRIST (Deemed
    to be University)
Bangalore, Karnataka, India

Mohamed Nassereddine
Electrical Department
University of Wollongong in Dubai,
    UAE

Rupendra Kumar Pachuri
Electrical and Electronics Engineering,
    Department, School of Engineering,
    University of Petroleum and Energy
Dehradun, India

Rishabh Pal
Ashoka Institute of Technology and
    Management
Varanasi, Uttar Pradesh, India

Rajneesh Panwar
Assistant Professor, School of Computer
    Science and Application, IIMT
    University Meerut
Uttar Pradesh, India

Sanjeev Patel
Department of Computer Science and
    Engineering, National Institute of
    Technology
Rourkela 769008, Odisha, India

Hanuman Prasad
Electrical Engineering Department,
    Model Institute of Engineering and
    Technology
Jammu, India

Ramasamy R.
Ramco Institute of Technology
Rajapalayam, Tamil Nadu, India

Kavitha Rajamohan
School of Sciences, CHRIST (Deemed
    to be University)
Bangalore, Karnataka, India

Sangeetha Rangasamy
School of Business and Management,
    CHRIST (Deemed to be University)
Bangalore, Karnataka, India

Kali Charan Rath
Associate Professor, Department of
    Mechanical Engineering, GIET
    University
Odisha 765022, India

Debanik Roy
Scientist, Division of Remote Handling
    and Robotics, Bhabha Atomic
    Research Centre; Professor,
    Homi Bhabha National Institute,
    Department of Atomic Energy
Govt. of India, Mumbai, India

Arul S.
Department of Electronics and
    Communication Engineering,
    Jeppiaar Institute of Technology
Chennai, Tamil Nadu, India

Geetha T. S.
Department of Electronics and
    Communication Engineering, Sriram
    Engineering College
Chennai, Tamil Nadu, India

Muktha D. S.
School of Sciences, CHRIST (Deemed
    to be University)
Bangalore,
Karnataka, India

Udhayanan S.
Vellore Institute of Technology
Vellore, Tamil Nadu, India

Raison Sabu
School of Sciences, CHRIST (Deemed
   to be University)
Bangalore, Karnataka, India

Shalini Sahay
SAGAR Institute of Research and
   Technology, Bhopal
Madhya Pradesh 462041, India

Aakansha Saxena
Rashtriya Raksha University,
   Ministry of Home Affairs,
   Lavad-Gandhinagar
Gandhinagar District, Gujarat,
   382305, India

Wasswa Shafik
Wasswa Shafik, School of Digital
   Science, Universiti Brunei
   Darussalam
Gadong, Brunei

Hinal Shah
Electrical Engineering Department,
   Indus University
Ahmedabad, India

Kewal Krishan Sharma
Professor, School of Computer Science
   and Application, IIMT University
   Meerut
Uttar Pradesh, India

Vikas Sharma
Assistant Professor, School of Computer
   Science and Application, IIMT
   University Meerut
Uttar Pradesh, India

Sarika Shrivastava
Ashoka Institute of Technology and
   Management
Varanasi, Uttar Pradesh, India

Neha Shrotriya
Department of Computer Engineering,
   Poornima College of Engineering
Jaipur, Rajasthan, India

Anita Shukla
BSH Department, Pranveer Singh
   Institute of Technology
Kanpur, India

Raghvendra Singh
ECE Department, Pranveer Singh
   Institute of Technology
Kanpur, India

Masoumeh Soleimani
Department of Mathematics and
   Statistical Sciences, Clemson
   University
South Carolina, USA

Morteza Soltani
Department of Industrial Engineering,
   Clemson University
South Carolina, USA

Vivek Sunnapwar
Department of Mechanical
   Engineering, Lokmanya Tilak
   College of Engineering, University
   of Mumbai
Mumbai, Maharashtra, India

Parthivi Thakore
UG-Computer Science and
   Engineering, Techno Engineering
   College Banipur
Habra, Kolkata, India

Poonam Tiwari
Department of Physical Sciences
Banasthali Vidyapith Rajasthan, India

Tarun Kumar Vashishth
Associate Professor, School of
    Computer Science and
Application, IIMT University Meerut
Uttar Pradesh, India

Rufai Yusuf Zakari
Skyline University Nigeria Location,
    Kano 700103, Nigeria; School of
    Digital Science, Universiti Brunei
    Darussalam
Gadong BE1410, Brunei Darussalam

Acknowledgments

This book *Advanced IoT Technologies and Applications in the Industry 4.0 Digital Economy* is based on the design and implementation of internet of things (IoT), artificial intelligence (AI), and applications in various industries.

Preparing and designing a book outline to introduce to readers around the world is the passion and noble mission of the editorial team. To be able to make ideas a reality and the success of this book, the biggest reward belongs to the efforts, experiences, enthusiasm, and trust of the contributors.

To all the reviewers with whom we have had the opportunity to collaborate and monitor their hard work remotely, we acknowledge their tremendous support and valuable comments not only for the book but also for future book projects.

We also express our deep gratitude for all the pieces of advice, support, motivation, sharing, collaboration, and inspiration we received from our faculty, contributors, educators, professors, scientists, scholars, engineers, and academic colleagues.

Last but not least, we are really grateful to our publisher CRC Press (Taylor & Francis Group) for the wonderful support in making sure the timely processing of the manuscript and bringing out this book to the readers soonest.

Thank you, everyone.

Editorial team: Alex Khang, Vugar Abdullayev, Vladimir Hahanov, Vrushank Shah

1 The Role of Internet of Things (IoT) Technology in Industry 4.0 Economy

Kali Charan Rath, Alex Khang, and Debanik Roy

INTRODUCTION

The manufacturing industry is a sector of the economy that is involved in the production of goods using labor, machinery, and materials. The manufacturing industry is responsible for creating a wide range of products, from simple household goods to complex industrial equipment. Manufacturing involves several stages of production, including design, prototyping, engineering, and production. The industry encompasses a wide range of sub-sectors, including aerospace, automotive, electronics, food and beverage, medical devices, and textiles, among others. Pricing, speed, quality, and innovation are some of the factors that manufacturing businesses compete on. To remain competitive, companies in the manufacturing industry must continually improve their processes, invest in research and development, and adapt to changing market conditions (Subhashini and Khang, 2023).

The manufacturing industry is always looking for ways to improve and innovate to remain competitive and efficient. Industrial revolutions are periods of rapid change and technological advancements that transform the way manufacturing is done. Modern economies have benefited from the expansion and development of the manufacturing sector. In recent years, the industry has undergone significant changes, including the adoption of advanced technologies such as automation, robotics, and digitalization. Here are some reasons why the manufacturing industry seeks industrial revolutions:

- **Increased Efficiency**: Industrial revolutions have historically been associated with the adoption of new technologies that significantly improve manufacturing efficiency. During the Second Industrial Revolution, the assembly line was invented, allowing for mass manufacturing and reduction in both production time and cost.
- **Cost Reduction**: The adoption of new technologies and innovations during an industrial revolution can lead to cost reductions in the manufacturing process. This can make manufacturing more affordable and accessible to a wider range of consumers.

DOI: 10.1201/9781003434269-1

- **Increased Production Capacity**: Industrial revolutions have the potential to significantly increase a company's production capacity. This allows manufacturers to meet demand more quickly and produce goods on a larger scale.
- **Improved Quality**: Industrial revolutions can also improve the quality of manufactured goods. For example, the third industrial revolution of the past century allowed for greater precision and accuracy in manufacturing, thanks to the use of computer-controlled automation.
- **Innovation**: Industrial revolutions can lead to the development of new products and industries with automation, which came to the front tire of the fourth industrial revolution known as Industry 4.0 (Kaushik & Singh, 2021).

GENERAL MANUFACTURING VS SMART MANUFACTURING SYSTEM

A general manufacturing system is a traditional manufacturing system that relies on manual labor and basic machinery to produce goods. This type of manufacturing system is often characterized by a segmented production process, with little real-time data or communication between different stages of production. In a general manufacturing system, the production process typically involves several steps, such as raw material handling, processing, assembly, and packaging. Workers operate machines, assemble parts, and perform quality control checks manually.

The efficiency and productivity of a general manufacturing system can be limited by factors such as the availability and skill level of workers, the speed and capacity of machines, and the complexity of the production process. While general manufacturing systems can be effective for producing small volumes of goods, they can be inefficient and costly for larger-scale production. Additionally, the lack of connectivity and automation can limit the ability to optimize and improve the production process.

The internet of things (IoT), artificial intelligence (AI), and machine learning (ML) are some of the cutting-edge technologies used in smart manufacturing. Smart manufacturing incorporates real-time data and communication between machines, equipment, and people, enabling the creation of a connected and intelligent manufacturing system (Bose & Liu, 2019).

A conventional manufacturing system and a smart manufacturing system have different degrees of automation, intelligence, and connection as shown in Figure 1.1. A general manufacturing system typically relies on manual labor and basic machinery, with limited automation and connectivity between machines and systems. The production process is often segmented, with little real-time data or communication between different stages of production. By integrating cutting-edge technologies like the internet of things (IoT), artificial intelligence (AI), and machine learning (ML), a smart manufacturing system can create a fully connected and intelligent production process, resulting in increased productivity, flexibility, and efficiency through real-time data and communication among people, machines, and other equipment.

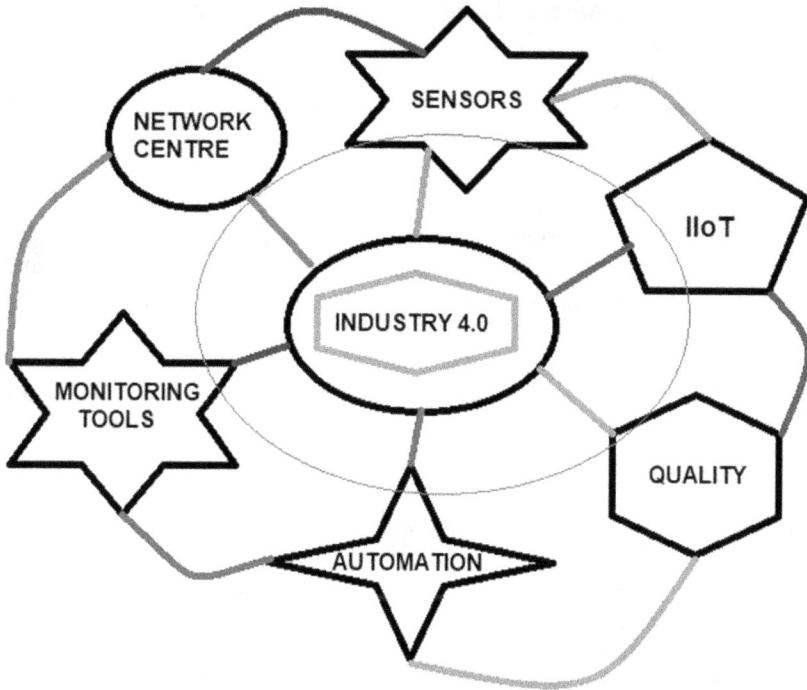

FIGURE 1.1   Industrial IoT (IIoT) in smart manufacturing

One of the goals of smart manufacturing is to reduce costs, waste, and environmental impact while enabling faster and more efficient production. By collecting and analyzing real-time data, smart manufacturing systems can optimize production processes, predict and prevent equipment failures, and improve quality control. Smart manufacturing can also enable mass customization, where products can be customized to meet specific customer needs without sacrificing efficiency or increasing costs. In a simple line, it can be stated that smart manufacturing represents a major shift in the way manufacturing is done, and has the potential to revolutionize many industries by enabling faster, more flexible, and more efficient production.

SCOPE FOR SMART MANUFACTURING INDUSTRY

The scope for the smart manufacturing industry is significant and continues to grow as more companies recognize the benefits of implementing smart manufacturing technologies. A few key areas where the industry has scope for growth and development include:

- **Integration with Industry 4.0**: Smart manufacturing technologies can be integrated with Industry 4.0 concepts, such as cyber-physical systems, to create even more efficient and responsive manufacturing processes.

- **Development of Advanced Analytics**: As more data is collected from sensors and other sources, there is scope for the development of advanced analytics tools that can provide deeper insights and drive further optimization.
- **Expansion to New Sectors**: While smart manufacturing is already being used in a variety of sectors, such as automotive, aerospace, and consumer electronics, there is potential for its expansion into new industries such as healthcare, food, and construction.
- **Increased Collaboration**: The smart manufacturing industry has scope for increased collaboration between manufacturers, technology providers, and other stakeholders to drive innovation and improve the adoption of these technologies.
- **Focus on Sustainability**: As sustainability becomes an increasingly important concern for companies and consumers, there is scope for smart manufacturing technologies to be used to create more sustainable and eco-friendly production processes.

SMART MANUFACTURING INDUSTRY AND IOT

The goal of smart manufacturing is to enhance quality control and decrease expenses by leveraging IoT technology to gather real-time data on product quality, thereby allowing manufacturers to identify defects and take corrective measures before delivering products to customers. By doing so, it can effectively minimize the occurrence of product recalls and enhance customer satisfaction, leading to a surge in sales and profitability.

RESEARCH AREAS IN THE SMART MANUFACTURING INDUSTRY

There are many research areas in the smart manufacturing industry, and some of the key interest research areas are:

- **Human–Machine Collaboration**: Smart manufacturing has also the potential to improve collaboration between humans and machines. Researchers might focus on developing new interfaces that allow humans and machines to work together more effectively or developing new training methods to help workers adapt to new technologies. Smart manufacturing systems are also designed to work alongside human operators, and researchers are exploring new ways to optimize human–machine collaboration. This includes developing new interfaces and tools that allow humans to work more effectively with smart machines, and improving the overall design of smart manufacturing systems to make them more intuitive and user-friendly.
- **Data Analytics and Machine Learning**: Several data are collected in smart manufacturing systems, and there will be a growing need for advanced data analytics tools to help manufacturers make sense of all the information. Researchers might focus on developing new machine learning algorithms

or improving the accuracy of existing analytics tools to help manufacturers gain insights and make data-driven decisions. One of the most critical areas of research in the smart manufacturing industry is data analytics and machine learning. Efforts are underway by researchers to design algorithms and tools capable of analyzing substantial amounts of data generated by smart manufacturing systems (Hu and Wang 2019). This would allow manufacturers to optimize production processes, improve product quality, and increase efficiency. By developing such technology, manufacturers would be able to enhance their productivity and efficiency, ultimately leading to better products and higher levels of customer satisfaction.

- **Predictive Maintenance**: Predictive maintenance is one of the most promising applications of IoT in smart manufacturing. It allows manufacturers to use sensors to monitor machines and equipment and predict when maintenance is required, resulting in the avoidance of costly downtime. Researchers are focusing on this area to develop more precise predictive models or enhance the efficiency of sensor networks that monitor equipment (Khang et al., 2024a). These advancements would enable manufacturers to optimize their maintenance practices, reduce costs, and improve their overall efficiency.
- **Smart Materials**: The use of smart materials in the manufacturing process is an area that researchers are exploring. Smart materials can respond to changes in the environment or external stimuli, resulting in the creation of products that are more adaptable, flexible, and responsive. By using these materials, manufacturers can create products that better meet the needs of their customers and are more efficient to produce. These advancements would enable manufacturers to create innovative products and stay competitive in the market.
- **Energy Efficiency**: Optimizing processes and minimizing waste are some of the ways smart manufacturing can improve energy efficiency. Researchers are looking into this area to develop new energy-saving technologies or improve the efficiency of existing processes through the use of sensors and data analysis. By doing so, manufacturers can significantly reduce their energy consumption and minimize their carbon footprint, leading to a cleaner and more sustainable manufacturing process. These advancements would enable manufacturers to become more environmentally responsible and cost-effective.
- **Quality Control**: Smart manufacturing leverages IoT sensors to monitor production processes in real time, thereby enhancing quality control. Researchers are exploring the development of new sensor technologies or improving the accuracy of existing sensors to ensure that products meet stringent quality standards (Hahanov et al., 2022). By using these advanced sensors, manufacturers can detect quality issues earlier and take corrective measures, ensuring that only high-quality products are delivered to customers. By using this strategy, firms can boost consumer happiness and their reputation while still competing in the market.

- **Supply Chain Optimization**: Smart manufacturing systems offer significant potential to optimize supply chain management in manufacturing (Huang et al., 2019). Researchers are actively working on developing new tools and technologies that can help manufacturers optimize their supply chains, reduce lead times, and enhance overall efficiency. IoT technology can also be leveraged to optimize supply chains in manufacturing, by using sensors to track the movement of goods and materials. This allows manufacturers to better manage inventory, reduce waste, and improve overall efficiency. Research in this area might explore new ways of utilizing sensor data to optimize supply chain logistics or developing more advanced machine learning algorithms to predict demand accurately. By implementing these advanced technologies and processes, manufacturers can improve the speed and reliability of their supply chains, leading to increased productivity and profitability.
- **Cybersecurity**: As smart manufacturing systems become more interconnected and digitized, cybersecurity is becoming increasingly important. As smart manufacturing and IoT become more widespread, cybersecurity will become an increasingly important concern. Researchers might focus on developing new security protocols or designing secure hardware and software systems to protect against cyber threats. To defend against online threats and to guarantee the security and integrity of smart manufacturing systems, researchers are attempting to create new cybersecurity solutions.

The smart manufacturing industry is a diverse field, and researchers are working on developing new solutions and technologies to improve efficiency, productivity, and overall performance throughout the entire manufacturing value chain. Smart manufacturing and IoT are rapidly growing areas of technology and research, and there are many exciting research areas where these two intersect. Some of these areas include predictive maintenance, energy efficiency optimization, supply chain management, and product quality control.

Researchers are also exploring the use of smart materials, which can create products that are more responsive, adaptable, and environmentally friendly. Additionally, the development of new algorithms and tools to process and analyze large volumes of data generated by smart manufacturing systems is a crucial research area. By combining these different technologies and research areas, manufacturers can transform their operations, leading to significant improvements in performance, sustainability, and competitiveness.

OBJECTIVE OF SMART MANUFACTURING INDUSTRY

The smart manufacturing industry aims to enhance production processes and efficiency by leveraging advanced technologies such as IoT, AI, and robotics. This industry strives to achieve various key goals including improving quality control, flexibility, agility, customization, and sustainability. Through the implementation of smart manufacturing, manufacturers can create a more automated and agile production environment, which can easily adapt to market changes and deliver high-quality products at reduced costs. To achieve this, smart manufacturing systems need to collaborate with IoT and IIoT for better monitoring and control. This book chapter aims to explain how IoT and IIoT can be integrated into smart manufacturing to improve efficiency and flexibility. By exploring research areas in this field, the chapter highlights the importance of IIoT and IoT applications in smart manufacturing systems, which can help reduce waste, increase productivity, and enhance sustainability in the manufacturing industry.

IOT, IIOT, AND INDUSTRY 4.0

IoT or the "internet of things" connects physical devices, vehicles, and various industry objects embedded with sensors, software, and connectivity to exchange data with other devices and systems over the internet. The technology enables a range of devices from simple sensors to complex systems that can control other devices, analyze data, and make decisions based on that data. Various industries such as manufacturing, transportation, healthcare, and agriculture use IoT to enhance efficiency, safety, and productivity.

In manufacturing, IoT collects real-time data from machines, processes, and products, which is then analyzed to identify patterns and insights that can inform business decisions. This technology allows for process optimization, reduced downtime, and increased efficiency. Additionally, IoT devices enable predictive maintenance, real-time inventory tracking, and enhanced safety in manufacturing plants (Kshetri & Voas, 2022). The integration of IoT and other technologies in Industry 4.0 is revolutionizing the way manufacturing and other industries operate, making them more responsive, productive, and efficient in meeting customer needs (Li et al., 2018).

IIoT focuses on using IoT technology in industrial environments, such as manufacturing plants, energy facilities, and transportation systems, with the goal of enhancing traditional processes through the integration of advanced technologies. Real-time monitoring of machines and equipment enabled by IIoT allows manufacturers to collect valuable data to optimize production, minimize downtime, and improve efficiency. Data gathered from sensors and other IoT devices can include information on temperature, humidity, equipment performance, and energy usage.

Industry 4.0 is a term used to describe the integration of advanced technologies like IoT, IIoT, artificial intelligence, and cloud computing (Miao et al., 2021). By utilizing IoT, IIoT, artificial intelligence, and cloud computing, manufacturers can gather and analyze real-time data, optimize production processes, and minimize downtime. This integration helps manufacturers to achieve better productivity and profitability while also improving the quality of their products. Overall, Industry 4.0 aims to revolutionize traditional manufacturing and industrial processes to create a more efficient, cost-effective, and innovative industry.

ADVANTAGES OF IoT AND IIoT IN INDUSTRY 4.0

In the context of Industry 4.0, data can be collected from different parts of the production process with the help of the internet of things. This data is analyzed in real time to provide insight into the production process (Yang et al., 2020). By using this data, manufacturers can identify inefficiencies and improve their operations.

IIoT leverages sensors and other devices embedded in machines and equipment to collect performance data that is analyzed to optimize their operations to get better efficiency and improve safety.

A few benefits of IoT and IIoT in application with Industry 4.0 are:

- Real-time monitoring of machines and processes for timely intervention and maintenance
- Predictive maintenance to reduce downtime and increase efficiency
- Improved supply chain management through real-time tracking of inventory and shipments
- Enhanced safety by monitoring environmental conditions and equipment performance in real time
- Improved quality control by monitoring product quality throughout the production process
- Increased efficiency through process automation, reducing the need for human intervention
- Enhanced customer experience through data collection on customer preferences and behavior
- Smart factories created through the integration of IoT devices, artificial intelligence, and other technologies
- Data analytics to identify patterns and trends that can inform business decisions
- Cost savings through improving efficiency, reducing downtime, and optimizing supply chains

DISADVANTAGES OF IoT AND IIoT IN INDUSTRY 4.0

While IoT and IIoT have many potential benefits in the context of Industry 4.0, there are also some disadvantages to consider (Hsiao et al., 2019). In the following, the main disadvantages of IoT and IIoT in Industry 4.0 are listed.

- **Security Concerns**: As IoT and IIoT systems become increasingly popular (Ries and Duan, 2022) and adopted in various industries, the risk of cyber-attacks and data breaches also rises, leaving these systems vulnerable to potential threats that can lead to sensitive information loss or control over critical systems. IIoT systems are at a higher risk due to the far-reaching consequences of a breach, impacting not just the affected company's operations but also the safety and well-being of its employees and customers. The use of IoT and IIoT systems in critical infrastructure also increases the potential for large-scale disruption and physical harm, making it crucial for companies to take proactive measures to secure their systems, including implementing strong authentication and access controls, updating software and firmware regularly, and conducting security audits and assessments.

Employee awareness and training on cybersecurity best practices are also vital to minimize human error and prevent data breaches and cyberattacks. As the use of IoT and IIoT systems continues to grow and evolve, companies must remain vigilant in their efforts to safeguard their critical systems and sensitive information (Khang et al., 2022b).

- **Complexity**: Designing, implementing, and maintaining IoT and IIoT systems can present challenges that require expertise across multiple disciplines, including hardware, software, and networking, which can be a daunting task for some businesses.
- **Cost**: For small- and medium-sized enterprises (SMEs) with limited funding, the cost of implementing IoT and IIoT systems can be a significant obstacle due to the high initial investment required.
- **Interoperability**: IoT and IIoT systems often rely on multiple devices and platforms, which can lead to compatibility issues and hinder interoperability between different systems.
- **Privacy Concerns**: Privacy concerns can be posed by the collection and transmission of sensitive data, such as personal information or proprietary data, in highly regulated industries.
- **Reliance on Connectivity**: IoT and IIoT systems require reliable connectivity to function properly. This can be a challenge in remote or harsh environments, where network connectivity may be limited or unreliable.

CLASSIFICATION OF IOT IN INDUSTRY 4.0

The classification of IoT in Industry 4.0 encompasses a broad range of applications, including real-time monitoring, predictive maintenance, worker safety, and supply chain management, which can aid in the identification of specific requirements and challenges for each type of IIoT system and help to categorize them (Khang et al., 2024b). The IIoT can be classified into several categories based on different criteria. Here are some common classification methods:

APPLICATION-BASED CLASSIFICATION

This classification is based on the application area of the IIoT system, such as smart manufacturing, predictive maintenance, asset tracking, supply chain management, and energy management. A brief description of case studies is given below.

Case Study 1: Smart Manufacturing

IIoT technologies can be used for smart manufacturing in the automobile industry to improve productivity, reduce downtime, and enhance product quality.

For instance, consider an IIoT system that is designed to optimize the production process of a car assembly line. Utilizing sensors and other devices, this system can gather data on various aspects of the production process, including robot performance, machinery status, and part quality, followed by the use of machine learning algorithms to analyze the data and detect patterns and trends. Based on this analysis, the system could optimize the production process by adjusting the performance of the robots, predicting maintenance needs, and flagging potential quality issues before they become serious (Shah et al., 2023).

By classifying this IIoT application based on application-based classification, we can better understand the specific requirements and challenges for implementing smart manufacturing in the automobile industry using IIoT technologies. We can also identify the potential benefits of using IIoT technologies in smart manufacturing, such as reducing downtime, improving product quality, and enhancing productivity (Khang et al., 2022a).

Case Study 2: Energy Management

IIoT technologies can be used for energy management in the energy sector to optimize energy consumption, reduce waste, and lower energy costs. For instance, consider an IIoT system that is designed to optimize energy consumption in a power generation facility. This system could use sensors and other devices to collect data on energy usage, such as fuel consumption, power output, and equipment performance. Employing machine learning algorithms, the system can analyze the data and identify patterns and trends to provide insights into the production process. Based on this analysis, the system could optimize the energy generation process by adjusting the performance of the equipment, predicting maintenance needs, and flagging potential issues before they become serious.

Case Study 3: Quality Control in the Mobile Manufacturing Sector

IIoT technologies can be used for quality control in the mobile manufacturing sector to ensure that the products meet the required quality standards. For instance, consider an IIoT system that is designed to optimize the quality control process of a mobile manufacturing company. The system can gather information on a range of production-related topics, such as tool performance, component quality, and test results. The gathered data can be assessed and potential quality issues identified by the system using machine learning, and alerts can be triggered to the relevant personnel, accompanied by suggestions for corrective measures.

Case Study 4: Predictive Maintenance

IIoT technologies can be used for predictive maintenance in the thermal power plant to reduce downtime and maintenance costs. For instance, consider an IIoT system that is designed to optimize the maintenance process of a thermal power plant. This system could use sensors and other devices to collect data on different aspects of the power plant, such as the performance of the turbines, the temperature and pressure of the steam, and the quality of the fuel. In order to schedule maintenance and repair work in advance, reduce the chance of unexpected downtime, and save money on maintenance, the system can use machine learning algorithms to evaluate the data and anticipate prospective equipment failures before they happen.

Case Study 5: Software Company

Application-based classification is a commonly used method in the software industry to categorize software based on their purpose or function. By using this method, software companies can effectively market their products to the right audience and generate more sales. Here are some examples of application-based classifications used by software companies:

- **Productivity Software**: Productivity software, comprising project management, time management, and collaboration tools, among others, aids users in enhancing their productivity.
- **Entertainment Software**: Video games, media players, and streaming services are software developed for leisure activities.
- **Health and Fitness Software**: This category includes software designed to help users track their health and wellness, such as diet tracking apps, workout tracking apps, and meditation apps.
- **Education Software**: This includes software designed to aid in the learning process, such as language learning software, educational games, and simulation software.
- **Finance Software**: This category includes software designed for financial management, such as accounting software, budgeting software, and investment management software.
- **Graphic Design Software**: This category includes software designed for graphic design, such as photo editing software, graphic design software, and desktop publishing software.
- **Communication Software**: This category includes software designed to help users communicate with others, such as email clients, instant messaging apps, and video conferencing software.

By classifying their software into these categories, software companies can tailor their marketing strategies to target the right audience. For example, a project management software company would want to target businesses and professionals, while a health and fitness app company would want to target fitness enthusiasts and people interested in wellness.

INDUSTRY-BASED CLASSIFICATION

This classification is based on the industry in which the IIoT system is deployed, such as manufacturing, healthcare, transportation, agriculture, and energy. Industry-based classification is a commonly used method in the IoT (internet of things) industry to categorize IoT solutions based on the industries they serve (Wang et al., 2020). By using this method, IoT companies can effectively market their products to the right audience and generate more sales. Few examples of industry-based classifications used in IoT are described below.

- **Smart Homes**: This includes IoT solutions designed for residential properties, such as smart thermostats, home security systems, and smart lighting systems.
- **Industrial IoT (IIoT)**: This includes IoT solutions designed for manufacturing, logistics, and other industrial applications, such as predictive maintenance systems, asset tracking systems, and industrial automation systems (Khang et al., 2023a).
- **Healthcare IoT**: This includes IoT solutions designed for healthcare applications, such as remote patient monitoring, medical device connectivity, and hospital asset tracking systems.
- **Agriculture IoT**: This includes IoT solutions designed for agriculture applications, such as soil monitoring systems, crop health monitoring systems, and precision farming solutions.
- **IoT-based Smart Cities**: This includes IoT solutions designed for urban infrastructure, such as smart traffic management systems, public safety systems, and waste management systems.
- **Retail IoT**: This includes IoT solutions designed for the retail industry, such as smart shelves, inventory tracking systems, and customer behavior tracking systems.

Each category represents a specific industry or application where IoT solutions can be used to enhance efficiency, improve safety, and streamline processes. For instance, IIoT solutions are designed for manufacturing and logistics companies to optimize processes, reduce downtime, and improve asset management. On the other hand, healthcare IoT solutions focus on remote patient monitoring, medical device connectivity, and hospital asset tracking systems.

By classifying their IoT solutions into industry-based categories, companies can better understand the needs of their target customers and offer solutions that meet their specific requirements. This approach helps companies differentiate themselves in the crowded IoT market and provides a clear value proposition to potential customers.

DEVICE-BASED CLASSIFICATION

Classifying devices is a crucial element of IoT and is especially significant in the smart manufacturing sector. This classification is based on the type of device or sensor used in the IoT system. Examples include temperature sensors, humidity sensors, motion sensors, pressure sensors, and gas sensors. Some examples of device-based classification in smart manufacturing:

- **Sensors**: Sensors are commonly used in smart manufacturing to monitor and collect data about the manufacturing process. Temperature sensors, pressure sensors, flow sensors, and humidity sensors are few typical sensor types utilized in smart manufacturing.
- **Actuators**: The control and movement of equipment, machinery, and robots based on data gathered from sensors is a crucial part of smart manufacturing.

- **Programmable Logic Controllers (PLCs)**: PLCs are a type of computer that is used to control and automate industrial processes. They can be programmed to monitor data from sensors and actuate machinery based on that data.
- **Industrial Robots**: Industrial robots are increasingly being used in smart manufacturing to automate tasks such as welding, painting, and assembly. These robots can be programmed to perform specific tasks and can be controlled by sensors and actuators.
- **Wearable Devices**: Wearable devices are used in Industry 4.0 to monitor the health and safety of workers. Examples include smart glasses that provide workers with augmented reality displays and smart watches that can monitor workers' vital signs.
- **Automated Guided Vehicles (AGVs)**: AGVs are vehicles that are used to transport materials and products around a factory or warehouse. They are typically equipped with sensors that allow them to navigate and avoid obstacles.
- **Smart Tools**: Smart tools are devices that are used to measure and control the manufacturing process. Examples include smart torque wrenches, which can measure the amount of torque applied to a bolt, and smart calipers, which can measure the dimensions of a part.
- **Asset Tracking Devices**: Asset tracking devices are used to monitor the location and status of equipment and inventory in a factory or warehouse. These devices can be equipped with sensors that can detect changes in temperature, humidity, and other environmental factors.

In Industry 4.0, device-based classification is important because it allows manufacturers to better understand and optimize their production processes. By collecting data from sensors and using actuators to control machinery, manufacturers can improve efficiency, reduce waste, and increase productivity.

TECHNOLOGY-BASED CLASSIFICATION

Smart manufacturing systems are designed to integrate advanced digital technologies into the manufacturing process to improve efficiency, productivity, and quality. Followings are some important technology-based classification in smart manufacturing systems.

- **Artificial Intelligence (AI)**: Artificial intelligence is used in smart manufacturing systems to improve production procedures, forecast equipment failures, and enhance product quality.
- **Internet of Things (IoT)**: The internet of things is a network of physical objects, vehicles, and other devices with software, sensors, and connections that allow them to exchange data with other devices and systems.
- **Industrial Internet of Things (IIoT)**: Using sensors, gadgets, and other cutting-edge technology for data gathering and exchange, the IIoT is

essential for monitoring and optimizing production processes in smart manufacturing systems.

- **Cloud Computing**: Cloud computing allows manufacturers to access real-time information and make data-driven decisions to improve production efficiency and quality using large volumes of data generated by smart manufacturing systems.
- **Cybersecurity**: Cybersecurity is critical in smart manufacturing systems to protect against data breaches and other security threats. This includes implementing secure network protocols, encrypting data, and controlling access to sensitive information.
- **Augmented Reality (AR)**: AR is used in smart manufacturing systems to provide workers with real-time information and guidance on production processes and equipment. This can improve worker efficiency, reduce errors, and increase product quality.
- **Digital Twin**: In the context of smart manufacturing systems, a digital twin is a computerized replica of a physical object or system, utilized for improving production processes, anticipating equipment failure, and enhancing product design (Han et al., 2020).
- **Robotics Technology**: Robotics involves the design, construction, and operation of robots to perform various tasks. Robotics is used in applications such as manufacturing, healthcare, and exploration.
- **Additive Manufacturing (3D Printing)**: In smart manufacturing systems, the utilization of additive manufacturing, such as 3D printing, enables the production of intricate components with higher efficiency and lower costs compared to conventional manufacturing techniques, leading to improved production flexibility and waste reduction.

The process of 3D printing involves adding materials layer by layer to create three-dimensional objects, and it is commonly utilized in product design, prototyping, and manufacturing applications.

In smart manufacturing systems, the utilization of additive manufacturing, such as 3D printing, enables the production of intricate components with higher efficiency and lower costs compared to conventional manufacturing techniques, leading to improved production flexibility and waste reduction.

Technology-based classification is important because it helps individuals and organizations to understand the different technologies available and how they can be used to solve specific problems. It also helps to identify the strengths and weaknesses of each technology, which can be useful in selecting the right technology for a particular application. In smart manufacturing systems, technology-based classification is important because it enables manufacturers to understand and leverage the various digital technologies available to them. By combining the correct technologies, production methods can be enhanced, costs can be cut, and product quality can be boosted by manufacturers.

DEPLOYMENT-BASED CLASSIFICATION

In the smart manufacturing industry, deployment-based classification can be used to categorize various technologies based on how they are deployed or implemented in the manufacturing environment. This classification is based on the deployment model of followings:

- **On-premises**: On-premises deployment of smart manufacturing technologies involves installing and operating hardware and software within the manufacturing facility itself. This can include sensors and other devices that collect data from machines and processes, as well as software applications that analyze and interpret this data. On-premises deployment offers greater control and security, but can be more costly and time-consuming to manage.
- **Cloud-based**: Cloud-based deployment of smart manufacturing technologies involves using a remote server or network of servers to store and process data. This can include cloud-based analytics platforms that analyze data collected from on-premises devices, as well as cloud-based software applications that automate manufacturing processes. Cloud-based deployment offers greater scalability and cost-effectiveness, but can be less secure and require a reliable internet connection.
- **Edge Computing**: In edge computing deployment of smart manufacturing technologies, computing resources are located nearer to where data is generated or processed, rather than being centralized in a data center or cloud. This can include edge computing devices that process data in real time, such as sensors that monitor machine performance and detect anomalies. Edge computing deployment is useful in smart manufacturing applications where real-time processing is critical.
- **Mobile**: Mobile deployment of smart manufacturing technologies involves deploying software applications on mobile devices such as tablets and smartphones. This can include mobile apps that allow operators to monitor and control machines and processes remotely, as well as mobile analytics platforms that provide real-time insights into manufacturing performance. Mobile deployment is useful in smart manufacturing applications where mobility is important, such as in field service and maintenance.
- **Hybrid**: Hybrid deployment of smart manufacturing technologies involves combining two or more deployment models, such as on-premises and cloud-based deployment, to achieve the benefits of both models. For example, a hybrid deployment might involve storing sensitive manufacturing data on-premises while using cloud-based analytics platforms to analyze this data.

Deployment-based classification in smart manufacturing is important because it helps organizations to understand the different deployment options available for smart manufacturing technologies and choose the most appropriate deployment model based on their needs and resources. It also helps to identify the advantages and disadvantages of each deployment model, which can be useful in making informed decisions about technology investments.

ARCHITECTURE-BASED CLASSIFICATION

Architecture-based classification is also relevant to smart manufacturing, where it can be used to categorize technologies based on their underlying architecture or design principles (Rana et al., 2021). This classification is based on the architecture of the IIoT system, such as centralized, decentralized, or hybrid. Some implemented architecture-based classifications in Industry 4.0 are:

- **Centralized Architecture**: Centralized architecture involves a single server or system that performs all the processing and storage functions. This architecture is commonly used in traditional manufacturing environments where there is limited connectivity and data exchange between machines. In Industry 4.0, centralized architecture is less common due to the need for real-time data exchange and distributed processing.
- **Distributed Architecture**: Distributed architecture involves multiple computing devices that work together to perform processing and storage functions. This architecture is commonly used in Industry 4.0 environments, where there is a need for real-time data exchange between machines and systems. Distributed architecture can include edge computing devices that process data in real-time and send it to cloud-based analytics platforms for further analysis.
- **Cloud Architecture**: Cloud architecture involves using a remote server or network of servers to store and process data. This architecture is commonly used in Industry 4.0 applications that require high scalability and cost-effectiveness, as it allows for easy access to computing resources and data storage.
- **Edge Computing Architecture**: Edge computing architecture involves deploying computing resources closer to the location where data is generated or processed, rather than in a centralized data center or cloud. This architecture is useful in Industry 4.0 applications where real-time processing is critical, as it allows for rapid decision-making based on real-time data.
- **Hybrid Architecture**: Hybrid architecture involves combining two or more architecture models, such as distributed and cloud architecture, to achieve the benefits of both models. For example, a hybrid architecture might involve storing sensitive manufacturing data on-premises while using cloud-based analytics platforms to analyze this data.

INTERNET OF PEOPLE (IOP)

The term "internet of people" (IoP) refers to the idea that the internet is not just a network of machines and devices, but also a network of people. It emphasizes the human-to-human connections and interactions that take place over the internet, rather than just the communication between machines. The IoP concept involves the use of technology to enhance social interactions, communication, and collaboration between individuals, as well as to create new opportunities for innovation and entrepreneurship. It aims to make the internet more accessible, inclusive, and user-friendly by putting people at the center of its design and development.

Advanced technologies such as sensors, artificial intelligence, and cloud computing are employed in Industry 4.0 to enhance human–machine interactions and improve productivity, efficiency, and quality in manufacturing. Internet of people (IoP) plays a crucial role in enabling the connection and collaboration between people, machines, and data. The IoP can help organizations to create smarter, more responsive, and more agile workplaces by providing workers with real-time access to data, insights, and feedback. This can improve decision-making, reduce errors, and optimize workflows, leading to increased productivity and cost savings.

For example, in a smart factory, workers can use wearable devices such as smart glasses or watches to receive real-time information and instructions about their tasks, access relevant data and documentation, and communicate with colleagues and machines. This can help them to perform their jobs more effectively, reduce downtime, and improve safety.

Moreover, the IoP can enable new forms of collaboration and innovation across different departments, functions, and organizations. By connecting people with diverse skills, knowledge, and perspectives, it can foster creativity, problem-solving, and continuous learning, leading to better products, services, and processes. However, the implementation of the IoP in Industry 4.0 also raises challenges related to data privacy, security, and ethical considerations. Organizations must ensure that the collection, use, and sharing of data are compliant with regulations and ethical standards, and that workers have the necessary skills, training, and support to use IoP technologies effectively and safely. IoP devices can be used to monitor worker health, track worker locations and activities, and provide real-time feedback to workers to help them perform their jobs more effectively.

INTERNET OF SYSTEMS (IOS)

The internet of systems (IoS) is a concept that describes the integration of different systems, devices, and technologies into a unified network of systems. It refers to the use of the internet to connect and communicate between different systems, enabling them to exchange data, collaborate, and coordinate their actions.

IoS aims to create a seamless and interoperable network of systems, in which data and information can flow seamlessly across different devices, platforms, and applications. IoS can enable the integration of different medical devices, electronic health records, and telemedicine platforms to create a connected and efficient healthcare system. In transportation, IoS can enable the integration of different systems such as traffic management, navigation, and vehicle-to-vehicle communication to create a safer and more efficient transportation system.

IoS also has implications for cybersecurity and privacy, as the integration of different systems can increase the potential attack surface and data breaches. Therefore, it is essential to ensure that IoS systems are designed and implemented with strong security and privacy controls, including encryption, authentication, and access controls.

The IoS is essential to Industry 4.0 because it unifies various systems, machines, and processes into a single network that facilitates information sharing, teamwork, and coordination. Building a linked and interoperable system that improves automation, flexibility, and productivity in manufacturing and other industries through real-time monitoring, control, and improved decision-making advances the Industry 4.0 goal.

By integrating various systems such as sensors, machines, Enterprise resource planning (ERP), and supply chain management systems, IoS can enable real-time monitoring and control of the entire production process in smart factories, allowing organizations to optimize production and improve product quality by reducing wastes. IoS also enables the use of data analytics and artificial intelligence to optimize industrial processes and identify new opportunities for innovation and value creation. By integrating data from different systems and processes, IoS can enable predictive maintenance, anomaly detection, and other advanced analytics applications, leading to better performance and cost savings.

VARIOUS LEVELS IN INDUSTRY 4.0

The integration of advanced technologies characterizes Industry 4.0, which can be distinguished by four implementation stages, each representing varying levels of technological complexity and integration. The four levels of Industry 4.0 implementation are:

- **Connectivity**: At this level, the focus is on connecting machines, sensors, and devices to enable real-time monitoring and data collection. This involves the deployment of IoT sensors and other devices to collect data from machines and processes, and the integration of this data into a centralized system for analysis.
- **Intelligence**: At this level, the focus is on leveraging the data collected from connected devices to enable intelligent decision-making. This involves the use of advanced analytics tools and AI algorithms.
- **Autonomy**: At this level, the focus is on enabling machines and processes to operate autonomously, with minimum involvement of work force. This adopts AI techniques and robotics to automate tasks such as maintenance, quality control, and production planning, leading to increased efficiency, productivity, and cost savings (Hajimahmud et al., 2022).
- **Cyber-physical Systems**: Digital twins, augmented reality, and virtual reality are used to facilitate real-time collaboration and decision-making between humans and machines, resulting in greater innovation and efficiency (Yi et al., 2019).

Various Levels in Amazon with IoT

Amazon, a multinational technology company headquartered in Seattle, Washington, was founded in 1994 by Jeff Bezos as an online bookstore but has since expanded to provide a diverse range of e-commerce products and services, including cloud computing, artificial intelligence, and digital streaming, making it one of the largest e-commerce platforms globally (Khanh and Khang, 2021). The company's e-commerce operations span multiple categories, including books, electronics, clothing, household goods, and more. Amazon has also expanded into digital media, with products like Amazon Prime Video, Amazon Music, and Amazon Kindle.

A range of cloud-based computing services, including data storage, processing, and machine learning, is provided by Amazon Web Services, making it a major player in the cloud computing industry in addition to its consumer-facing businesses. Amazon's use of IoT can be viewed on multiple levels, including:

- **Consumer Level**: At the consumer level, Amazon uses IoT to enhance the shopping experience for its customers. For example, the Amazon Dash button allows customers to quickly reord household items by pressing a button, which triggers an order through the Amazon website.
- **Operational Level**: At the operational level, Amazon uses IoT to optimize its logistics and supply chain operations. For example, Amazon uses IoT sensors and Radio-frequency identification (RFID) tags to track inventory and monitor the condition of goods in its warehouses and during transportation.
- **Industrial Level**: At the industrial level, Amazon uses IoT to enhance its manufacturing and production processes. For example, Amazon's use of robots in its warehouses to move inventory and fulfill orders is enabled by IoT technology.
- **Cloud Services Level**: At the cloud services level, Amazon's Web Services (AWS) IoT platform provides a suite of tools and services that enable businesses to build, deploy, and manage IoT applications and devices. This includes device management, data storage and analytics, machine learning, and more (Rani et al., 2023).

Amazon's use of IoT is in their Amazon Go stores, which use a combination of sensors, cameras, and machine learning algorithms to create a seamless checkout experience for customers. Here's how it works: when a customer enters an Amazon Go store, they scan their Amazon Go app to gain access. Once inside, they can browse and select items as they normally would in any other store. But instead of having to go through a traditional checkout process, customers simply walk out of the store with their items.

A system that tracks the items shoppers put in their bags is supported by a network of sensors and cameras placed throughout the store. As customers leave the store, the system automatically charges their Amazon account for the items they took. The system is able to accurately track items and identify which customer picked them up by using machine learning algorithms that analyze the data from the sensors and cameras. This technology has allowed Amazon to eliminate the need for cashiers and create a frictionless shopping experience for their customers. Overall, Amazon's use of IoT spans multiple levels and has enabled the company to create new business models, streamline operations, and enhance the customer experience.

Various Levels in Healthcare with IoT

The potential transformation of healthcare through the improvement of patient outcomes, reduction of costs, and increase in efficiency is enabled by IoT technology, which can be viewed at various levels.

- **Wearables and Personal Health Monitoring**: At the consumer level, IoT devices such as wearables, health monitors, and home health monitoring systems can provide patients with real-time monitoring of their vital signs and health metrics. This can enable patients to track their health status and provide doctors with valuable data to inform their treatment plans.
- **Medical Equipment and Asset Tracking**: At the operational level, IoT sensors can be used to monitor and track medical equipment and supplies, which can help healthcare organizations reduce waste, improve efficiency, and ensure that equipment is being used effectively.
- **Smart Hospitals and Healthcare Facilities**: At the institutional level, IoT technology can be used to create smart hospitals and healthcare facilities. This can involve the integration of IoT sensors and devices throughout the facility to provide real-time monitoring of patients, staff, and equipment, enabling healthcare providers to quickly identify and respond to issues as they arise.
- **Healthcare Supply Chain and Logistics**: Ensuring that medical supplies and equipment are delivered to the right location at the right time is one of the benefits of using the internet of things.
- **Healthcare Data Analytics**: Large volumes of healthcare data can be collected and analyzed using internet of things technology, which can provide insights into patient health trends, treatment effectiveness, and more. This can enable healthcare providers to make data-driven decisions and improve patient outcomes.

Various Levels in Car Manufacturing Company with IoT

IoT technology has the potential to transform the automotive industry by improving supply chain efficiency, reducing costs, and increasing productivity. There are several ways to look at how IoT is used in the supply chain for making cars, including:

- **Supply Chain Management**: The utilization of internet of things (IoT) sensors at the supply chain level can track the movement of parts and components throughout the manufacturing process, resulting in improved inventory management, waste reduction, and ensuring the availability of parts when required for car manufacturers.

- **Manufacturing Operations**: Car manufacturers can optimize their manufacturing processes, reduce downtime, and increase productivity by utilizing IoT sensors at the operational level to monitor the performance of manufacturing equipment, track the movement of assets such as tools and machinery, and predict when maintenance is needed.
- **Parts and Component Tracking**: In order to improve inventory control, cut waste, and guarantee the availability of parts when required, automakers can use the internet of things to track the movement of parts and components during the production process.
- **Asset Tracking and Monitoring**: IoT sensors can be used to track the whereabouts and motion of machinery, tools, and equipment in real-time. This can aid producers in maximizing asset use and avoiding equipment breakdowns.
- **Inventory Management**: By using the internet of things, manufacturers can gain real-time visibility into the movement of raw materials and finished goods, enabling them to improve inventory management, reduce inefficiencies, and ensure the timely availability of critical supplies.
- **Equipment Tracking**: IoT sensors have the potential to enhance the efficiency of production and minimize the downtime of equipment for automotive manufacturers through proper tracking of various types of equipment.
- **Predictive Maintenance**: Automotive firms can leverage the power of the internet of things to gain valuable insights into the operational performance of their production equipment. IoT sensors can be deployed to monitor various parameters such as temperature, vibration, and energy consumption. This data can be analyzed to identify patterns and detect anomalies that could indicate potential equipment failures. By proactively detecting maintenance needs, automotive firms can schedule repairs and maintenance activities in a timely manner, thus minimizing equipment downtime. This approach can prevent unexpected breakdowns, which can cause significant production delays and lead to losses. Furthermore, it can also help automakers reduce maintenance costs and avoid the need for expensive emergency repairs.
- **Quality Control**: IoT sensors can play a crucial role in maintaining the quality of components and finished vehicles at the manufacturing level. By deploying sensors at various stages of the production process, car manufacturers can capture real-time data on critical parameters such as temperature, humidity, pressure, and more. This data can be analyzed to ensure that the components and vehicles meet the required quality standards. By monitoring the quality of components and finished vehicles using IoT sensors, car manufacturers can detect quality issues early on and take corrective action before they become more serious. This approach can prevent defective products from reaching customers, reducing the risk of product recalls and the associated costs. Furthermore, IoT sensors can also help

car manufacturers optimize their production processes by identifying areas where quality improvements can be made. By analyzing the data collected from the sensors, manufacturers can gain insights into the root causes of quality issues and take steps to improve their processes accordingly.

- **Worker Safety**: At the operational level, the deployment of IoT sensors can be a valuable tool for enhancing worker safety in real time. By monitoring workers' activity, manufacturers can capture real-time data on workers' movements and identify potentially hazardous conditions. This data can be analyzed to detect safety risks and provide timely alerts to workers to take appropriate action. Using IoT sensors, manufacturers can track workers' location and movements, ensuring that they are following safety protocols and operating in designated areas. Sensors can also detect unauthorized entry to restricted areas or the operation of machinery without proper training. In such cases, the sensors can immediately alert the workers and their supervisors to take corrective action, preventing accidents and improving safety. Moreover, IoT sensors can also detect hazardous conditions, such as high temperatures, gas leaks, and machine malfunctions. In the event of such occurrences, sensors can automatically shut down the equipment and alert workers to evacuate the area, preventing potential harm.
- **Manufacturing Analytics**: By leveraging IoT technology to collect and analyze data from the entire manufacturing process, including supplier performance, production efficiency, and other critical factors, manufacturers can gain valuable insights to make data-driven decisions and optimize their manufacturing processes at the analytics level.
- **Vehicle Telemetry**: IoT sensors installed in vehicles can provide real-time monitoring and analysis of performance factors such as safety systems, fuel efficiency, and engine performance, enabling car manufacturers to make data-driven decisions to enhance vehicle design and performance.
- **Customer Experience**: At the customer level, IoT technology can be used to create connected cars that provide customers with a range of features and services, such as real-time traffic information, remote vehicle monitoring, and mobile app integration. This can enhance the overall customer experience and drive customer loyalty.

IoT (internet of things) is a technology that connects various devices and sensors to the internet to collect and exchange data. In the automotive industry, Ford Motor Company is one of the companies that implemented IoT to improve their operations. Ford uses IoT sensors to monitor the performance of their assembly line equipment, which enables them to detect any issues or malfunctions quickly and take corrective actions.

Moreover, Ford also uses IoT sensors to track the movement of their vehicles through the supply chain, from manufacturing to delivery. This provides Ford with real-time visibility into their logistics process, which allows them to optimize their operations and reduce delays. As a result, Ford has experienced improved efficiency, reduced downtime, and better quality control, which ultimately leads to increased customer satisfaction.

VARIOUS LEVELS IN AUTOMATION OF WELDING INDUSTRIES

The automation of welding procedures uses cutting-edge technology such as robotic, artificial intelligence, and other similar technologies to carry out welding operations with the least amount of human involvement (Figure 1.2). The ways in which welding processes can be automated are:

- **Robotic Welding**: The use of robotic arms and advanced welding technology in welding systems allows for highly precise and consistent performance of welding tasks, and the ability to program them to operate 24 hours a day enhances their productivity and efficiency.
- **Welding Positioners**: Welding positioners are used to rotate and position the workpiece during welding, reducing the need for manual adjustment and improving weld quality.
- **Welding Wire Feeders**: Welding wire feeders are used to automatically feed welding wire to the welder, reducing the need for manual adjustment and improving weld consistency.
- **Welding Sensors**: Real-time monitoring of welding processes can be achieved with the use of welding sensors, which provide constant feedback on the weld's quality.
- **Welding Power Sources**: Advanced welding power sources can be used to automatically adjust welding parameters based on the material being welded and the welding position, improving the quality and consistency of the weld. The welding process monitoring by IoT can help to ensure consistent weld quality, detect defects as they occur, and improve the overall efficiency of the welding process. By using IoT sensors and data analytics, operators can monitor the welding process in real time and take corrective

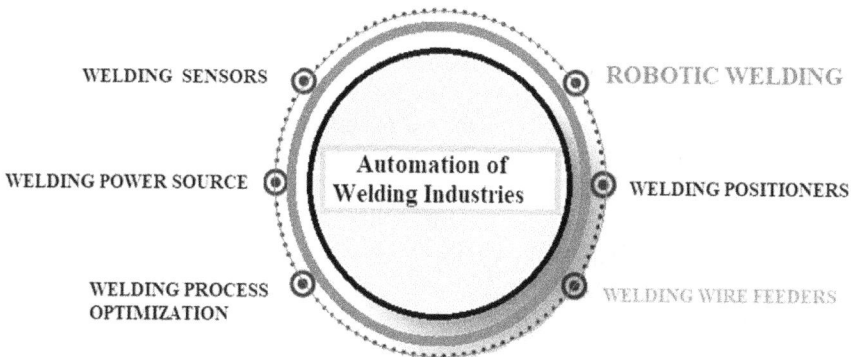

WELDING SENSORS

ROBOTIC WELDING

WELDING POWER SOURCE

Automation of Welding Industries

WELDING POSITIONERS

WELDING PROCESS OPTIMIZATION

WELDING WIRE FEEDERS

FIGURE 1.2 Automation of welding industries

action as needed, reducing the risk of defects and improving overall productivity. The following are some key points to note about welding process monitoring by IoT.

- **Welding Process Optimization**: Real-time monitoring of critical factors such as welding parameters, gas flow, and wire feed speed is possible with the implementation of the internet of things. Here are some ways in which IoT is being used in welding.
- **Welding Process Monitoring**: Welding process monitoring is an important aspect of ensuring consistent weld quality and improving the efficiency of the welding process. By using IoT sensors and data analytics, welding process monitoring can provide real-time monitoring of welding parameters, allowing operators to adjust settings on the fly to ensure consistent quality. Additionally, automated defect detection systems can be used to detect defects in welds, allowing operators to take corrective action quickly.

Manufacturers will be able to pinpoint the primary sources of problems and take corrective action by employing data analytics to spot patterns and trends in weld quality. Additionally, traceability systems can be implemented to ensure compliance with regulatory requirements and provide documentation for quality control purposes. IoT sensors are used to monitor various parameters such as welding voltage, current, and wire speed in real time. This data can be analyzed to optimize the welding process and reduce the risk of defects.

- **Quality Control**: Quality control in welding industries is a critical process that involves monitoring and inspecting welding operations to ensure that welds meet the required standards. IoT technology can play an important role in quality control in welding industries by providing real-time monitoring, defect detection, automated inspection, data analytics, and traceability. The implementation of IoT sensors for real-time monitoring of welding parameters and subsequent analysis can guarantee uniformity in welding procedures and timely detection of defects, with the aid of automated inspection through machine vision systems further increasing consistency and decreasing the need for human intervention, while analysis of IoT data may highlight patterns and trends in weld quality, ultimately allowing manufacturers to identify and rectify the root causes of defects and to ensure comprehensive traceability of the entire welding process, from materials used to inspection outcomes.
- **Automation of Welding Processes**: Advanced automation in the welding industry is augmented by the implementation of IoT technology to automate welding processes, which not only boosts productivity but also minimizes human intervention, rendering it particularly advantageous for high-volume manufacturing, while an array of sensors is utilized for real-time tracking of weld quality and rapid detection of defects.
- **Data Analytics and Optimization**: The gathering and analysis of internet of things data can help with the welding process, increase productivity,

and lower costs. By integrating IoT technology with welding processes, manufacturers can improve the efficiency and quality of welding while also enhancing worker safety. By monitoring welding parameters in real time and predicting when maintenance is needed, welding equipment can be operated more efficiently, reducing the risk of equipment failure and downtime. By automating the welding process and using data analytics to optimize the process, productivity can be improved while costs are reduced.

IMPLEMENTATION OF INDUSTRY 4.0 IN INDIA

India is adopting Industry 4.0. The country recognizes the potential of this technology in transforming its industries, increasing productivity, and improving competitiveness on a global level. The following are some of the initiatives taken by India toward implementing Industry 4.0:

- **National Policy on Industry 4.0**: The adoption of new technologies, such as IoT, artificial intelligence, and big data analytics, in India's manufacturing sector is being promoted through the introduction of a National Policy on Industry 4.0 by the government of India (Bhambri et al., 2022).
- **Smart Cities Mission**: An aim to develop smart cities across India by incorporating technology and digital infrastructure to improve citizens' quality of life and promote sustainable economic growth was launched by the Indian government.
- **Digital India Program**: The National Optical Fiber Network is one of the initiatives of India's Digital India program. The aim of the program is to provide broadband to all areas of the country.
- **Atmanirbhar Bharat Abhiyan**: The Atmanirbhar Bharat Abhiyan is a self-reliance campaign launched by the Indian government to promote domestic manufacturing and reduce dependence on imports. The campaign is focused on increasing investment in research and development, promoting innovation, and developing a skilled workforce.
- **Center of Excellence in Industry 4.0**: The Indian government has established several Centers of excellence in Industry 4.0. The aim of these centers is to promote research and development in emerging technologies.

CONCLUSION

Incorporating the IoT as a crucial element, Industry 4.0 has transformed various sectors by enabling a variety of smart devices to connect and communicate with each other. This technology has increased productivity, efficiency, and safety in numerous industries. Smart factories, which are highly automated and networked industrial facilities, are made possible by IoT technology and are indispensable in Industry 4.0. In smart factories, IoT devices monitor and enhance manufacturing processes, anticipate maintenance requirements, and improve product quality. Additionally, logistics, inventory, and supply chain management processes are made more efficient by the use of IoT technology.

Gaining insights into consumer behavior, market trends, and manufacturing methods by gathering and analyzing massive amounts of data from interconnected devices is one of the primary advantages of IoT in Industry 4.0. This data can be utilized by businesses to make better decisions and operate their operations more efficiently. However, implementing IoT in Industry 4.0 also comes with challenges, including the need for robust cybersecurity measures to protect sensitive data, the high cost of implementation, and the potential for job displacement due to increased automation.

FUTURE SCOPE OF WORK IN INDUSTRY 4.0

While Industry 4.0 presents exciting opportunities for growth and innovation, there are still research problems that need to be addressed to fully realize its potential. One such problem is the lack of standardization in IoT devices and data analytics, which can result in compatibility issues and hinder seamless integration across various systems. Another research problem is the ethical and legal considerations surrounding the use of AI, particularly in decision-making processes. Additionally, there is a need for further research on the social and economic impacts of Industry 4.0, particularly in terms of job displacement and the changing nature of work. Addressing these research problems will be crucial for the continued development and success of Industry 4.0.

REFERENCES

Bhambri, P., Rani, S., Gupta, G., & Khang, A. (2022). *Cloud and Fog Computing Platforms for Internet of Things*. CRC Press. https://doi.org/10.1201/9781003213888

Bose, I., & Liu, L. (2019). "Industry 4.0 and big data analytics for supply chain management: A review." *International Journal of Production Research*, 57(15–16), 4980–4993. https://doi.org/10.1080/00207543.2018.1568985

Hahanov, V., Khang, A., Litvinova, E., Chumachenko, S., Hajimahmud, V. A., & Alyar, V. A. (2022). "The key assistant of smart city - sensors and tools." *AI-Centric Smart City Ecosystems: Technologies, Design and Implementation* (1st ed.), 17(10). CRC Press. https://doi.org/10.1201/9781003252542-17

Hajimahmud, V. A., Khang, A., Hahanov, V., Litvinova, E., Chumachenko, S., & Alyar, V. A. (2022). "Autonomous robots for smart city: Closer to augmented humanity." *AI-Centric Smart City Ecosystems: Technologies, Design and Implementation* (1st ed.) (12). CRC Press. https://doi.org/10.1201/9781003252542-7

Han, X., Guo, J., & Fang, Z. (2020). "A digital twin-driven approach to energy-efficient production planning and control for smart manufacturing." *Journal of Cleaner Production*, 245, 118860. https://www.sciencedirect.com/science/article/pii/S0959652621043407

Hu, L., & Wang, S. (2019). "Big data-driven smart manufacturing system design and optimization for Industry 4.0." *Journal of Industrial Information Integration*, 13, 49–58. https://ieeexplore.ieee.org/abstract/document/8079310/

Huang, G. Q., Zhang, Y. F., Mak, K. L., & Maropoulos, P. G. (2019). "Big data analytics for advanced manufacturing systems: Trends, challenges, and perspectives." *Journal of Manufacturing Systems*, 49, 194–206. https://www.sciencedirect.com/science/article/pii/S0278612521000601

Hsiao, Y. C., Chang, C. H., & Wu, T. Y. (2019). "IoT and industry 4.0: A literature review." *Journal of Industrial Integration and Management*, 4(3), 1950003. https://doi.org/10.1142/s2424862219500037

Kaushik, S., & Singh, R. (2021). "Industry 4.0 and digital transformation: A review of literature and future research directions." *Journal of Enterprise Information Management*, 34(2), 361–387. https://doi.org/10.1108/jeim-10-2019-0411

Khang, A., Abdullayev, V., Hahanov, V., & Shah, V. (2024b). *Advanced IoT Technologies and Applications in the Industry 4.0 Digital Economy* (1st ed.). CRC Press. https://doi.org/10.1201/9781003434269

Khang, A., Gupta, S. K., Rani, S., & Karras, D. A. (Eds.). (2023a). *Smart Cities: IoT Technologies, Big Data Solutions, Cloud Platforms, and Cybersecurity Techniques.* CRC Press. https://doi.org/10.1201/9781003376064

Khang, A., Hahanov, V., Abbas, G. L., & Hajimahmud, V. A. (2022). "Cyber-physical-social system and incident management." *AI-Centric Smart City Ecosystems: Technologies, Design and Implementation* (1st ed.), 2(15). CRC Press. https://doi.org/10.1201/9781003252542-2

Khang, A., Misra, A., Abdullayev, V., & Litvinova, E. (Eds.). (2024a). *Machine Vision and Industrial Robotics in Manufacturing: Approaches, Technologies, and Applications.* CRC Press. https://doi.org/10.1201/ 9781003438137

Khang, A., Rani, S., & Sivaraman, A. K. (2022a). *AI-Centric Smart City Ecosystems: Technologies, Design and Implementation* (1st ed.). CRC Press. https://doi.org/10.1201/9781003252542

Khanh, H. H., & Khang, A. (2021). "The role of artificial intelligence in blockchain applications." *Reinventing Manufacturing and Business Processes through Artificial Intelligence*, 2, 20–40. CRC Press. https://doi.org/10.1201/9781003145011-2

Kshetri, N., & Voas, J. (2022). "A literature review of blockchain technology in Industry 4.0 and the IIoT." *International Journal of Information Management*, 63, 102503. https://doi.org/10.1016/j.ijinfomgt.2021.102503

Li, X., Xu, X., Xu, L. D., & Wang, L. (2018). "Industry 4.0: State of the art and future trends." *International Journal of Production Research*, 56(8), 2941–2962. https://www.tandfonline.com/doi/abs/10.1080/00207543.2018.1444806

Miao, C., Wang, J., & Shi, W. (2021). "A review of industry 4.0 technologies and applications." *International Journal of Advanced Manufacturing Technology*, 112(9–10), 2809–2827. https://doi.org/10.1007/s00170-020-06409-7

Rana, G., Khang, A., Sharma, R., Goel, A. K., & Dubey, A. K. (Eds.). (2021). *Reinventing Manufacturing and Business Processes through Artificial Intelligence*. CRC Press. https://doi.org/10.1201/9781003145011

Rani, S., Bhambri, P., Kataria, A., Khang, A., & Sivaraman, A. K. (2023). *Big Data, Cloud Computing and IoT: Tools and Applications* (1st ed.). Chapman and Hall/CRC. https://doi.org/10.1201/9781003298335

Ries, J., & Duan, Y. (2022). "Industry 4.0, IIoT and the digital transformation of manufacturing: A review and future research agenda." *International Journal of Production Economics*, 243, 108505. https://doi.org/10.1016/j.ijpe.2022.108505

Shah, V., Jani, S., & Khang, A. (Eds.). (2023). "Automotive IoT: Accelerating the automobile industry's long-term sustainability in smart city development strategy." *Smart Cities: IoT Technologies, Big Data Solutions, Cloud Platforms, and Cybersecurity Techniques*. CRC Press. https://doi.org/10.1201/9781003376064-9

Subhashini, R., & Khang, A. (Eds.). (2023). "The role of Internet of Things (IoT) in smart city framework." *Smart Cities: IoT Technologies, Big Data Solutions, Cloud Platforms, and Cybersecurity Techniques*. CRC Press. https://doi.org/10.1201/9781003376064-3

Wang, W., Jiang, P., & Wan, J. (2020). "A survey on industrial Internet of Things: A rising paradigm for Industry 4.0." *Journal of Industrial Information Integration*, 17, 100116. https://doi.org/10.1016/j.jii.2019.100116

Yang, Y., Zhang, L., & Sun, L. (2020). "Industry 4.0 and industrial Internet of Things (IIoT)-based smart manufacturing: A case study." *International Journal of Production Research*, 58(18), 5598–5610. https://doi.org/10.1080/00207543.2020.1719214

Yi, W., Zhang, X., & Tao, F. (2019). "Industrial big data analytics and cyber-physical systems for future maintenance & service innovation." *Journal of Industrial Information Integration*, 15, 53–67. https://www.sciencedirect.com/science/article/pii/S2212827115008744

2 Role of Artificial Intelligence (AI)-aided Internet of Things (IoT) Technologies in Business and Production

Navneet Kaur, Shalini Sahay, and Shruti Dixit

INTRODUCTION

The internet of things (IoT) was created in 2008 (Evans, 2011), and since then, it has grown tremendously and now has a position in many households and companies. The IoT is difficult to define because it has changed and evolved since it was first introduced, but it is best described as a network of digital and analog equipment and computer devices that have been given unique identifiers (UIDs) and the capability to exchange data without the involvement of a person (Rouse, 2020). This typically takes the form of a user interacting with a central hub device or application, which is frequently a smartphone app, before sending information and instructions to one or more periphery IoT devices (Linthicum, 2019). If necessary, the fringe devices can carry out tasks and send data back to the hub device or application so that a person can examine it. In terms of device connectivity, the IoT concept has improved the world's accessibility, integrity, availability, scalability, confidentiality, and interoperability (Lu and Xu, 2019). However, because of their numerous attack surfaces, lack of security standards and regulations, and newness, IoTs are susceptible to cyber-attacks. Business operations have been revolutionized by IoT machine learning, which converts massive volumes of data into useful insights and decision-making tools.

The technological era is continually changing, with new innovations coming practically daily. The fusion of IoT with machine learning (ML) is one such area that has shown tremendous growth in recent years. This cutting-edge combination of technology is creating new economic opportunities and is positioned to have a significant impact on how the world develops in the future. IoT machine learning offers organizations a fresh and exciting way to harness the power of big data and get

DOI: 10.1201/9781003434269-2

a competitive edge in the market in a world that is becoming more and more data-based. Because of its numerous uses and endless possibilities, IoT machine learning is set to be one of the key drivers of innovation and growth in the years to come (Rana et al., 2021).

The systems are made predictive, prescriptive, and autonomous, thanks to the combination of AI and IoT. The nature of developing applications is changing as a result of the confluence of AI and IoT, moving from being helped to augment and ultimately to autonomous intelligence. All sectors of the economy, including manufacturing, retail, healthcare, telecommunications, and transportation, will be impacted by this continuum (Khang et al., 2023c). IoT sensors will enable the gathering of enormous amounts of data, while AI can assist in generating intelligence for creating smarter apps for a smarter world. Additionally, the developing 5G environment offers a base for maximizing the potential of AI-powered IoT. The massive connection provided by 5G combined with ultra-low latency capability will pave the way for innovative applications in many industries (Bhambri et al., 2022).

Artificial intelligence (AI)-aided internet of things technology combines AI technologies and IoT infrastructure. The purpose of AI-assisted IoT technology is to improve human–machine interactions, make IoT operations more efficient, and improve data management and analytics (Rani et al., 2021).

Natural language processing, speech recognition, and machine vision are three common applications of AI, which is the emulation of human intelligence processes by machines, particularly computer systems. The IoT is a network of interconnected computing devices, mechanical and digital machines, and objects with the capacity to send data across a network without the need for human or computer-to-human interaction. A thing in the IoT can be any device that can be given an internet protocol address and transmit data across a network, such as a person's implanted heart monitor or a car with built-in sensors that warn the driver when tire pressure is low (Yongjun et al., 2021).

AI-assisted IoT technologies are game-changing and mutually advantageous for both types of technology, as AI adds value to IoT through connectivity, signaling, and data sharing, while IoT adds value to AI through machine learning capabilities and enhanced decision-making processes. By generating greater value out of the data supplied by the IoT, it can enhance businesses and their services. Using AI, an IoT device better analyses, learns from, and makes decisions using accumulated massive data without human assistance.

IoT networks are used to connect many infrastructure parts, such as chipsets and software that make up AI-assisted IoT devices. Next, an application programming interface (API) is employed to guarantee that all hardware, software, and platform components can function and communicate with one another without the end user having to do any effort (Tedeschi et al., 2019).

When in use, IoT devices generate and collect data, which AI then analyses to produce insights and increase productivity. AI gains insights by employing techniques like data learning.

APPLICATIONS AND EXAMPLES OF AI-AIDED IOT TECHNOLOGY

The following lists numerous instances of the wider application of AI-aided IoT technology (Syed et al., 2021).

- **Smart Cities**: Data is gathered using smart technologies including sensors, lights, and meters in order to increase operational effectiveness, spur economic growth, and enhance resident quality of life (Khang et al., 2022a).
- **Smart Retail**: Businesses employ smart cameras to identify customers' faces and determine if they used the self-checkout to scan their products before leaving the store.
- **Smart Home**: Through human interaction and response, smart appliances learn. IoT devices with AI assistance can also save and learn from user data to comprehend user habits and offer tailored support.
- **Business and Industrial**: Smart chips are used by manufacturers to identify when a piece of equipment needs to be repaired or replaced.
- **Social Media and Personnel (HR)**: Social media and platforms for HR professionals can be connected with AI-assisted IoT solutions to provide an AI decision as a service function.
- **Autonomous Vehicles**: Several camera and sensor systems are used by autonomous vehicles to collect information about other vehicles in the area, monitor road conditions, and search for pedestrians.
- **Automated Delivery Robots**: Sensors acquire information about the robot's surroundings, such as a warehouse, and then use AI to generate judgments based on traversal.
- **Healthcare**: Real-time health data, such as heart rate, is collected and monitored by medical devices and wearables, and irregular heartbeats can be eliminated (Mathad and Khang, 2023).

AI and IoT applications in advanced manufacturing technology have recently emerged as a result of the industrial era 4.0's rapid development. As a result, IoT systems can boost automation in the manufacturing process, which will help to improve the operational effectiveness and efficiency of the smart manufacturing processes. A deep understanding of cutting-edge technologies that are incorporated into manufacturing processes, such as AI, IoT, big data analytics, machine learning and deep learning, virtual reality, and autonomous robots, is necessary for smart manufacturing (MohdJavaid et al., 2022).

The ways that organizations used to operate are being redefined by AI and IoT. IoT has increased the scope of communication between devices and humans through effective intelligent technology, while AI has opened the way for smarter task execution with real-time analysis and greater interaction between humans and machines. The applications of artificial intelligence and the internet of things are enhanced by their interactions. The entire corporate landscape has undergone a profound upheaval as a result of AI and IoT. The combination of IoT and AI can assist managers in making accurate judgments. AI aids in the assimilation and evaluation of the massive volumes of data that IoT gathers through device access to the internet (Rani et al., 2023).

IoT devices use machine learning, a kind of artificial intelligence, to recognize patterns and spot errors in data collection using very sophisticated sensors. Over time, this technology derives intrinsic phenomena including air stimulation, temperature, humidity, pollution, sound, vibrations, lights, etc. IoT and machine learning, as opposed to conventional technology, accelerate operational estimates by 20× while improving accuracy. This explains why companies using AI technology experience an increase in their sales figures. The IoT revolution's use of artificial intelligence has resulted in a huge income increase and the sale of more linked products.

There is no doubt that IoT demand is increasing. Every company, whether an established company or a startup, is currently asking for IoT and AI. As neither of these skills has a cap, businesses want to unleash their potential and utilize them to the fullest. Figure 2.1 shows a comparison of various technologies and demonstrates that are now in vogue (Bourechak et al., 2023).

ADVANTAGES OF AI AND IOT IN BUSINESS

IoT and AI working together are inexorable technological forces. To take advantage of the many advantages that IoT and AI offer, the two can be merged. The following provides additional details (Ghosh et al., 2018).

DATA COLLECTION, DISSEMINATION, AND FORMULATING USER PERCEPTIONS

The development and expansion of a firm depend greatly on data collection. A company with an IoT strategy is aware of how technology might reduce data compulsion by facilitating easier access to customer data. Dealing with information is made simpler by AI in business. IoT devices offer a special feature that allows users and device interaction patterns to be tracked, recorded, and observed. Companies develop better ways to improve customer experiences using the data they have collected.

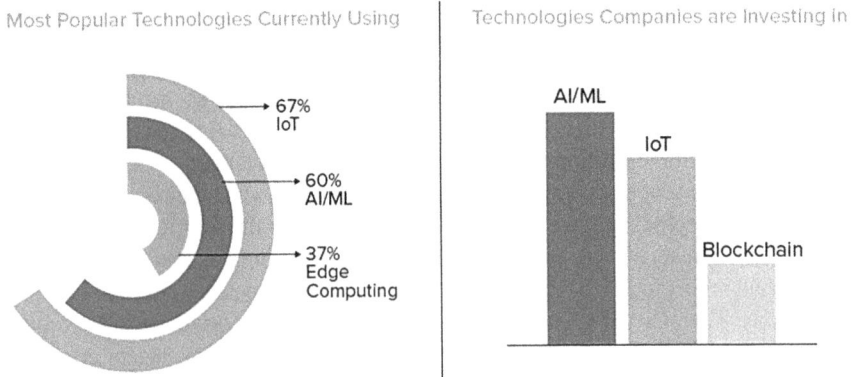

FIGURE 2.1   Popular technologies

REDUCTION OF IDLE TIME

Heavy machinery used by oil and gas manufacturing companies is susceptible to unforeseen or unplanned breakdowns. Downtime brought on by this can result in significant losses. Predictive maintenance is made possible by having an IoT platform with AI capabilities. Analytics helps to foresee equipment failures and breakdowns in advance so that you can prepare a course of action and avoid having your operations affected: a decrease in maintenance planning time of 20–50%; a maintenance expense reduction of 5–10%; and an increase in equipment uptime and availability of 10–20%.

STRENGTHENING SECURITY MEASURES

Security and safety are top concerns for a corporation because of the recent increase in data breaches and theft of private information. AI-powered IoT technology offers tenacious protection for your private information and forbids intrusion by outside parties. Some firms are facilitating machine-to-machine communication to detect incoming threats and send out automated answers to hackers. A typical application would be in the banking industry, where IoT sensors detect suspicious activity in ATMs and promptly alert police enforcement (Hahanov et al., 2022).

OPERATIONAL EFFICIENCY WITH AUTOMATION

IoT adoption simplifies operations and aids in creating precise predictions, both of which are vital for enhancing the effectiveness of the company. Spending money on internet of things investments is essential in the modern era because the technology also enables you to gain insight into time-consuming and redundant tasks. Google's reduction in cooling costs, which they could accomplish with AI and IoT, is a nice example. You can determine which of your operational tasks need some fine-tuning, much like Google, to maintain efficiency.

HELPS IN PROCESSING BUSINESS ANALYSIS

The relationship between supply and demand must be carefully balanced. AI assists in enhancing inventory management and relieving pressure on the stock by enabling you to foresee when you need to refill. Retailers can benefit from this since they sometimes stockpile too many things only to learn later that they couldn't all be sold. This demonstrates the accuracy compared to manual approaches. They can gather data and do analyses for stock maintenance with the aid of IoT technologies.

IMPROVED RISK MANAGEMENT

We have talked about how IoT platforms and AI work together to preserve cybersecurity. The two deal with issues and provide rapid reactions with ease when it comes to risk management, which includes handling financial loss, employee safety, and cyber threats. This helps to prevent scenarios like this from occurring. For instance, using data gathered from wearable devices and AI, the Japanese IT equipment and service provider Fujitsu ensures worker safety.

POSSIBILITY OF DEVELOPING NEW AND BETTER GOODS AND SERVICES

The technique known as Natural Language Processing (NLP), which strives to enhance communication through speech, text, or gestures, has improved the flow of information between people and machines. Monitoring and inspection take on a whole new meaning, thanks to AI-powered robotics, especially with the previously unheard-of use of drones in the construction sector. It assists in retrieving data that a person could never physically be able to do. This demonstrates how bright the future of IoT and AI is. Monitoring all measurable information aids in fleet management for commercial vehicles. Rolls Royce is a fantastic illustration of IoT use cases enabled by AI. Plans call for implementing AI technology to address IoT-enabled aircraft engine maintenance requirements.

CHALLENGES AND BENEFITS OF AI-AIDED IOT TECHNOLOGIES

The internet of things produces bulk volumes of data from millions of devices. IoT machine learning is led by data and generates insight from it. Machine learning utilizes past behavior to identify patterns and builds models that help in predicting future behavior and events. It can be used to assign future trends, find variance, and enhance intelligence by using images, videos, and audio. The IoT has made the world more sensitive and smarter by integrating the digital and physical world together. At last, the two combined will open the way to new solutions and experiences that transform businesses across numerous industries, creating new opportunities altogether (Pande and Padwalkar, 2014).

BENEFITS OF MACHINE LEARNING INFERENCE FOR IOT

Data formulation and data collection, one of the key components of an organization, is the development and collection of data. IoT and AI are skilled at organizing voluminous amounts of data for easy comprehension (Khang et al., 2024a). IoT has the capacity to observe and document how users interact with devices, as well as their behavior patterns. Most firms gather and store this data in order to enhance the digital consumer experience.

- **Security Enhancement**: Due to the attempts made by hackers, malware, and cybercriminals to steal sensitive data, safety and security are concerns for every firm. The integration of AI and IoT prevents unauthorized access to the data and promptly alerts the operator to any suspicious activity.
- **Automated Operations**: IoT increases efficiency by making critical decisions ahead of time. It streamlines the process, saves time, and analyses areas that require modification in order to achieve the desired result.

- **Chain Management**: AI and IoT make asset tracking simple and keep a central database of all inventory. Additionally, it enables firms to more effectively manage their inventories and assets while enhancing their cash flow.
- **Improved Risk Management**: Through automatic reactions or notifications, organizations are able to safeguard themselves, their staff, and customers from cyber dangers and financial losses. It keeps track of the patterns employed by ATM banks and foresees any trouble in advance.

CHALLENGES IN AI AND IoT

When IoT and AI are combined, there are additional, more complex difficulties that arise. A few examples include the following.

Safety

Due to the fact that AI and IoT are gathering sensitive and critical data from their users or clients, it is imperative to ensure that the data is secure and in trustworthy hands. Security is the biggest problem with any technology since we don't know when hackers will steal our sensitive and vital data.

Compatibility and Complexity

As we all know, the IoT is the merging of numerous devices with numerous different technologies. This may result in various challenges after connecting all of the devices into one.

Artificial Stupidity

Artificial stupidity means the inability of an AI program to perfectly perform basic tasks, AI systems, and the algorithms. AI uses need to be well developed to understand and interpret data so more accurate and sensible decisions can be made. This is a work in progress.

Lack of Confidence

Knowing that IoT is a new and developing technology, both consumers and businesses are gravely concerned about security and have limited faith in the ability to safeguard IoT devices and the integrity of the data produced.

Cloud Attacks

It is not surprising that dangerous attacks by viruses have gained unwanted attention due to the rapid development of cloud computing technology. IoT requires a lot of data, which is kept on the cloud, increasing the danger of data security (Rani et al., 2023).

Technology

Instead of having many more difficulties of IoT technology in the future, we may say that this challenge involves bringing rivalry to all technologies. It is not an easy task to have this challenge and give competition to every AI-based technology. Therefore, we can find probable future directions and potential growth areas for AI-aided IoT in smart business and production.

The combination of artificial intelligence (AI) and the internet of things (IoT) can transform how businesses operate and produce goods. AI-aided IoT has already revolutionized the way businesses and production facilities operate by providing valuable insights and automation capabilities. Here are some probable future directions and potential growth areas for AI-aided IoT in smart business and production (Khang et al., 2023a):

- **Edge Computing**: As more and more devices are connected to the internet, edge computing will become essential in AI-aided IoT. This technology enables data to be processed locally at the edge of the network, closer to the source, reducing latency, and improving response times.
- **Predictive Maintenance**: Predictive maintenance is an application of AI-aided IoT that uses data analysis to predict when a machine is likely to fail. IoT sensors can gather data on equipment performance, and AI algorithms can analyze sensor data from machines and predict when they are likely to fail, enabling businesses to schedule maintenance proactively, reduce downtime, extend the life of their equipment, save costs, and increase efficiency. For example, in the manufacturing industry, IoT sensors can monitor the condition of machines and detect any anomalies in their performance. The AI algorithms can then analyze this data to predict when maintenance is required, enabling the manufacturer to take corrective action before any breakdown occurs (Bhanu et al., 2022) (Ashish et al., 2018).
- **Autonomous Systems**: Autonomous systems, such as robots and drones, are already in use in manufacturing and logistics, and we can expect to see more widespread adoption of these technologies in the future. AI-aided IoT will enable these systems to make real-time decisions and adapt to changing conditions.
- **Intelligent Supply Chain**: AI algorithms can analyze data from sensors and other sources to optimize inventory management and improve logistics efficiency. It can be used to optimize the supply chain by providing real-time visibility into inventory levels, shipping status, and delivery times. This can help businesses to reduce waste, improve efficiency, and respond to customer demands faster.
- **Smart agriculture**: Smart agriculture is an emerging field that uses IoT and AI to optimize crop yields and reduce waste (Figure 2.2). IoT sensors can be used to monitor crops and soil conditions, and AI algorithms can analyze this data to optimize irrigation, fertilizer usage, and pest control (Khang et al., 2023b). The AI algorithms can then analyze this data to adjust the irrigation system accordingly, resulting in increased crop yields and reduced water waste (Qazi et al., 2022).
- **Cyber Security**: As the number of connected devices continues to grow, so does the risk of cyber-attacks. AI-aided IoT can help to detect and prevent cyber-attacks by analyzing network traffic, identifying anomalies, and responding to threats in real time (Khang et al., 2022b).

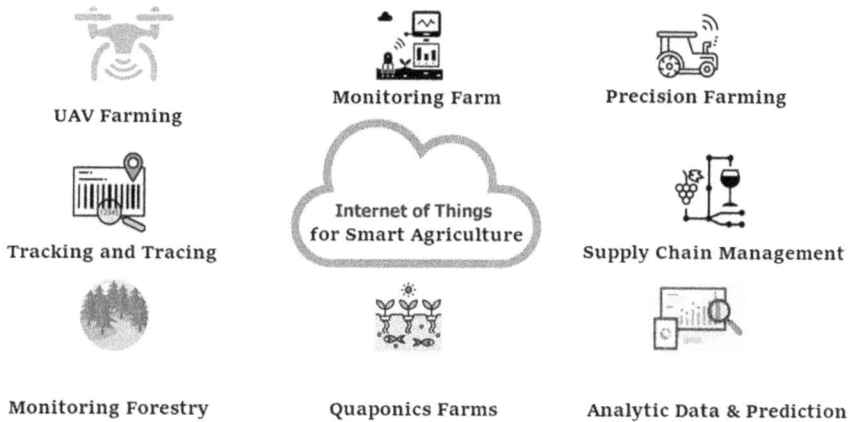

FIGURE 2.2 IoT-based agriculture

- **Energy Management**: Energy management is another potential application of AI-aided IoT. By optimizing energy consumption in factories and other production facilities, businesses can save on energy costs and reduce their carbon footprint. With AI-aided IoT, businesses can analyze sensor data to optimize energy consumption and reduce energy waste (Siguo et al., 2022).
- **Autonomous Robotics**: By integrating AI and IoT, businesses can create autonomous robots that can perform a range of tasks, from picking and packing in warehouses to assembly in factories. Autonomous robotics can help businesses to be more efficient, cost-effective, and flexible (Tzafestas, 2018).
- **Industrial Automation**: Industrial automation is another application of the IoT and AI blend. IoT sensors and AI algorithms can be used to automate various industrial processes, from assembly lines to warehouse management. This can help to increase efficiency, reduce waste, and improve safety (Figure 2.3). AI algorithms can also optimize the assembly line process to reduce production time and waste (Hajimahmud et al., 2022).
- **Smart Asset Tracking**: IoT sensors can be used to track assets, such as tools and equipment, throughout a production facility. AI algorithms can analyze this data to optimize asset utilization, reduce loss, and increase efficiency.
- **Customer Experience**: AI-aided IoT can be used to create personalized experiences for customers, from smart fitting rooms in retail stores to personalized recommendations in online shopping.
- **Retail**: AI and IoT can be used to personalize the shopping experience for customers. IoT sensors can collect data on customer behavior, such as their browsing history and purchase history. AI algorithms can then use this data to recommend products that are relevant to the customer's interests and preferences.

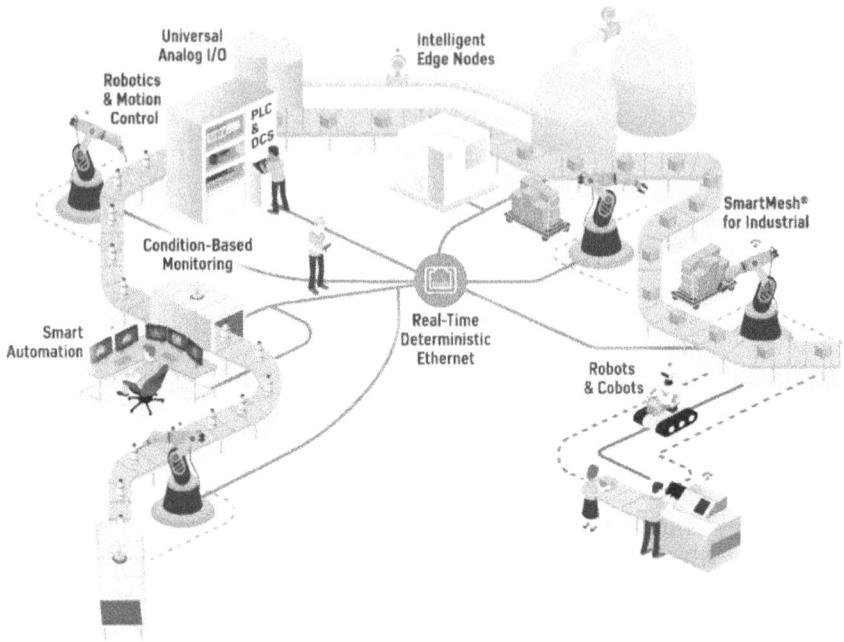

FIGURE 2.3   Industrial automation

- **Logistics**: AI and IoT can be used to optimize logistics operations. IoT sensors can be used to track the location and condition of goods in transit, providing real-time information on the status of shipments. AI algorithms can then use this data to optimize delivery routes, reduce delivery times, and improve customer satisfaction.
- **Manufacturing**: AI and IoT can be used to optimize the manufacturing process. IoT sensors can collect data on machine performance, and AI algorithms can analyze this data to predict when maintenance is required, reducing downtime and increasing efficiency. AI-powered robots can also be used to automate repetitive tasks, reducing the need for human labor and increasing production capacity.
- **Healthcare**: AI and IoT can be used to monitor patient health remotely. Wearable devices and other IoT sensors can collect data on patient vitals, which can be analyzed by AI algorithms to detect potential health issues before they become a problem. This can help to progress patient outcomes and lower healthcare costs (Khang et al., 2023d).
- **Smart Library**: An AI-based IoT smart library is a library system that integrates artificial intelligence (AI) and internet of things (IoT) technologies to enhance the efficiency, accessibility, and overall user experience of a library. Some key features and components that could be included in such a system are: smart bookshelves, intelligent book recommendations, smart

lighting and climate control, digital signage, and many more (Zheng and Zhang, 2022).

- **Smart Homes**: AI and IoT can be used to create smart homes that are more energy efficient and secure. IoT sensors can be used to monitor energy usage and detect unusual activity, such as the presence of intruders. AI algorithms can then use this data to adjust energy settings and send real-time alerts to homeowners (Zheng and Zhang, 2022).

CONCLUSION

People will wear intelligent devices in the future, ingest intelligent capsules that assess the effects of medications on the body, live in intelligent dwellings, and so on. Though it may sound like science fiction, this is the topic of all current research. Everything will be internet-connected and intelligent. All scientific disciplines will work together to produce something extremely valuable. A smart cyber revolution will occur. The question of whether or not we are moving toward creative destruction is still up for debate.

In the future, the internet of things (IoT) and artificial intelligence (AI) will be crucial. The increasing need for technologies in businesses and governments is fueled by a variety of sources. Technologists, engineers, and scientists have already begun implementing it at various levels. For example, machines are increasingly able to perform fewer regular activities, and this transition is taking place at a time when many workers are already having difficulty. However, with the right policies, we can have automation without a severe unemployment problem. Human cleverness eventually alters the function of productive activity. The promotion of educational possibilities will result in more skilled labor and the re- and up-skilling of existing workers. We will be compelled to reevaluate the implications of such automation on the circumstances of human life as we continue to introduce AI models into the real world.

Despite the numerous advantages of these systems, they come with some inherent risks, such as privacy invasion, and codifying serious repercussions will result from any hardware or software malfunctions or flaws. Even a power outage can be very inconvenient. In order to track the location of this AI-enabled IoT at all times, we might need to add another AI system on top of it. One day, we could require such democratic systems that can restrain themselves from acting irrationally. Technology will continue to dominate our life and become our only source of everything. Whatever the case, humans should still rule over all artificial intelligence. Only then we will be able to control this revolution without becoming its slaves.

REFERENCES

Bhambri, P., S. Rani, G. Gupta, and A. Khang. 2022. *Cloud and Fog Computing Platforms for Internet of Things.* CRC Press. https://doi.org/ 10.1201/9781003213888.

Bi, Siguo, Cong Wang, Jilong Zhang, Wutao Huang, Bochun Wu, Yi Gong, and Wei Ni. 2022. "A survey on artificial intelligence aided internet-of-things technologies in emerging smart libraries." *Sensors.* https://doi.org/10.3390/s22082991; https://www.mdpi.com/journal/sensors.

Bourechak, Amira, Mohamed NadjibKouahla, OuardaZedadra, Antonio Guerrieri, Hamid Seridi, and Giancarlo Fortino. 2023. "At the confluence of artificial intelligence and edge computing in IoT-based applications: A review and new perspectives." *Sensors*, vol. 23, no. 3, p. 1639. https://doi.org/10.3390/s23031639.

Chander, Bhanu, Souvik Pal, Debashis De, and Rajkumar Buyya. 2022. *Artificial Intelligence-based Internet of Things for Industry 5.0*. Springer. https://link.springer.com/chapter/10.1007/978-3-030-87059-1_1.

Evans, D. 2011. *The Internet of Things: How the Next Evolution of the Internet Is Changing Everything*. Cisco Internet Business Solutions Group: Cisco. https://media.telefonicatech.com/telefonicatech/uploads/2021/1/126528_Internet_of_Things_IoT_IBSG_0411FINAL.pdf.

Ghosh, Ashish, Debasrita Chakraborty, and Anwesha Law. 2018. "Artificial intelligence in Internet of things." ISSN 2468. *CAAI Transactions on Intelligence Technology*, vol. 3, no. 4, pp. 208–218. https://10.1049/trit.2018.1008.

Hajimahmud, V. A., A. Khang, V. Hahanov, E. Litvinova, S. Chumachenko, and V. A. Alyar. 2022. "Autonomous robots for smart city: Closer to augmented humanity." *AI-Centric Smart City Ecosystems: Technologies, Design and Implementation* (1st ed.), 7 (12). CRC Press. https://doi.org/10.1201/9781003252542-7.

Hahanov, V., A. Khang, E. Litvinova, S. Chumachenko, V. A. Hajimahmud, and V. A. Alyar. 2022. "The key assistant of smart city - sensors and tools." *AI-Centric Smart City Ecosystems: Technologies, Design and Implementation* (1st ed.), 17 (10). CRC Press. https://doi.org/10.1201/9781003252542-17.

Khang, A., G. Rana, R. K. Tailor, and V. A. Hajimahmud. (Eds.). 2023. *Data-Centric AI Solutions and Emerging Technologies in the Healthcare Ecosystem*. CRC Press. https://doi.org/10.1201/9781003356189.

Khang, A., S. Rani, and A. K. Sivaraman. 2022b. *AI-Centric Smart City Ecosystems: Technologies, Design and Implementation* (1st ed.). CRC Press. https://doi.org/10.1201/9781003252542.

Khang, A., S. K. Gupta, V. Shah, and A. Misra. (Eds.). 2023a. *AI-aided IoT Technologies and Applications in the Smart Business and Production*. CRC Press. https://doi.org/10.1201/9781003392224.

Khang, A., S. K. Gupta, V. A. Hajimahmud, J. Babasaheb, and G. Morris. 2023b. *AI-Centric Modelling and Analytics: Concepts, Designs, Technologies, and Applications* (1st ed.). CRC Press. https://doi.org/10.1201/9781003400110.

Khang, A., V. Abdullayev, V. Hahanov, and V. Shah. 2024a. *Advanced IoT Technologies and Applications in the Industry 4.0 Digital Economy* (1st ed.). CRC Press. https://doi.org/10.1201/9781003434269.

Khang, A., V. Hahanov, E. Litvinova, S. Chumachenko, H. V. A. Triwiyanto, R. N. Ali, A. V. Alyar, and P. T. N. Anh. 2023d. "The analytics of hospitality of hospitals in healthcare ecosystem." *Data-Centric AI Solutions and Emerging Technologies in the Healthcare Ecosystem* (P 4). CRC Press. https://doi.org/10.1201/9781003356189-4.

Khang, A., V. Hahanov, G. L. Abbas, and V. A. Hajimahmud. 2022c. "Cyber-physical-social system and incident management." *AI-Centric Smart City Ecosystems: Technologies, Design and Implementation* (1st ed.), 2 (15). CRC Press. https://doi.org/10.1201/9781003252542-2

Linthicum, D. "App nirvana: When the internet of things meets the API economy." https://techbeacon.com/app-dev-testing/app-nirvana-wheninternet-things-meets-api-economy. Accessed 15 November 2019.

Lu, Y., and L. D. Xu. 2019. "Internet of Things (IoT) cybersecurity research: A review of current research topics." *IEEE Internet of Things Journal*, vol. 6, no. 2, pp. 2103–2105. https://ieeexplore.ieee.org/abstract/document/8462745/.

Mathad, K., and A. Khang. 2023. "Hospital 4.0: Capitalization of health and healthcare in industry 4.0 economy." *Data-Centric AI Solutions and Emerging Technologies in the Healthcare Ecosystem* (P 14). (1st ed.). CRC Press. https://doi.org/10.1201 /9781003356189-19.

MohdJavaid, A. Haleem, Ravi Pratap Singh, and Rajiv Suman. 2022. "Artificial intelligence applications for industry 4.0: A literature-based study." *Journal of Industrial Integration and Management*, vol. 7, no. 01, pp. 83–111. https://www.worldscientific .com/doi/abs/10.1142/S2424862221300040.

Pande, P., and A. R. Padwalkar. 2014, February. "Internet of things - A future of internet: A survey." *International Journal of Advance Research in Computer Science and Management Studies*, vol. 2, no. 2, pp. 354–361. https://link.springer.com/chapter/10 .1007/978-981-16-6210-2_2.

Qazi, S., B. A. Khawaja, and Q. U. Farooq. 2022. "IoT-equipped and AI-enabled next generation smart agriculture: A critical review, current challenges and future trends." *IEEE Access*. https://ieeexplore.ieee.org/abstract/document/9716089/.

Rani, S., M. Chauhan, A. Kataria, and A. Khang (Eds.). 2021. "IoT equipped intelligent distributed framework for smart healthcare systems." *Networking and Internet Architecture*, vol. 2, p. 30. https://doi.org/10.48550/arXiv.2110.04997.

Rouse, M. "What is IoT (Internet of Things) and how does it work? IoT Agenda, TechTarget." Accessed 11 February 2020. http://www.internetofthingsagenda.techtarget.com/defnition/Internet-of-Things-IoT.

Syed, Abbas Shah, Daniel Sierra-Sosa, Anup Kumar, and Adel Elmaghraby. 2021. "IoT in smart cities: A survey of technologies, practices and challenges." *Smart Cities*, vol. 4, no. 2, pp. 429–475. https://doi.org/10.3390/smartcities4020024.

Tedeschi, Stefano, Christos Emmanouilidis, Jörn Mehnen, and Rajkumar Roy. 2019. "A design approach to IoT endpoint security for production machinery monitoring." *Sensors*, vol. 19, no. 10, p. 2355. https://doi.org/10.3390/s19102355.

Tzafestas, Spyros G. 2018. "Synergy of IoT and AI in modern society: The robotics and automation case." *Robotics& Automation Engineering Journal*, vol. 3, no. 5: RAEJ. MS.ID.5555621.. https://pdfs.semanticscholar.org/eb1a/6afb510603dea5185249342 9c03f860919df.pdf.

Xu, Yongjun, Xin Liu, and Xin Cao. 2021. "Artificial intelligence: A powerful paradigm for scientific research." *The Innovation*, vol. 2, no. 4, p. 100179. ISSN: 2666-6758. https:// www.cell.com/article/S2666-6758(21)00104-1/fulltext.

Zheng, Zhi-Xian, and Fuquan Zhang. 2022. "Image real-time detection using LSE-Yolo neural network in artificial intelligence-based internet of things for smart cities and smart homes." *Wireless Communications and Mobile Computing, Hindawi*. Article ID 2608798, 8 pages. https://doi.org/10.1155/2022/2608798.

3 Exploring the Transformative Power of Internet of Things (IoT) Technologies in the Age of Industry 4.0
Unleashing the Potential in the Digital Economy

Bhupendra Kumar, Vikas Sharma,
Tarun Kumar Vashishth, Rajneesh Panwar,
Kewal Krishan Sharma, and Sachin Chaudhary

INTRODUCTION

Industry 4.0, also known as the Fourth Industrial Revolution, is characterized by the integration of advanced digital technologies such as the internet of things (IoT), artificial intelligence (AI), and robotics in the industrial sector. This integration is transforming the way industries operate, leading to new business models, increased productivity, and improved efficiency (Figure 3.1).

IoT technologies are a key driver of Industry 4.0, enabling the creation of smart factories and connected supply chains. This chapter explores the transformative impact of IoT technologies in the era of Industry 4.0 digital economy. It analyses the key drivers, challenges, and opportunities for unlocking the potential of IoT technologies in the industrial sector. The chapter also examines the role of IoT in smart manufacturing, supply chain management, and predictive maintenance. Furthermore, it discusses the potential impact of IoT technologies on workforce and job creation (Khang & Rani et al., 2023).

They are transforming industries, creating new business models, and unlocking untold potential. This exploration aims to delve into the transformative impact of IoT technologies in the era of Industry 4.0 and the digital economy. We will examine how IoT technologies are changing the face of industries, such as manufacturing,

DOI: 10.1201/9781003434269-3

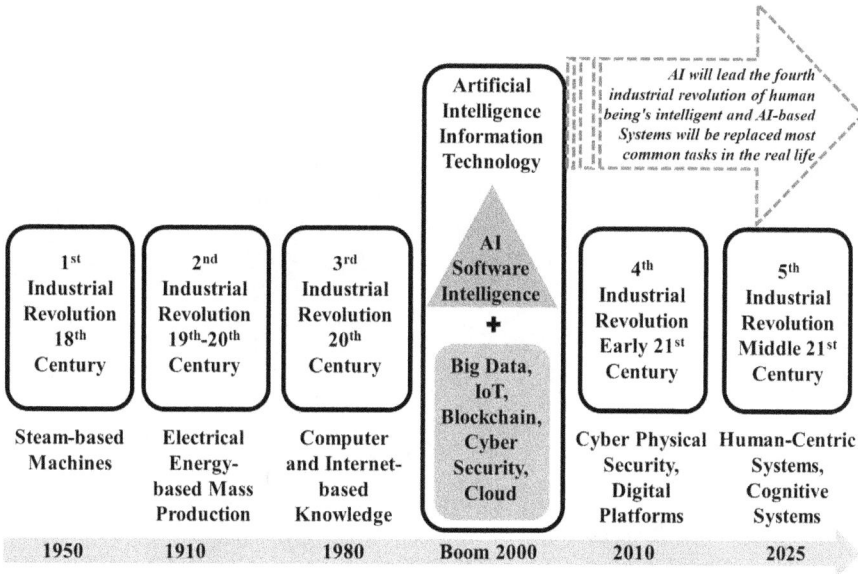

FIGURE 3.1 Evolution of humankind and industrialization (Khang, 2021)

healthcare, transportation, and agriculture (Khang et al., 2023). We will also explore the challenges that come with implementing IoT technologies and the strategies that can be employed to overcome them.

In addition, we will examine the role of data analytics, artificial intelligence, and machine learning in the IoT ecosystem. We will explore how these technologies are enhancing the capabilities of IoT devices and enabling the development of new, innovative applications. This exploration seeks to provide a comprehensive overview of the IoT landscape in the era of Industry 4.0 and the digital economy. We hope that it will inspire new ideas, spark discussions, and ultimately help organizations unlock the full potential of IoT technologies.

The IoT technologies have emerged as a key enabler of Industry 4.0, the Fourth Industrial Revolution that is reshaping the manufacturing and production landscape. With the rise of the digital economy, IoT technologies are transforming the way businesses operate, creating new opportunities for innovation, efficiency, and competitiveness. IoT technologies refer to a network of connected devices and sensors that can collect and exchange data in real time, enabling businesses to monitor, analyze, and optimize their operations (Figure 3.2).

Some of the key IoT technologies used in Industry 4.0 include:

- **Sensors and Actuators**: These are devices that can detect changes in the physical environment and respond to them. They are used to monitor equipment, production processes, and supply chain operations, enabling real-time monitoring and control.

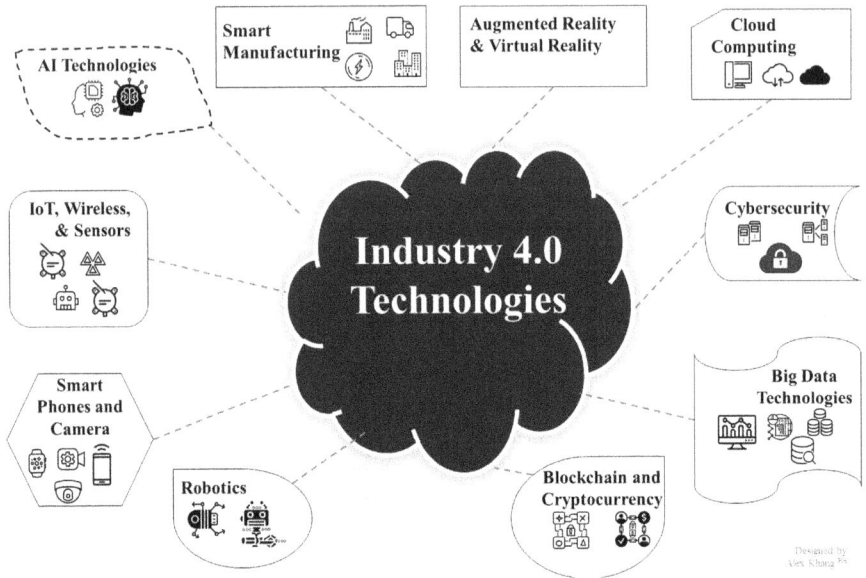

FIGURE 3.2   What is Industry 4.0 (Khang, 2021)?

- **Cloud Computing**: Cloud computing provides a scalable and flexible platform for storing and analyzing large amount of data generated by IoT devices. It enables businesses to process and analyze data in real time, providing insights into operational performance and identifying opportunities for optimization.
- **Artificial Intelligence (AI)**: AI technologies such as machine learning and deep learning are used to analyze large amount of data generated by IoT devices. They enable businesses to identify patterns and trends in the data, predict future outcomes, and optimize operations.
- **Blockchain**: Blockchain technology provides a secure and transparent platform for tracking and verifying transactions in the supply chain. It enables businesses to ensure the authenticity of products and reduce the risk of fraud (Khang & Chowdhury et al., 2022).
- **Edge Computing**: Edge computing involves processing data at the edge of the network, closer to the source of the data. It enables businesses to reduce latency and improve the speed of data processing, making it ideal for real-time applications.

KEY DRIVERS OF IOT IN INDUSTRY 4.0

The key drivers of IoT in Industry 4.0, as explored in this chapter, can be summarized as follows:

- **Connectivity and Data Exchange**: The advancement of IoT technologies enables seamless connectivity and data exchange between various devices and systems. This interconnectedness allows for real-time communication and collaboration, fostering a highly connected ecosystem in Industry 4.0.
- **Automation and Efficiency**: IoT technologies play a pivotal role in automating processes and increasing operational efficiency. By integrating sensors, actuators, and IoT devices into industrial systems, tasks can be automated, reducing manual intervention, minimizing errors, and optimizing resource allocation.
- **Data-Driven Decision-Making**: IoT generates vast amount of data from sensors and devices, providing valuable insights for data-driven decision-making. Real-time data analytics and predictive algorithms enable businesses to make informed decisions, optimize operations, and enhance overall performance.
- **Predictive Maintenance and Asset Management**: IoT facilitates predictive maintenance by continuously monitoring and analyzing equipment data. This proactive approach helps identify potential faults or maintenance needs, allowing for timely interventions and reducing unplanned downtime. Additionally, IoT enables efficient asset management through real-time tracking, monitoring, and utilization of resources.
- **Supply Chain Optimization**: IoT technologies enable end-to-end visibility and transparency across the supply chain. By integrating sensors and IoT devices into logistics operations, businesses can track inventory, monitor shipment conditions, optimize routes, and streamline logistics processes, leading to improved efficiency and customer satisfaction.
- **Enhanced Customer Experience**: IoT technologies allow for personalized and customized customer experiences. By collecting and analyzing data from IoT devices, businesses can gain insights into customer behavior, preferences, and usage patterns, enabling them to deliver tailored products, services, and experiences.
- **Innovation and New Business Models**: IoT opens up new opportunities for innovation and the development of new business models. By leveraging IoT technologies, businesses can create innovative products, services, and business processes that drive growth and competitive advantage in the digital economy.
- **Collaboration and Integration**: IoT promotes collaboration and integration among various stakeholders in the industrial ecosystem. By connecting different devices, systems, and partners, businesses can achieve seamless integration, foster collaboration, and enable cross-functional data sharing, leading to enhanced productivity and innovation.

These key drivers of IoT in Industry 4.0 collectively contribute to the transformative power of IoT technologies in unleashing their potential in the digital economy. By harnessing these drivers, businesses can leverage the benefits of IoT to drive efficiency, innovation, and growth in the era of Industry 4.0.

LITERATURE REVIEW

This literature review aims to explore the transformative power of internet of things (IoT) technologies within the context of Industry 4.0 and its impact on the digital economy. With the rapid advancement of IoT technologies, there has been a paradigm shift in various industries, enabling unprecedented connectivity, data exchange, and automation. This review synthesizes existing research and provides insights into the potential of IoT technologies to revolutionize industrial processes, enhance productivity, optimize resource allocation, and drive innovation. Additionally, it discusses the challenges and considerations associated with adopting IoT technologies in the digital economy. The findings of this literature review contribute to a deeper understanding of the transformative impact of IoT technologies in the era of Industry 4.0. Some of the studies are as follows:

- Kali et al.'s (2024) article examines the applications, investments, and challenges of the IoT for enterprises. The authors argue that the IoT can enable businesses to create new revenue streams, increase efficiency, and improve customer experiences, but they also highlight the challenges of data security, privacy, and interoperability.
- Khang and Misra et al. (2023) review the technologies, applications, and challenges of the Industrial IoT (IIoT), which is a subset of the IoT focused on industrial processes and systems. The authors discuss the potential benefits of IIoT, including predictive maintenance, real-time monitoring, and remote control, as well as the challenges related to data integration, standardization, and security.
- Rani et al. (2022) examine the development and impact of Industry 4.0, which refers to the Fourth Industrial Revolution characterized by the integration of digital technologies into manufacturing processes. The authors argue that Industry 4.0 can lead to increased productivity, flexibility, and customization, but they also acknowledge the risks of job displacement, skills gaps, and ethical issues.
- Shah & Jani (2024) investigate the impact of Industry 4.0 on the automotive industry, which is a major adopter of digital technologies. The authors analyze the changes in business models, supply chains, and customer expectations brought by Industry 4.0 and provide recommendations for companies to remain competitive.
- Debosree et al. (2023) propose an information framework for creating a smart city through IoT, which involves the integration of various sensors, devices, and platforms to enable data-driven decision-making and improve urban services. The authors discuss the challenges of data management, privacy, and interoperability, and provide examples of smart city applications in energy, transportation, and healthcare.
- Zanella et al. (2014) focus on the IoT for smart cities and provide a comprehensive review of the technologies, architectures, and challenges involved in building smart city systems. The authors emphasize the importance of

interoperability, scalability, and sustainability in designing IoT solutions for smart cities.

- Nassereddine et al. (2024) present a survey of the IoT, covering its history, definitions, architectures, and applications. The authors highlight the potential benefits of the IoT in various domains, such as healthcare, agriculture, and logistics, but also raise concerns about the privacy, security, and standardization of IoT devices and networks.

- Schuh et al. (2017) introduce the Industry 4.0 Maturity Index (I4.0MI), which is a framework for assessing the digital maturity of companies and guiding their digital transformation strategies. The authors argue that the I4.0MI can help companies identify their strengths and weaknesses in terms of digital capabilities, and provide a roadmap for improving their competitiveness.

- Khang & Hajimahmud et al. (2024) discuss the concept of the digital twin, which is a virtual representation of a physical product or process that can be used for simulation, optimization, and monitoring. The authors argue that the digital twin is a key enabler of Industry 4.0, as it allows companies to leverage real-time data and analytics for decision-making and innovation.

- Rana et al. (2021) provide an overview of Industry 4.0, including its definition, components, and government initiatives. The authors discuss the role of digital technologies, such as artificial intelligence, cloud computing, and IoT, in enabling Industry 4.0 and highlight the policy implications and challenges of its implementation.

This literature review provides a comprehensive examination of the transformative power of IoT technologies within the context of Industry 4.0 and their potential impact on the digital economy. By synthesizing existing research, it highlights the various benefits and challenges associated with the adoption of IoT technologies in industrial processes, supply chains, and customer experiences. The findings of this review contribute to a deeper understanding of how IoT technologies can unleash the potential in the digital economy, ultimately paving the way for innovation and economic growth in the era of Industry 4.0.

APPLICATIONS OF IOT TECHNOLOGIES IN INDUSTRY 4.0

Internet of things (IoT) technologies have become increasingly important in the era of Industry 4.0 digital economy applications. With the proliferation of connected devices and the growth of data analytics and artificial intelligence, IoT technologies are being used to improve the efficiency, productivity, and safety of industrial operations. Some of the key applications are as follows:

- **Predictive Maintenance**: IoT sensors can be used to monitor machines and equipment, collect data on their performance, and identify potential problems before they occur. This enables proactive maintenance and reduces downtime.

- **Supply Chain Optimization**: IoT technologies can be used to track products, materials, and shipments throughout the supply chain, providing real-time visibility and enabling better inventory management, logistics, and transportation.
- **Smart Manufacturing**: IoT sensors can be used to monitor and optimize production processes, improving efficiency and reducing waste. This can also enable better quality control and real-time adjustments to production schedules.
- **Energy Management**: IoT technologies can be used to monitor and control energy usage in industrial facilities, reducing costs and improving sustainability.
- **Safety and Security**: IoT sensors can be used to monitor and detect potential safety hazards, such as leaks, fires, and equipment malfunctions. This can improve workers' safety and prevent accidents.

CHALLENGES OF IOT TECHNOLOGIES IN INDUSTRY 4.0

While IoT technologies offer numerous benefits in Industry 4.0, there are also several challenges that need to be addressed to fully realize their potential (Figure 3.3).

Figure 3.3 depicts some of the key challenges that include the following:

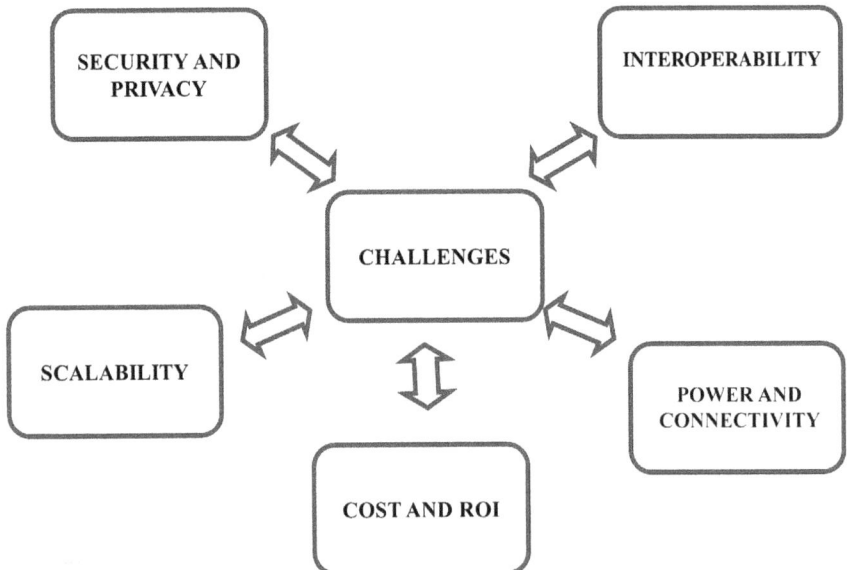

FIGURE 3.3 Key challenges of Industry 4.0

- **Security and Privacy**: IoT devices can be vulnerable to cyber-attacks, which can compromise sensitive data and potentially cause physical harm. Ensuring the security and privacy of IoT networks is essential for their widespread adoption.
- **Interoperability**: IoT devices and systems may use different protocols and standards, making it difficult to integrate them into existing systems. Interoperability standards need to be established to ensure seamless integration of IoT technologies.
- **Scalability**: As the number of connected devices and data generated by IoT systems grows, managing and processing this data can become a significant challenge. Scalable infrastructure and data management solutions are needed to handle this data effectively.
- **Power and Connectivity**: Many IoT devices are powered by batteries and may have limited connectivity, which can affect their reliability and functionality. Advances in battery technology and connectivity solutions are needed to overcome these limitations.
- **Cost and Return on Investment (ROI)**: Implementing IoT technologies can be costly, and the ROI may not always be clear. Organizations need to carefully assess the costs and benefits of IoT solutions to ensure their viability.

Overall, addressing these challenges is essential for the widespread adoption and success of IoT technologies in Industry 4.0. As the technology continues to evolve, it is important to stay up to date with emerging trends and best practices to overcome these challenges and fully realize the potential of IoT in industry.

PROPOSED METHODOLOGY

IoT technologies are rapidly transforming the industrial landscape by enabling the seamless integration of physical devices and systems with digital technologies. In the era of Industry 4.0, where digital transformation is driving the economy, the importance of IoT technologies cannot be overstated. However, the deployment and integration of these technologies can be complex and challenging (Figure 3.4).

The proposed methodology involves the following steps:

- **Business Process Analysis**: The first step in deploying IoT technologies in the era of Industry 4.0 is to analyze the existing business processes and identify areas where IoT technologies can be used to optimize and streamline these processes. This analysis should consider the potential benefits of IoT technologies such as increased efficiency, reduced costs, and improved quality.
- **Technology Selection**: The next step is to identify the IoT technologies that are best suited for the identified business processes. This involves evaluating the available technologies based on their capabilities, compatibility with existing systems, and cost-effectiveness.

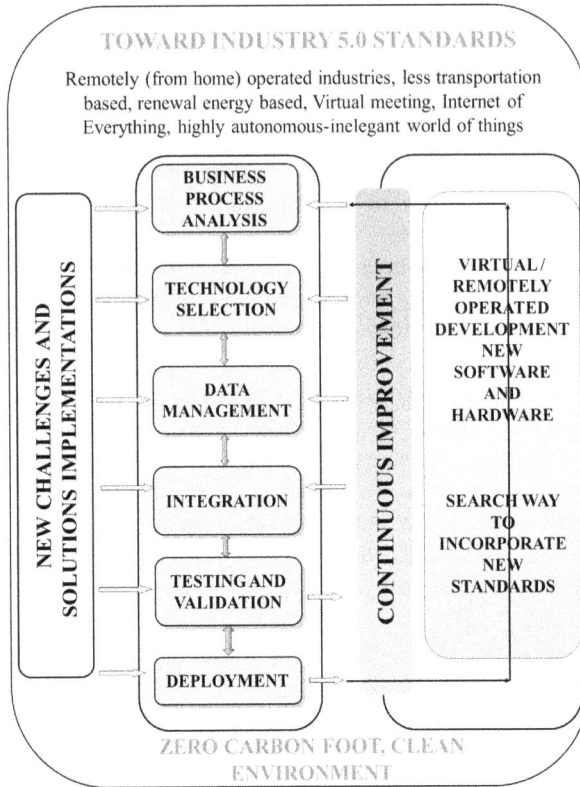

FIGURE 3.4   Integration of various factors in Industry 4.0

- **Data Management**: IoT technologies generate vast amount of data, and it is crucial to have a robust data management system in place to handle this data effectively. This involves identifying the data that is essential for decision-making, establishing data governance policies, and implementing data security measures.
- **Integration**: IoT technologies must be integrated with existing systems to ensure seamless communication between devices and systems. This involves identifying the interfaces required for integration, developing APIs, and establishing communication protocols.
- **Testing and Validation**: Before deploying IoT technologies in a live environment, it is essential to conduct rigorous testing to ensure that the system works as intended. This involves testing the system under different scenarios and validating the results against expected outcomes.
- **Deployment**: Once the system has been tested and validated, it can be deployed in a live environment. This involves training the users and establishing procedures for monitoring and maintaining the system.

- **Continuous Improvement**: IoT technologies are constantly evolving, and it is essential to continuously evaluate and improve the system to ensure that it remains effective and efficient. This involves monitoring the system's performance, identifying areas for improvement, and implementing changes as required.

The proposed methodology provides a structured approach to deploying IoT technologies in the era of Industry 4.0 digital economy. By following this methodology, organizations can maximize the benefits of IoT technologies while minimizing the risks and challenges associated with their deployment.

CASE STUDIES

Case studies of this chapter include the following examples:

CASE STUDY 1: IoT IMPLEMENTATION IN MANUFACTURING

- **Scope**: Smart Factory Optimization
- **Company**: Zolta Manufacturing
- **Overview**: Zolta Manufacturing, a leading automotive parts manufacturer, implemented IoT technologies to optimize their factory operations and improve overall productivity.

Implementation Details

- **Sensor Integration**: IoT sensors were deployed across the manufacturing floor to monitor machine performance, energy consumption, and environmental conditions such as temperature and humidity.
- **Real-time Data Collection**: The sensors collected real-time data on machine status, production output, and energy usage. The data was transmitted to a central IoT platform for analysis and visualization.
- **Predictive Maintenance**: By analyzing the collected data, machine performance patterns were identified, enabling predictive maintenance. Maintenance alerts were automatically generated when any machine showed signs of potential failure, allowing timely interventions to prevent breakdowns and reduce downtime.
- **Machine Optimization**: With real-time monitoring, Zolta Manufacturing gained insights into production inefficiencies and bottlenecks. The data-driven analysis enabled them to optimize machine settings, adjust production schedules, and improve overall equipment effectiveness (OEE).
- **Inventory Management**: IoT sensors were integrated into the warehouse to track inventory levels and provide real-time visibility into stock availability. This allowed for accurate demand forecasting, streamlined material flow, and reduced inventory holding costs.

Results and Benefits

- **Increased Productivity**: IoT-enabled machine optimization and predictive maintenance reduced unplanned downtime, resulting in increased production output and improved overall productivity by 15%.
- **Cost Savings**: By optimizing machine settings and adjusting production schedules based on real-time data, Zolta Manufacturing achieved energy savings of 20% and reduced maintenance costs by 25%.
- **Enhanced Quality Control**: Real-time monitoring of machine performance and environmental conditions ensured consistent product quality, reducing defects and customer complaints.
- **Efficient Inventory Management**: IoT-enabled inventory tracking and demand forecasting reduced stock-outs, minimized excess inventory, and improved overall inventory management efficiency by 30%.
- **Data-driven Decision-Making**: The availability of real-time data and analytics empowered Zolta Manufacturing to make data-driven decisions, enabling continuous improvement and proactive problem-solving (Khang et al., 2024).

This case study highlights how IoT implementation in manufacturing can transform traditional factories into smart, connected environments. By leveraging IoT technologies, Zolta Manufacturing optimized their production processes, improved quality control, reduced costs, and enhanced overall operational efficiency. The success of their IoT implementation serves as a testament to the transformative power of IoT technologies in the manufacturing sector within the context of Industry 4.0.

CASE STUDY 2: IoT APPLICATIONS IN TRANSPORTATION AND LOGISTICS

- **Scope**: Fleet Management Optimization
- **Company**: Creation Logistics
- **Overview**: Creation Logistics, a global logistics company, implemented IoT technologies to optimize their fleet management operations and improve efficiency in transportation and logistics.

Implementation Details

Vehicle Tracking and Monitoring: IoT-enabled tracking devices were installed in the company's fleet of trucks, providing real-time visibility into the location, speed, and route of each vehicle. The devices also collected data on fuel consumption, engine performance, and driver behavior (Khanh & Khang et al., 2021).

- **Route Optimization**: By integrating GPS data with real-time traffic information, Creation Logistics used IoT analytics to optimize route planning and delivery schedules. This reduced transportation time, minimized fuel consumption, and improved on-time delivery performance.
- **IoT Tracking Devices**: IoT tracking devices were deployed on shipping containers, vehicles, and packages to provide real-time location tracking throughout the supply chain.

- **Real-time Monitoring**: The tracking devices transmitted location data, temperature, humidity, and other relevant information to a centralized IoT platform. This enabled real-time monitoring and visibility of shipments at each stage of the logistics process.
- **Route Optimization**: Using IoT data, Creation Logistics optimized transportation routes by analyzing traffic conditions, weather forecasts, and historical data. The IoT platform provided real-time recommendations for route adjustments, allowing for efficient delivery schedules and reduced transit times.
- **Condition Monitoring**: IoT sensors were integrated into containers and packages to monitor environmental conditions such as temperature, humidity, and shock/vibration. Any deviations from predefined thresholds triggered alerts, enabling proactive actions to mitigate risks of damage or spoilage.

Results and Benefits

- **Customer Notifications**: Real-time shipment status updates and delivery notifications were sent to customers via mobile apps or web portals. This enhanced transparency and improved customer satisfaction by providing accurate and timely information.
- **Enhanced Operational Efficiency**: Route optimization based on real-time IoT data reduced transportation costs and improved delivery efficiency, resulting in a 20% reduction in transit times and a 15% decrease in fuel consumption.
- **Enhanced Quality Control**: Real-time monitoring of machine performance and environmental conditions ensured consistent product quality, reducing defects and customer complaints.
- **Risk Mitigation**: IoT-enabled condition monitoring ensured the integrity of sensitive shipments, reducing the risk of product damage or spoilage. This led to improved product quality and customer satisfaction.
- **Proactive Issue Resolution**: Early detection of potential delivery delays or deviations from predefined conditions allowed Creation Logistics to take proactive measures, such as alternative transportation arrangements or rerouting, minimizing service disruptions.
- **Customer Satisfaction**: Real-time shipment status updates and delivery notifications improved customer communication and satisfaction. Customers had access to accurate and timely information, reducing inquiries and enhancing their overall experience.

This case study exemplifies the transformative impact of IoT technologies in transportation and logistics. By leveraging IoT for supply chain visibility, route optimization, condition monitoring, and customer engagement, Creation Logistics was able to streamline operations, reduce costs, mitigate risks, and enhance customer satisfaction. The successful implementation of IoT in transportation and logistics showcases the immense potential of IoT technologies in optimizing and transforming the industry in the era of Industry 4.0.

CASE STUDY 3: IoT-ENABLED SMART CITIES

- **Scope**: Sustainable Energy Management
- **Organization**: New Delhi Smart City
- **Overview**: New Delhi Smart City implemented IoT technologies to optimize energy management, reduce carbon footprint, and enhance sustainability within the city.

Implementation Details

- **Smart Grid and Energy Monitoring**: IoT devices were integrated into the city's power grid to monitor energy consumption, distribution, and demand in real time. Smart meters were installed in residential and commercial buildings to track individual energy usage.
- **Energy Efficiency Solutions**: IoT-enabled systems were implemented to control and optimize energy usage in public lighting, buildings, and other infrastructure. Automated lighting systems adjusted brightness based on natural light levels, occupancy, and traffic patterns.
- **Renewable Energy Integration**: IoT technologies facilitated the integration of renewable energy sources, such as solar panels and wind turbines, into the power grid. Real-time monitoring and analytics enabled efficient utilization of renewable energy and seamless grid integration.
- **Demand Response Programs**: IoT devices allowed for demand response programs, where energy consumers could adjust their usage based on peak load periods or renewable energy availability. Consumers received real-time pricing and incentives to encourage energy conservation during high-demand periods.

Results and Benefits

- **Energy Efficiency**: IoT-enabled energy management systems resulted in significant energy savings by optimizing consumption and reducing waste. Smart lighting systems reduced energy usage by 30%, while smart metering helped consumers reduce their energy consumption by 15%.
- **Renewable Energy Utilization**: Integration of renewable energy sources through IoT technologies increased the share of renewable energy in the city's power supply, reducing reliance on fossil fuels and lowering carbon emissions by 20%.
- **Cost Savings**: Efficient energy management and demand response programs led to cost savings for both the city and consumers. Residents and businesses benefited from reduced energy bills, while the city reduced infrastructure maintenance costs and improved overall budget management.
- **Sustainability and Environmental Impact**: New Delhi Smart City IoT initiatives contributed to the city's sustainability goals by promoting renewable energy, reducing carbon emissions, and enhancing environmental stewardship. The city became a role model for sustainable urban living.

- **Citizen Engagement**: Real-time energy consumption data and personalized feedback empowered citizens to actively participate in energy conservation efforts. Consumers had access to energy usage insights and were motivated to adopt more sustainable behaviors.

This case study highlights how IoT technologies can transform cities into smart, sustainable, and energy-efficient environments. Through the integration of IoT devices, New Delhi Smart City optimized energy management, reduced carbon footprint, and engaged citizens in sustainable practices. The successful implementation of IoT-enabled smart city initiatives demonstrates the potential of IoT technologies to create livable, eco-friendly urban spaces in the era of Industry 4.0. These case studies illustrate the transformative impact of IoT technologies in different sectors, showcasing how they revolutionize industrial processes, optimize resource utilization, and improve overall efficiency. They demonstrate the potential of IoT to drive innovation, cost savings, and enhanced customer experiences, highlighting the transformative power of IoT technologies in the age of Industry 4.0.

CONCLUSION

In conclusion, this review has delved into the transformative power of IoT technologies in the context of Industry 4.0 and their potential to unleash a significant impact on the digital economy. Through the synthesis of existing research, several key findings have emerged.

First, IoT technologies enable enhanced connectivity and communication, paving the way for seamless data exchange and collaboration across various industrial sectors. This connectivity drives automation and smart manufacturing, leading to improved productivity, streamlined processes, and optimized resource allocation. The availability of real-time data and advanced analytics empowers data-driven decision-making, further enhancing operational efficiency and enabling predictive maintenance. Moreover, IoT technologies play a pivotal role in supply chain optimization by facilitating end-to-end visibility, inventory management, and demand forecasting. The ability to track and monitor assets in real time enables more efficient logistics and inventory control, reducing costs and enhancing customer satisfaction. Additionally, IoT enables personalized customer experiences by capturing and analyzing user data, allowing businesses to tailor their offerings and deliver targeted services.

Despite the transformative potential, challenges must be addressed. Security and privacy concerns associated with IoT implementation need to be effectively mitigated through robust encryption, authentication, and data protection mechanisms. Interoperability and standardization are critical for seamless integration and collaboration among IoT devices and platforms. Adequate data management and governance frameworks are essential to ensure data integrity, quality, and compliance with regulations. Equipping the workforce with the necessary skills and knowledge to leverage IoT technologies is crucial to fully capitalize on their potential (Khang, 2024).

Looking ahead, the future scope lies in advancing IoT security, integrating AI into IoT systems, exploring edge and fog computing, promoting standardization and interoperability, addressing sustainability and ethical implications, and understanding the economic and societal impact of IoT technologies. Additionally, regulatory frameworks, human–machine collaboration, and cross-domain applications of IoT warrant further exploration. By embracing these future directions, we can unlock the full transformative potential of IoT technologies in the age of Industry 4.0. This will pave the way for innovation, economic growth, and a digital economy that is connected, efficient, and responsive to the evolving needs of businesses and society at large.

The potential of IoT technologies in the era of Industry 4.0 digital economy is vast and transformative. The seamless integration of physical devices and systems with digital technologies is enabling organizations to optimize their business processes, improve efficiency, reduce costs, and enhance the overall quality of their products and services. However, realizing the full potential of IoT technologies requires a structured approach that considers the unique needs of each organization. By following a methodology that includes business process analysis, technology selection, data management, integration, testing, deployment, and continuous improvement, organizations can unlock the transformative impact of IoT technologies and gain a competitive advantage in the digital economy. As we move forward, it is clear that IoT technologies will continue to play a critical role in shaping the future of Industry 4.0 and driving the digital economy forward.

FUTURE SCOPE

With the help of IoT technologies, cities will become more intelligent, efficient, and sustainable. Sensors will monitor everything from traffic flow to air quality, and data will be used to optimize city services and infrastructure. There is a wide scope of research in this area. IoT technologies will drive significant changes in manufacturing from predictive maintenance to supply chain management. Smart factories will be able to optimize production processes and reduce waste, while real-time data analysis will enable manufacturers to make informed decisions about production schedules and inventory management.

1. **Advancing IoT Security**: As IoT technologies continue to expand, ensuring robust security measures will be crucial. Future research can focus on developing innovative security solutions, encryption techniques, and authentication methods to protect IoT devices and networks from cyber threats.
2. **Integration of Artificial Intelligence (AI)**: Combining IoT with AI can amplify the transformative potential of both technologies. Future studies can explore the integration of AI algorithms and machine learning techniques to enhance data analytics, predictive maintenance, and autonomous decision-making in IoT-enabled systems.
3. **Edge Computing and Fog Computing**: With the increasing volume of data generated by IoT devices, there is a need for efficient data processing

and analysis at the network edge. Future research can investigate the implementation of edge computing and fog computing architectures to reduce latency, enhance real-time processing capabilities, and alleviate bandwidth constraints in IoT systems.

4. **Standardization and Interoperability**: The lack of standardization and interoperability among IoT devices and platforms hinders seamless integration and data exchange. Future efforts can focus on developing industry-wide standards and protocols that promote interoperability, compatibility, and ease of integration across various IoT ecosystems.

5. **Sustainability and Green IoT**: With growing environmental concerns, there is a need to explore sustainable practices in IoT deployments. Future research can focus on developing energy-efficient IoT devices, optimizing resource consumption, and leveraging IoT technologies for environmental monitoring and conservation.

6. **Ethical and Social Implications**: As IoT technologies become more pervasive, it is essential to address the ethical and social implications they raise. Future studies can examine the impact of IoT on privacy, data ownership, social equity, and the potential risks of algorithmic bias to ensure responsible and inclusive IoT deployments.

7. **Business Models and Economic Implications**: The transformative power of IoT technologies can reshape business models and economic systems. Future research can explore the evolving landscape of IoT-enabled business models, the economic impact of IoT in different sectors, and the potential for new revenue streams and market opportunities.

8. **Regulatory Frameworks and Policies**: As IoT technologies continue to evolve, there is a need for comprehensive regulatory frameworks and policies to address privacy, security, data governance, and liability issues. Future studies can analyze the legal and regulatory aspects of IoT deployments and propose frameworks that strike a balance between innovation and protection.

9. **Human–Machine Collaboration**: The role of humans in the IoT ecosystem is evolving, moving from passive users to active participants. Future research can explore the dynamics of human–machine collaboration, the impact on workforce skills, and the design of user-friendly interfaces to enhance human interaction with IoT devices and systems.

10. **Cross-Domain Applications**: IoT technologies have the potential to create synergies across various domains, including healthcare, agriculture, transportation, and energy. Future studies can investigate the cross-domain applications of IoT, exploring how IoT can solve complex societal challenges and improve quality of life (Khang & Hajimahmud et al., 2024).

By focusing on these future research directions, we can further unlock the transformative power of IoT technologies in the age of Industry 4.0, unleashing their full potential to drive innovation, efficiency, and economic growth in the digital economy (Khang & Gupta et al., 2023).

REFERENCES

Debosree, G., & Khang, A. (2023), "Quantum Computing Internet of Things: Unlocking New Frontiers for Connectivity, Security, and Efficiency in Data Processing," *Applications and Principles of Quantum Computing.* IGI Global Press. https://doi.org/10.4018/979 -8-3693-1168-4-ch017

Khang, A. (2021), "Material4Studies," *Material of Computer Science, Artificial Intelligence, Data Science, IoT, Blockchain, Cloud, Metaverse, Cybersecurity for Studies.* https:// www.researchgate.net/publication/370156102_Material4Studies

Khang, A. (Ed.) (2024), *AI-Oriented Competency Framework for Talent Management in the Digital Economy: Models, Technologies, Applications, and Implementation.* CRC Press. https://doi.org/10.1201/9781003440901

Khang, A., Hajimahmud, V. A., Hrybiuk, Olena, & Shukla, A. K. (2024), *Computer Vision and AI-Integrated IoT Technologies in Medical Ecosystem* (1st ed.). CRC Press. https:// doi.org/10.1201/9781003429609

Khang, A., Gujrati, R., Uygun, H., Tailor, R. K., & Gaur, S. S. (2024), *Data-driven Modelling and Predictive Analytics in Business and Finance.* CRC Press. https://doi.org/10.1201 /9781032600628

Khang, A., Gupta, S. K., Rani, S., & Karras, D. A. (Eds.) (2023), *Smart, Cities: IoT Technologies, Big Data Solutions, Cloud Platforms, and Cybersecurity Techniques.* CRC Press. https://doi.org/10.1201/9781003376064

Khang, A., Misra, A., Abdullayev, V. A., & Litvinova, E. (Eds.) (2024), *Machine Vision and Industrial Robotics in Manufacturing: Approaches, Technologies, and Applications.* CRC Press. https://doi.org/10.1201/ 9781003438137

Khang, A., Misra, A., Gupta, S. K., & Shah, V. (Eds.) (2023), *AI-aided IoT Technologies and Applications in the Smart Business and Production.* CRC Press. https://doi.org/10.1201 /9781003392224

Khang, A., Ragimova, N. A., Hajimahmud, V. A., & Alyar, V. A. (2022), "Advanced technologies and data management in the smart healthcare system," *AI-Centric Smart City Ecosystems: Technologies, Design and Implementation* (1st ed.), 16(10). CRC Press. https://doi.org/10.1201/9781003252542-16

Khang, A., Rani, S., Gujrati, R., Uygun, H., & Gupta, S. K. (Eds.) (2023), *Designing Workforce Management Systems for Industry 4.0: Data-Centric and AI-Enabled Approaches.* CRC Press. https://doi.org/10.1201/9781003357070

Khanh, H. H., & Khang, A. (2021), "The role of artificial intelligence in blockchain applications," *Reinventing Manufacturing and Business Processes through Artificial Intelligence*, 2, 20–40. CRC Press. https://doi.org/10.1201/9781003145011-2

Nassereddine, M., & Khang, A. (2024), "Applications of Internet of Things (IoT) in smart cities," *Advanced IoT Technologies and Applications in the Industry 4.0 Digital Economy* (1st ed.). CRC Press. https://doi.org/10.1201/9781003434269-6

Rana, G., Khang, A., Sharma, R., Goel, A. K., & Dubey, A. K. (Eds.) (2021), *Reinventing Manufacturing and Business Processes through Artificial Intelligence.* CRC Press. https://doi.org/10.1201/9781003145011

Rani, S., Bhambri, P., Kataria, A., & Khang, A. (2022), "Smart city ecosystem: Concept, sustainability, design principles and technologies," *AI-Centric Smart City Ecosystems: Technologies, Design and Implementation* (1st ed.), 1(20). CRC Press. https://doi.org/10 .1201/9781003252542-1

Rath, K. C., Khang, A., & Roy, D. (2024), "The role of Internet of Things (IoT) technology in industry 4.0," *Advanced IoT Technologies and Applications in the Industry 4.0 Digital Economy* (1st ed.). CRC Press. https://doi.org/10.1201/9781003434269-1

Schuh, G., Anderl, R., Gausemeier, J., & Schmitt, R. (2017), *Industrie 4.0 Maturity Index – Managing the Digital Transformation of Companies.* Fraunhofer Institute for Industrial Engineering IAO. https://epub.fir.de/frontdoor/index/index/docId/1112

Shah, V., Jani, S., & Khang, A. (Eds.) (2023), "Automotive IoT: Accelerating the automobile industry's long-term sustainability in smart city development strategy," *Smart Cities: IoT Technologies, Big Data Solutions, Cloud Platforms, and Cybersecurity Techniques.* CRC Press. https://doi.org/10.1201/9781003376064-9

Shah, V., Jani, S., & Khang, A. (2024), "Automotive IoT: Accelerating the Automobile Industry's Long-Term Sustainability in Smart City Development Strategy," *Advanced IoT Technologies and Applications in the Industry 4.0 Digital Economy* (1 Ed.). CRC Press. https://doi.org/10.1201/9781003434269-16

Zanella, A., Bui, N., Castellani, A., Vangelista, L., & Zorzi, M. (2014), "Internet of Things for smart cities," *IEEE Internet of Things Journal*, 1(1), 22–32. https://ieeexplore.ieee.org/abstract/document/6740844/

4 Internet of Things (IoT) Technology
A Critical Component of Industry 4.0

Rachit Jain, Poonam Tiwari, Pooja Jain,
Ramasamy R., Javeed S. Imran, and Udhayanan S.

INTRODUCTION

Industry 4.0, often known as the fourth industrial revolution, is a paradigm shift in manufacturing and other industries that is being driven by advances in digital technologies. It includes physical and digital system integration, automation, data interchange, and the application of emerging technologies such as artificial intelligence, the internet of things (IoT), cloud computing, and big data analytics. This section defines Industry 4.0 in detail and emphasizes its importance in altering industries around the world.

Industry 4.0 denotes a new era of industrial change marked by the convergence of cyber-physical systems, IoT, and advanced digital technologies. It reflects the marriage of physical production processes and digital technologies, allowing for more automation, real-time data processing, and intelligent decision-making. The term "Industry 4.0" was coined in Germany in 2011 as part of a high-tech strategy aimed at fostering the digital transformation of manufacturing (Khang et al., 2024).

KEY PRINCIPLES OF INDUSTRY 4.0

Several essential principles distinguish Industry 4.0 from earlier industrial revolutions, which are as follows:

- **Interconnectivity**: Through IoT technology, Industry 4.0 emphasizes the seamless connectivity and integration of machines, devices, and systems. It provides data and information transmission across the entire value chain, allowing for real-time monitoring, control, and optimization.
- **Information Transparency**: For informed decision-making, Industry 4.0 relies on the availability of comprehensive, real-time data. Relevant

DOI: 10.1201/9781003434269-4

information is accessible at all stages of the manufacturing process, thanks to IoT-enabled sensors and data analytics, enhancing transparency and efficiency.

- **Technical Support and Decentralized Decision-Making**: Industry 4.0 employs modern technologies such as artificial intelligence (AI), machine learning, and robotics to assist human workers and enable decentralized decision-making. Smart systems and robots may conduct activities autonomously, analyze data, and deliver actionable insights, thereby empowering workers and increasing total productivity.
- **Flexibility and Customization**: Industry 4.0 allows for flexible and adaptable manufacturing processes to fit the needs of unique customers. Manufacturers can quickly adjust to changes in demand, optimize resource allocation, and achieve mass customization with IoT-enabled systems.

SIGNIFICANCE OF INDUSTRY 4.0

Industry 4.0 holds significant implications for industries and economies worldwide. Its importance stems from various factors:

- **Increased Productivity and Efficiency**: Industry 4.0 technologies offer streamlined and automated operations, lowering operational costs and increasing productivity. IoT connectivity and real-time data analytics enable predictive maintenance, resource optimization, and effective supply chain management.
- **Improved Quality and Customization**: Industry 4.0 enables greater quality control, real-time monitoring, and faster identification of faults or abnormalities through the integration of modern sensors and data analytics. Product customization and personalization become more practicable, allowing for the fulfillment of particular consumer preferences and enhancing customer satisfaction.
- **Time-to-Market and Accelerated Innovation**: Industry 4.0 promotes innovation by enabling quick prototyping, iterative design, and agile manufacturing. Companies can use digital technology and IoT connections to reduce product development cycles, bring innovations to market more quickly, and adapt quickly to changing market demands.
- **Job Transformation and Workforce Skill Development**: Industry 4.0 changes the nature of work, causing existing job positions to change and new skill requirements to emerge. To capitalize on the potential of Industry 4.0 technology, workers must adapt and improve digital literacy, data analysis, and problem-solving abilities.
- **Global Competitiveness and Sustainable Growth**: For sectors and economies that embrace digital transformation, Industry 4.0 provides considerable competitive advantages. It has the potential to drive long-term growth by optimizing resource use, lowering environmental impact, and enabling more (Agarwal et al., 2021; Anbalagan and Moreno-Garcia, 2021; Asemani et al., 2019; Asghar et al., 2015; Crnjac et al., 2017).

INTRODUCTION TO IOT TECHNOLOGY AND ITS ROLE IN INDUSTRY 4.0

The internet of things (IoT) has emerged as a critical technology in the Industry 4.0 era, revolutionizing industries and propelling digital transformation. The internet of things (IoT) is a network of interconnected physical objects, sensors, and actuators that are embedded with technology for data collection and sharing. This section introduces IoT technology, its major components, and its critical role in realizing the vision of Industry 4.0 (Forcina and Falcone, 2021; Georgios et al., 2019).

REQUIREMENTS FOR A SYSTEM

What Is IoT Technology?

The internet of things (IoT) technology enables the seamless connectivity and communication of objects, machines, and systems, facilitating data interchange and intelligent decision-making. It is a network of physical things that have sensors, software, and connectivity to gather and transmit data via the internet. These objects might range from simple consumer electronics to sophisticated industrial machines (Ghosh et al., 2022).

Key Components of IoT Technology

- **Sensors and Actuators**: Sensors, which collect real-time data from the physical world, are at the heart of IoT technology. They can measure temperature, pressure, humidity, and location, among other things. Actuators, on the other hand, enable physical responses or actions depending on sensor data.
- **Connectivity**: To communicate data to other devices or centralized systems, IoT devices rely on numerous connectivity choices. Wi-Fi, Bluetooth, Zigbee, and cellular networks such as 4G and 5G are examples of wireless protocols. Devices may communicate and share data in real-time, thanks to connectivity.
- **Data Analytics**: IoT generates huge amount of data, which must be processed and analyzed in order to yield relevant insights. Data analytics techniques like machine learning and artificial intelligence are critical in extracting actionable information from acquired data. This allows for more informed decision-making and process optimization.
- **Cloud Computation**: Cloud systems offer scalable and secure IoT data storage and computation capabilities. They allow for the seamless integration and management of data from various sources, allowing for real-time data processing, storage, and analysis (Figure 4.1). Cloud computing also makes it possible to install IoT apps and services (Ghosh et al., 2022; Hassoun et al., 2022).

ROLE OF IOT IN INDUSTRY 4.0

IoT technology plays a critical role in realizing the vision of Industry 4.0, transforming industries and revolutionizing the way businesses operate as shown in Figure 4.1.

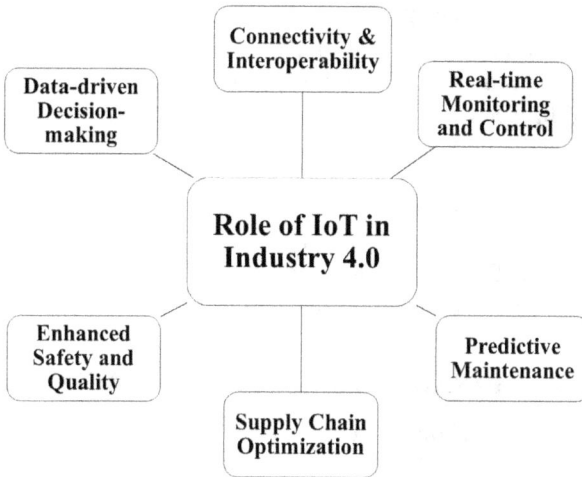

FIGURE 4.1   Role of IoT in Industry 4.0 (Khang, 2021)

The integration of IoT within Industry 4.0 enables the following (Hayat et al., 2023; Khan and Javaid, 2023).

Connectivity and Interoperability

The internet of things enables the seamless connection and integration of machines, devices, and systems across the whole value chain. This allows for the real-time flow of data and information, encouraging interoperability among diverse components and enabling end-to-end visibility.

Real-time Monitoring and Control

The internet of things (IoT) allows for real-time monitoring and control of physical assets, processes, and activities. Data is collected, analyzed, and used to optimize manufacturing processes, improve efficiency, and detect probable breakdowns or bottlenecks using sensors and connectivity.

Predictive Maintenance

IoT sensors and data analytics enable predictive maintenance, in which machines and equipment can be continuously monitored for symptoms of malfunction or wear. This allows for preventive maintenance, which reduces downtime and improves overall equipment efficacy (Kali et al., 2024).

Supply Chain Optimization

End-to-end visibility and traceability in the supply chain are made possible by IoT technology. It enables real-time product tracking, inventory management, and demand forecasting. This aids in logistical optimization, cost reduction, and inventory management efficiency.

Enhanced Safety and Quality

The internet of things allows for real-time monitoring of environmental conditions, worker safety, and product quality. Sensors can identify anomalies, monitor safety parameters, and trigger warnings or automatic responses to assure product quality and compliance with safety regulations.

Data-driven Decision-making

The internet of things creates massive volumes of data that may be analyzed to yield meaningful insights. This enables data-driven decision-making, process optimization, resource allocation optimization, and finding opportunities for improvement.

INDUSTRY 4.0: A PARADIGM SHIFT IN MANUFACTURING

AI's potential and applicability in a wide range of human life sectors are growing all the time, despite the fact that it is still classified as an emerging or middle-stage technology (Figure 4.2). Global technology corporations are investing in developing computers that understand and behave more like humans using techniques such as machine learning and neural networks (Kumar et al., 2019).

OVERVIEW OF THE CONCEPT OF INDUSTRY 4.0

Industry 4.0 is a game-changing concept that will transform the manufacturing landscape. It is propelled by the integration of modern digital technologies, automation, and data interchange, which results in the development of smart factories and connected production systems. The confluence of physical and digital systems in

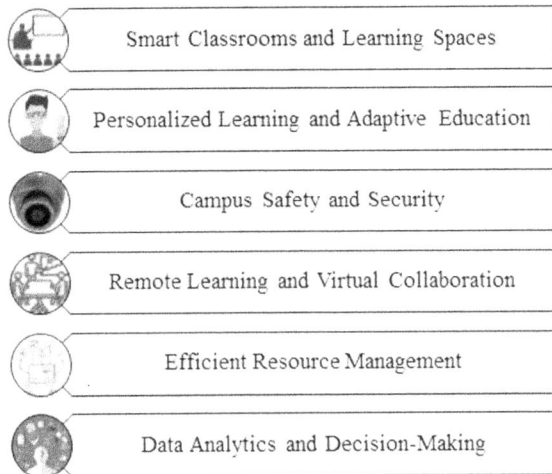

FIGURE 4.2   Industry 4.0 flexible and customized production processes

Industry 4.0 enables real-time data analysis, process optimization, and intelligent decision-making.

KEY PRINCIPLES AND OBJECTIVES OF INDUSTRY 4.0

Industry 4.0 is guided by several key principles that define its objectives and implementation (Figure 4.3):

- **Interconnectivity**: Through IoT technology, Industry 4.0 emphasizes the seamless connectivity and integration of machines, devices, and systems. Because of this interconnection, data and information can be sent across the whole value chain, allowing for real-time monitoring, control, and optimization.
- **Transparency of Information**: To enable informed decision-making, Industry 4.0 relies on the availability of comprehensive, real-time data. Data is captured at various phases of the manufacturing process using IoT-enabled sensors, enabling visibility and transparency. This enables accurate process and resource analysis and optimization.
- **Technical Support and Decentralized Decision-Making**: Industry 4.0 employs modern technology such as artificial intelligence, machine learning, and robotics to assist human workers and enable decentralized decision-making. Smart systems and robots may conduct activities autonomously, analyze data, and deliver actionable insights, thereby empowering workers and increasing total productivity (Hajimahmud et al., 2022).

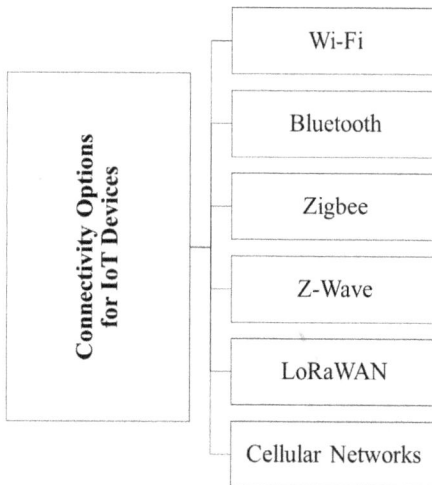

FIGURE 4.3   Key components of IoT technology (Khang, 2021)

INTEGRATION OF PHYSICAL AND DIGITAL SYSTEMS

The seamless integration of physical and digital systems is an important feature of Industry 4.0. Manufacturing processes have traditionally worked in silos, with physical operations segregated from digital information systems. Industry 4.0, on the other hand, breaks down these barriers by combining the physical and digital realms to create a connected and intelligent industrial ecosystem. Machinery, equipment, robots, and other tangible assets employed in manufacturing are examples of physical systems. Sensors, actuators, and IoT-enabled devices supplement these systems, capturing real-time data from the physical world. This information is subsequently transferred and integrated into digital systems, allowing for real-time monitoring, analysis, and control.

Data analytics, cloud computing, artificial intelligence, and machine learning are all examples of digital systems. These systems collect and interpret data from physical systems, offering vital insights for decision-making, optimization, and predictive capacities. In Industry 4.0, the integration of physical and digital systems allows for a bidirectional flow of information. Physical systems' real-time data is used to optimize operations and make informed decisions, while digital systems supply instructions and feedback to physical systems, enabling adaptive and autonomous control.

EMERGENCE OF CYBER-PHYSICAL SYSTEMS (CPS)

Cyber-physical systems (CPS) are at the foundation of Industry 4.0. The integration of computing algorithms, communication networks, and physical processes is referred to as CPS. It brings together the cyber world of software, data, and communication with the physical world of machinery, equipment, and manufacturing processes.

In Industry 4.0, CPS allow for the seamless integration of physical and digital systems, resulting in a synergistic relationship. Sensors and actuators are used in these systems to monitor physical processes, collect data, and interact with other systems. Real-time data analysis and processing enable intelligent decision-making and adaptive control.

Predictive maintenance, real-time optimization, self-monitoring, and self-optimization are all possible with CPS.

UNDERSTANDING IOT TECHNOLOGY

DEFINITION AND KEY COMPONENTS OF IOT

The internet of things (IoT) is a massive network of interconnected physical devices, objects, and systems with integrated sensors, actuators, and communication. These internet of things devices may gather, exchange, and process data, allowing them to interact with the real environment and communicate with one another via the internet. The core concept of IoT is to enable seamless communication and data sharing among diverse devices and systems, resulting in smart and intelligent functions (Sadeeq et al., 2021).

KEY COMPONENTS OF IoT

- **Sensors**: Sensors are critical components of IoT devices. They are in charge of acquiring data from their surroundings by detecting changes in temperature, humidity, pressure, motion, light, and a variety of other characteristics. These sensors turn physical signals into electrical signals that the IoT system can handle and analyze.
- **Actuators**: Actuators are mechanisms in IoT devices that perform physical actions depending on sensor data or instructions from the central IoT system. They have the ability to control and manipulate physical things or systems, such as opening and closing valves, turning on and off switches, and altering motor speeds.
- **Connectivity**: To transfer data to other devices or central systems, IoT devices rely on a variety of connectivity options. Wi-Fi, Bluetooth, Zigbee, Z-Wave, LoRaWAN, cellular networks (e.g., 4G and 5G), and satellite communication are common IoT networking technologies. The connectivity option chosen is determined by characteristics such as range, power consumption, data rate, and network coverage.
- **Data Processing and Storage**: Because of their compact size and power constraints, IoT devices may have limited processing capabilities. As a result, data processing and storage are frequently performed in cloud or edge computing servers. Cloud computing offers scalable and centralized storage and computational capacity, but edge computing brings data processing closer to the source, lowering latency and bandwidth use.

ROLE OF SENSORS AND ACTUATORS IN IoT SYSTEMS

Sensors and actuators are the primary enablers of IoT functionality, working in tandem to create an interconnected ecosystem. Here's how they play their roles:

- **Sensing and Data Collection**: Sensors are in charge of collecting data from the physical environment. A temperature sensor in a smart thermostat, for example, measures room temperature, whereas a motion sensor in a security camera detects movement in a monitored area. The data is collected and delivered to the central IoT system for analysis and action.
- **Data Transmission**: After collecting data, IoT devices employ connectivity options such as Wi-Fi or Bluetooth to send the data to the central IoT platform or other devices in the network. Depending on the application and use case, this data transmission can be in real-time or at periodic intervals.
- **Data Processing and Analysis**: To draw relevant insights, the central IoT system processes and analyses the incoming data. Machine learning algorithms, for example, are frequently employed in data analytics to find patterns, trends, and anomalies that can be used for predictive maintenance, optimization, and decision-making.

- **Actuation and Control**: Based on sensor data analysis, the IoT system delivers instructions to actuators to conduct physical actions. An IoT-enabled irrigation system, for example, may utilize actuators to manage the flow of water depending on soil moisture levels, whereas a smart lighting system may use actuators to alter brightness based on ambient lighting conditions (Figure 4.3).

CONNECTIVITY OPTIONS FOR IoT DEVICES

- **Wi-Fi**: Wi-Fi is a common networking option for IoT devices that require high data rates and have access to LANs or the internet. Wi-Fi enables dependable and secure wireless connectivity, making it ideal for smart homes, workplaces, and industrial environments. However, when compared to alternative low-power options, Wi-Fi may appear power-hungry.
- **Bluetooth**: Bluetooth is a short-range wireless technology that is frequently used to link IoT devices to smartphones, tablets, and other nearby devices. It is appropriate for low-power communication and short-range connection applications such as wearable devices, health monitoring, and home automation systems.
- **Zigbee**: Zigbee is a low-power, low-data-rate wireless communication protocol intended for applications involving a large number of devices attached in a mesh network. It is commonly used in home automation systems, building automation systems, and industrial monitoring systems. Zigbee offers dependable connectivity with low power usage and long-range transmission.
- **Z-Wave**: Z-Wave is a wireless communication protocol comparable to Zigbee that was developed primarily for home automation applications. It operates at sub-GHz frequencies, allowing for more range and penetration through walls than Wi-Fi or Bluetooth. Z-Wave is well known for its low power consumption and interoperability among devices from various manufacturers.
- **LoRaWAN**: Long Range Wide Area Network (LoRaWAN) is a wireless communication system that allows IoT devices to connect across long distances. It is appropriate for smart cities, agricultural, and industrial monitoring applications when devices are dispersed across a large region. LoRaWAN has low data speeds but high long-range coverage and minimal power consumption (Khang et al., 2023a).
- **Cellular Networks**: IoT devices can connect to cellular networks like 4G LTE and 5G. Cellular connectivity provides broad coverage and dependable connectivity, making it ideal for IoT applications that demand mobility or are located in remote locations. However, cellular communication is often more expensive and consumes more power than alternative options.

INTRODUCTION TO IoT PROTOCOLS AND STANDARDS

Interoperability, safe connection, and fast data transmission between IoT devices and systems are all dependent on IoT protocols and standards. Here's an overview of several popular IoT protocols and standards:

- **Message Queuing Telemetry Transport (MQTT)**: MQTT is a lightweight publish-subscribe communications protocol intended for internet of things (IoT) applications. It is designed for low-bandwidth, unstable networks and runs on top of TCP/IP. Because MQTT is asynchronous and event-driven, it is appropriate for resource-constrained devices and applications requiring real-time data exchange.
- **Confined Application Protocol (CoAP)**: CoAP is a lightweight protocol designed for IoT devices with limited resources and confined networks, such as low-power wireless networks. CoAP has a request–response model similar to HTTP, allowing IoT devices to interface with web services using RESTful methods.
- **Hypertext Transfer Protocol (HTTP)**: HTTP is a popular protocol for transferring data between web clients and servers. HTTP can be used for device management, data retrieval, and control in the context of IoT. Through application programming interfaces (APIs), it allows IoT devices to reveal their capabilities and connect with web-based applications and services.
- **Open Platform Communications Unified Architecture (OPC UA)**: OPC UA is a machine-to-machine communication protocol that is widely used in industrial automation and internet of things (IoT) applications. It provides a secure and dependable communication platform for device and system interoperability across manufacturers. OPC UA supports a wide range of data models and allows for standardized data interchange in industrial environments.

Thread is an IPv6-based networking protocol targeted for smart home and IoT applications. Its primary goal is to establish a dependable and secure mesh network of devices within a home or business. Thread is excellent for home automation and IoT scenarios due to its low-power operation, simple network setup, and secure communication. These protocols and standards serve as the cornerstone for IoT communication, providing device and system compatibility, security, and fast data flow across varied IoT contexts (Paul et al., 2021; Peralta et al., 2017).

IOT IN INDUSTRY 4.0: KEY COMPONENTS AND ARCHITECTURE

ROLE OF IOT IN THE DIGITAL TRANSFORMATION OF INDUSTRIES

The internet of things (IoT) is a critical enabler of digital transformation in a variety of businesses. It transforms old business models, improves operational efficiency, and fuels innovation. Here are some critical roles of IoT in industry digital transformation:

- **Connectivity and Data Collection**: The internet of things connects physical items, devices, and systems, allowing for the seamless collection of data from multiple sources. Sensors and actuators on IoT devices collect real-time data on characteristics such as temperature, pressure, location, and machine condition. This information provides useful insights into operations, performance, and customer behavior.

- **Real-time Monitoring and Control**: The internet of things allows for real-time monitoring of assets, processes, and environments. Industries can remotely monitor equipment, track inventories, optimize supply chains, and automate operations using IoT-enabled sensors and actuators. Real-time data helps detect anomalies, predict maintenance needs, and optimize resource allocation, leading to increased productivity and reduced downtime.
- **Predictive Analytics and Maintenance**: Industries can apply predictive analytics and maintenance strategies by exploiting IoT data. IoT devices generate data, which is then analyzed by machine learning algorithms to predict equipment failures or performance degradation. This enables preventive maintenance while avoiding costly unplanned downtime, optimizing asset utilization and increasing asset lifespan.
- **Supply Chain Optimization**: The internet of things (IoT) plays a critical role in revolutionizing supply chains. It provides complete visibility and traceability of commodities, assets, and inventory across the supply chain. IoT devices and sensors give real-time information about the location, condition, and status of items, allowing for more efficient inventory management, demand forecasting, and logistical optimization.
- **Enhanced Customer Experience**: IoT enables enterprises to provide personalized and engaging consumer experiences. Smart gadgets and connected devices enable personalized product suggestions, remote device control, and real-time customer support. IoT data assists in understanding client preferences, behavior, and usage patterns, resulting in customized offerings, increased customer happiness, and loyalty.
- **Automation and Efficiency**: The internet of things enables process automation, decreasing manual intervention and increasing operational efficiency. Devices and systems that are linked can communicate with one another, triggering actions based on predetermined rules or machine learning algorithms. This automation eliminates errors, streamlines workflows, and frees up human resources to focus on more important duties.
- **Safety and Sustainability**: IoT helps to improve industrial safety and sustainability. IoT sensors can monitor environmental conditions, detect dangerous circumstances, and verify that safety rules are followed. Furthermore, IoT allows for energy control, resource optimization, and waste reduction, hence aiding sustainability programs and decreasing environmental impact.
- **Business Model Innovation**: The internet of things (IoT) creates new prospects for business model innovation. Based on IoT data and insights, industries can shift from selling items to providing value-added services. Equipment makers, for example, can transition from selling machinery to delivering equipment-as-a-service, with consumers paying depending on usage and performance.

Overall, IoT plays a revolutionary role in allowing connectivity, data-driven decision-making, automation, and customer-centricity in the digital transformation of sectors. It drives operational excellence, improves efficiency, and fosters innovation, positioning industries for success in the digital era (Suleiman et al., 2022; Ungurean et al., 2014).

OVERVIEW OF THE IoT ARCHITECTURE IN INDUSTRY 4.0

The IoT architecture in Industry 4.0 consists of interconnected components that work together to enable seamless data flow and communication. These components are as follows:

- **Sensors and Actuators**: Sensors capture data from the physical environment and play an important role in IoT-enabled devices. They capture data on characteristics like temperature, pressure, humidity, vibration, and other variables. Actuators, on the other hand, receive IoT system commands and conduct physical actions based on the data received. Sensors and actuators work together to collect and act on data in real-time.
- **Connectivity**: To transfer data to other devices or central systems, IoT devices rely on numerous connectivity methods. This includes traditional connections like Ethernet as well as wireless technologies like Wi-Fi, Bluetooth, Zigbee, and cellular networks. IoT devices can communicate with one another, as well as with cloud platforms or edge computing infrastructure, thanks to connectivity.
- **Edge Computing**: Edge computing is a distributed computing concept that moves computational power and storage closer to the data source. Edge computing in the context of IoT helps process and analyze data near devices or at the network edge, decreasing latency, bandwidth utilization, and reliance on cloud services. In IoT systems, edge computing improves real-time decision-making, data filtering, and local control.
- **Cloud Computing**: Cloud computing is a critical component of Industry 4.0's IoT architecture. Cloud platforms offer scalable storage, computing power, and powerful data analytics. Data can be sent from IoT devices to the cloud for long-term storage, historical analysis, and processing. Cloud-based services provide advanced analytics, machine learning algorithms, and predictive models to generate insights and aid decision-making (Xu et al., 2018).

SENSOR NETWORKS AND DATA COLLECTION IN IoT-ENABLED SYSTEMS

Sensor networks serve as the foundation for IoT-enabled systems in Industry 4.0. These networks are made up of interconnected sensors that are placed in industrial settings to collect data from machines, equipment, and production processes (Hajimahmud et al., 2022). Sensors detect changes in a variety of parameters and send the collected data to a central IoT system for further analysis and action. The following steps are involved in the acquisition of sensor data:

- **Data Collection**: Sensors gather information by monitoring physical characteristics like temperature, pressure, vibration, or machine condition. They convert physical signals into electrical signals that the IoT system can process.
- **Data Transmission**: The collected data is forwarded to the core IoT system or edge computing infrastructure via wired or wireless communication channels. Depending on the application's needs, this data transmission might be done in real-time or on a regular basis.
- **Data Preprocessing**: When data arrives at the IoT system, it is preprocessed to remove noise, outliers, and useless information. To prepare the data for future analysis, preprocessing procedures such as data filtering, normalization, or feature extraction may be used (Nassereddine and Khang, 2024).
- **Data Storage and Analysis**: For long-term storage and analysis, processed data is kept in cloud-based or edge-based storage systems. Statistical analysis, machine learning algorithms, and predictive modeling are examples of data analytics techniques that can be used to obtain actionable insights and help decision-making (Yu et al., 2018).

CLOUD COMPUTING AND EDGE COMPUTING IN IoT

Cloud and edge computing are two complementary technologies that play critical roles in the internet of things (IoT) ecosystem. They allow for efficient data processing, storage, and analysis, but in separate parts of the IoT architecture.

Cloud computing is the distribution of on-demand computer resources such as storage, processing power, and services via the internet. Cloud computing, in the context of IoT, provides a centralized infrastructure for storing, managing, and analyzing the massive amounts of data created by IoT devices. The following are the most important characteristics of cloud computing in IoT:

- **Scalability**: Cloud computing provides virtually limitless scalability, allowing IoT applications to handle vast volumes of data while meeting ever-increasing demands. The cloud infrastructure can dynamically allocate resources based on the requirements of the IoT system.
- **Storage and Data Processing**: Cloud systems offer storage services, allowing IoT data to be safely kept for long-term access and analysis. Furthermore, cloud-based data processing and analytics tools allow advanced data analysis, machine learning, and artificial intelligence algorithms to extract useful insights from IoT data.
- **Cost-Effectiveness**: Cloud computing eliminates the need for on-premises infrastructure, lowering hardware, maintenance, and upgrade expenses. It provides a pay-as-you-go paradigm, allowing IoT applications to expand resources in response to demand, resulting in cost savings.
- **Collaboration and Integration**: Cloud-based IoT platforms make it easier for multiple stakeholders, such as device manufacturers, application developers, and data analysts, to collaborate and integrate. They provide

standardized APIs, development tools, and services that allow IoT devices, apps, and data to be seamlessly integrated (Zawra et al., 2017).

EDGE COMPUTING

Edge computing brings computational power and data storage closer to the source of data generation, enabling real-time processing and localized decision-making. It involves deploying computing resources at or near the IoT devices or at the edge of the network. Here's how edge computing enhances IoT systems:

- **Low Latency**: By processing data at or near the edge devices, edge computing reduces latency in IoT applications. Real-time decision-making can occur without relying on cloud-based services, resulting in faster response times and improved operational efficiency.
- **Bandwidth Optimization**: Edge computing reduces the need for sending raw data to the cloud for processing, which optimizes bandwidth usage. Only relevant data or processed information is transmitted, reducing network congestion and lowering communication costs.
- **Offline Operation**: Edge computing enables IoT devices to operate autonomously even when network connectivity is unavailable or inconsistent. In cases when cloud connectivity is poor, local processing and decision-making capabilities provide ongoing operation and resilience.
- **Privacy and Security**: Privacy and security are addressed by edge computing by processing sensitive data locally, closer to the source. It lowers data exposure to external networks, improving data privacy and security (Khang et al., 2022b).

Edge computing enables IoT devices to operate autonomously even when network connectivity is unavailable or inconsistent. In cases when cloud connectivity is poor, local processing and decision-making capabilities provide ongoing operation and resilience. Privacy and security are addressed by edge computing by processing sensitive data locally, closer to the source. It lowers data exposure to external networks, improving data privacy and security (Zheng et al., 2018).

BENEFITS OF IOT IN INDUSTRY 4.0

The internet of things (IoT) integration in Industry 4.0 provides several benefits to enterprises across industries. Here are some of the most significant benefits of adopting IoT technology in the context of Industry 4.0:

REAL-TIME DATA ANALYSIS IMPROVES OPERATIONAL EFFICIENCY

IoT enables the collection of real-time data from a variety of sources, including machines, devices, and sensors. This information can be analyzed and used to improve operations and decision-making processes. Businesses can use real-time information to identify bottlenecks, optimize resource allocation, and increase overall operational efficiency.

IMPROVED PRODUCTIVITY AND COST SAVINGS

IoT allows for automation and streamlines procedures, resulting in increased productivity. Connected devices and systems can communicate and collaborate, decreasing the need for manual intervention and human error. This automation, together with real-time data analysis, assists in identifying cost-saving opportunities, eliminating waste, and optimizing resource utilization.

PREDICTIVE MAINTENANCE AND PROCESS OPTIMIZATION

IoT provides predictive maintenance by analyzing data from connected devices to find patterns and anomalies. This helps firms to predict equipment faults and undertake maintenance tasks before they occur. Businesses may cut expenses and maintain ongoing operations by minimizing unplanned downtime and optimizing maintenance plans.

QUALITY CONTROL AND TRACEABILITY

IoT technology allows for real-time monitoring and control of production processes, ensuring product quality and regulatory compliance. Connected sensors can continuously monitor characteristics like temperature, humidity, and vibration, providing insights into product quality and allowing for early fault detection. Furthermore, IoT allows firms to follow products along the supply chain, assuring accountability and transparency.

NEW BUSINESS MODELS AND REVENUE STREAMS

The internet of things allows enterprises to create new business models and revenue streams. Businesses can provide value-added services like predictive maintenance, remote monitoring, and personalized experiences by exploiting IoT data and insights. This move from selling items to offering services generates new revenue sources while also strengthening client relationships.

IMPROVED SAFETY AND EMPLOYEE WELL-BEING

In industrial environments, IoT technology promotes safety and employee well-being. Connected sensors can monitor and detect dangerous conditions, allowing for quick intervention and risk reduction. Wearables and personal protective equipment (PPE) with IoT capabilities can track vital signs and send real-time alerts, assuring worker safety and well-being.

SUPPLY CHAIN OPTIMIZATION

The internet of things allows for end-to-end visibility and optimization of the supply chain. Connected devices and sensors deliver real-time data on inventory levels, demand, and logistics, allowing organizations to enhance inventory management, reduce stockouts, and increase order fulfillment. IoT also improves supply chain collaboration by allowing stakeholders to share data and better coordinate actions.

Sustainability and Environmental Impact

Energy management, resource optimization, and waste reduction are all made possible by IoT technology. Energy consumption can be monitored via connected devices and sensors, which can also suggest areas for improvement and automate energy-saving solutions. IoT also allows for more efficient resource allocation, reducing waste and environmental impact.

Finally, IoT technology in Industry 4.0 provides several advantages such as increased operational efficiency, increased productivity, predictive maintenance, quality control, new business models, and sustainability. Businesses may optimize processes, cut costs, and open up new prospects for development and innovation by leveraging IoT (Zhi-hong and Yi-jun, 2012).

CHALLENGES AND CONSIDERATIONS IN IOT SYSTEMS

Security and Privacy Concerns in IoT Systems

In IoT systems, security and privacy are key challenges. There is an increased risk of cyberattacks and data breaches with so many connected devices and sensors gathering and transmitting data. Unauthorized access, data tampering, or abuse of sensitive information can all result from inadequate security measures. To secure IoT systems and data privacy, organizations must employ strong security mechanisms such as encryption, authentication, and access control.

Interoperability Concerns and Efforts at Standardization

Interoperability refers to the capacity of various IoT devices, platforms, and systems to communicate and share data in real-time. However, the lack of standardization in protocols, data formats, and interfaces makes interoperability difficult. Incompatible systems and devices impede IoT solution integration and scalability. Standardization efforts, such as the development of common protocols and frameworks, are essential to ensure seamless interoperability and enable widespread adoption of IoT technology.

Data Management, Storage, and Governance

IoT systems generate massive volumes of data, which organizations must efficiently manage, store, and analyze in order to draw actionable insights. Data quality, data integration, and data lifecycle management are key difficulties in data management. Data governance practices, such as data privacy, consent, and compliance with rules such as the General Data Protection Regulation (GDPR), must also be considered by organizations. For generating value from IoT data, it is necessary to establish data management policies and install adequate data storage and analytics infrastructure.

Workforce Training and Skill Requirements

IoT technology adoption necessitates a professional workforce with experience in a variety of fields, including IoT architecture, data analytics, cybersecurity, and system integration. Organizations must engage in workforce training and development programs to close the skill gap and ensure that staff have the knowledge needed to effectively implement, manage, and use IoT technologies. To stay up with the growing IoT landscape, organizations should foster a culture of continuous learning and give chances for upskilling and reskilling.

Scalability and Infrastructure Requirements

Organizations encounter scalability and infrastructure requirements difficulties when IoT systems expand in size. Scaling IoT installations across several locations, managing a large number of connected devices, and dealing with an ever-increasing flow of data necessitate a strong infrastructure that includes network capacity, cloud resources, and edge computing capabilities. To avoid bottlenecks and enable seamless expansion of their IoT ecosystems, organizations must prepare for scalability from the beginning of IoT adoption.

Implications for Ethics and Society

The broad usage of IoT technology presents ethical and social issues that must be carefully considered. Data ownership, permission, and the possibility of surveillance are all examples of ethical problems. Furthermore, the influence of IoT on employment, privacy, and social dynamics should be assessed to guarantee that IoT benefits are balanced with ethical concerns and societal well-being.

It is critical to address these issues and factors for the successful implementation and adoption of IoT technology. To maximize the promise of IoT in Industry 4.0 while limiting related dangers, organizations must build complete strategies that include security measures, interoperability standards, data management practices, workforce training, and ethical principles (Zhou et al., 2015).

FUTURE OUTLOOK AND EMERGING TRENDS IN IOT TECHNOLOGY AND INDUSTRY 4.0

Evolving Trends in IoT Technology and Industry 4.0

IoT technology is constantly growing, and various factors are influencing IoT's future in the context of Industry 4.0. Below are some examples:

- **Edge Intelligence**: The migration of data processing and analytics closer to the edge devices is referred to as edge intelligence. By processing sensitive data locally, this trend minimizes latency, enables real-time decision-making, and improves data privacy.

- **Digital Twins**: Virtual representations of physical assets or systems are known as digital twins. They enable physical object monitoring, analysis, and modeling in real-time. Digital twins aid in the optimization of performance, the prediction of maintenance requirements, and the simulation of scenarios for better decision-making.
- **Swarm Intelligence**: Swarm intelligence is the coordination and collaboration of multiple internet of things devices to achieve a shared purpose. This method mimics natural swarm behavior and can be used to solve complicated issues, optimize resource allocation, and enable distributed decision-making.
- **Autonomous Systems**: In Industry 4.0, autonomous systems powered by IoT, AI, and robotics are gaining traction. These systems are capable of operating autonomously, making intelligent decisions, and collaborating with human personnel. They provide for greater productivity, efficiency, and adaptability in industrial processes.

INTEGRATION OF IoT WITH EMERGING TECHNOLOGIES SUCH AS AI, ML, AND BLOCKCHAIN

IoT technology is being merged with other developing technologies to open up new opportunities and improve capabilities. The combination of IoT and AI/ML provides sophisticated analytics, predictive modeling, and intelligent automation. For proactive decision-making and optimization, AI algorithms can analyze IoT data, find patterns, and make accurate predictions (Rana et al., 2021).

Blockchain technology is also being implemented into IoT devices to improve security, privacy, and trust. Blockchain provides safe data sharing, smart contracts, and decentralized and immutable record-keeping. It has the potential to improve data integrity, enable secure transactions, and permit trustworthy interactions between IoT devices and stakeholders (Khanh and Khang, 2021).

THE ROLE OF 5G CONNECTIVITY IN ENABLING IoT APPLICATIONS

5G network deployment is predicted to have a transformational influence on IoT applications. In comparison to earlier generations of cellular networks, 5G provides much higher bandwidth, decreased latency, and expanded network capacity. This offers seamless connectivity, faster data transfer, and real-time communication, allowing IoT applications to be widely adopted.

By connecting a huge number of devices at the same time, 5G connection enables vast IoT deployments. It enables mission-critical IoT applications that require ultra-low latency and great reliability, such as autonomous vehicles and remote surgery. 5G's enhanced network capacity and speed also makes it easier to deploy IoT solutions across a variety of industries, including smart cities, healthcare, agriculture, and manufacturing (Khang et al., 2023b).

POTENTIAL IMPACT OF EDGE COMPUTING AND FOG COMPUTING

In IoT systems, edge computing and fog computing are developing paradigms that supplement cloud computing. Edge computing is the processing of data at or near the network edge, whereas fog computing expands this concept by dispersing computing resources closer to the network edge.

Edge computing and fog computing have a number of advantages in IoT applications. They reduce latency by processing data locally, allowing for real-time decisions and faster response times. They also cut network capacity needs by filtering and analyzing data at the edge and transferring only relevant data to the cloud. This improves network utilization and reduces communication expenses.

Furthermore, edge computing and fog computing improve data privacy and security by limiting exposure to external threats by keeping sensitive data within local networks. These paradigms also support offline operation, allowing IoT systems to function autonomously even when cloud access is unavailable.

Finally, developing themes such as edge intelligence, digital twins, swarm intelligence, and autonomous systems will shape the future of IoT in Industry 4.0. Integration of IoT with AI, machine learning, and blockchain improves capabilities and security (Khang et al., 2022a).

CONCLUSION

In the end, this research chapter offered a thorough description of IoT technology as a vital component of Industry 4.0. We looked at the definition and main components of IoT, its significance in digital transformation of industries, and IoT integration in Industry 4.0 design. We've also talked about the advantages of IoT in Industry 4.0, such as increased operational efficiency, increased productivity, predictive maintenance, quality control, and the introduction of new business models and income streams.

We also looked at the obstacles and considerations connected with IoT adoption, such as security and privacy concerns, interoperability issues, data management, and workforce training. Understanding and tackling these issues are critical for successful IoT adoption and utilization in Industry 4.0.

Emerging developments in IoT technology and Industry 4.0 point to a transition toward edge intelligence, digital twins, swarm intelligence, and autonomous systems in the future. Integration of IoT with new technologies such as AI, ML, blockchain, and the deployment of 5G networks will expand IoT's capabilities and effect across industries (Hussain et al., 2022).

IoT technology has far-reaching consequences in Industry 4.0. It allows firms to improve operational efficiency, productivity, and cost savings. The internet of things enables real-time data analysis, predictive maintenance, quality control, and traceability, resulting in better decision-making and consumer satisfaction. Furthermore, IoT creates new business prospects, supports innovation, and promotes environmentally friendly practices.

Businesses and industries should examine the recommendations for future research and implementation to fully realize the promise of IoT in Industry 4.0. In order to protect IoT systems and data privacy, organizations should prioritize the adoption of effective security solutions. To reduce security threats, encryption, authentication, and access control techniques must be implemented. Below are some examples:

- **Interoperability and Standardization Should be Promoted**: Efforts should be made to define standardized protocols, data formats, and interfaces to improve interoperability among IoT devices, platforms, and systems. Initiatives to standardize will assist the seamless integration and scalability of IoT systems.
- **Creating Comprehensive Data Management Strategies**: Organizations should prioritize the creation of comprehensive data management strategies in order to efficiently collect, store, analyze, and govern IoT data. This includes ensuring data quality, integration, and regulatory compliance.
- **Investing in Workforce Training and Development**: Organizations should engage in training programs to provide staff with the skills and knowledge needed to properly deploy, manage, and use IoT technologies. Initiatives to promote continuous learning and upskilling will assure a competent workforce in the future.
- **Embracing Upcoming Technologies and Connectivity**: Businesses should investigate the integration of IoT with developing technologies such as AI, machine learning, and blockchain (Vrushank et al., 2023), as well as the capabilities provided by 5G networks. Advanced analytics, greater security, and faster, more dependable communication will be enabled by these technologies.

Moreover, IoT technology is a vital component of Industry 4.0, revolutionizing industries worldwide. Its influence on operational efficiency, productivity, cost reductions, and the development of new business models cannot be overstated. Businesses may fully harness the promise of IoT in Industry 4.0 and promote sustainable growth and innovation by resolving issues, embracing emerging trends, and following the recommended tactics.

REFERENCES

Agarwal V., S. Tapaswi, and P. Chanak, "A Survey on Path Planning Techniques for Mobile Sink in IoT-Enabled Wireless Sensor Networks," *Wireless Personal Communications*, Mar. 2021, https://doi.org/10.1007/s11277-021-08204-w.

Anbalagan A. and C. F. Moreno-Garcia, "An IoT Based Industry 4.0 Architecture for Integration of Design and Manufacturing Systems," *Materials Today: Proceedings*, vol. 46, pp. 7135–7142, 2021, https://doi.org/10.1016/j.matpr.2020.11.196.

Asemani M., F. Abdollahei, and F. Jabbari, "Understanding IoT Platforms: Towards a Comprehensive Definition and Main Characteristic Description," *IEEE Xplore*, Apr. 01, 2019, https://ieeexplore.ieee.org/abstract/document/8765259 (accessed Aug. 11, 2020).

Asghar M. H., A. Negi, and N. Mohammadzadeh, "Principle Application and Vision in Internet of Things (IoT)," *IEEE Xplore*, May 01, 2015, https://ieeexplore.ieee.org/abstract/document/7148413 (accessed Nov. 09, 2020).

Crnjac M., I. Veža, and N. Banduka, "From Concept to the Introduction of Industry 4.0," *International Journal of Industrial Engineering and Management (IJIEM)*, vol. 8, no. 1, pp. 21–30, 2017, http://ijiemjournal.org/images/journal/volume8/ijiem_vol8_no1_3.pdf.

Forcina A. and D. Falcone, "The Role of Industry 4.0 Enabling Technologies for Safety Management: A Systematic Literature Review," *Procedia Computer Science*, vol. 180, pp. 436–445, 2021, https://doi.org/10.1016/j.procs.2021.01.260.

Georgios S. K. and A. Theofylaktos, "Internet of Things in the Context of Industry 4.0: An Overview," *dspace.vsp.cz*, vol. 7, no. 1, Jun. 2019, http://dspace.vsp.cz/handle/ijek/103.

Ghosh R. K., A. Banerjee, P. Aich, D. Basu, and U. Ghosh, "Intelligent IoT for Automotive Industry 4.0: Challenges, Opportunities, and Future Trends," *Internet of Things*, pp. 327–352, 2022, https://doi.org/10.1007/978-3-030-81473-1_16.

Hajimahmud V. A., A. Khang, V. Hahanov, E. Litvinova, S. Chumachenko, and V. A. Alyar, "Autonomous Robots for Smart City: Closer to Augmented Humanity," *AI-Centric Smart City Ecosystems: Technologies, Design and Implementation* (1st ed.), 7(12). CRC Press, 2022, https://doi.org/10.1201/9781003252542-7.

Hassoun A. et al., "The Fourth Industrial Revolution in the Food Industry—Part I: Industry 4.0 Technologies," *Critical Reviews in Food Science and Nutrition*, pp. 1–17, Feb. 2022, https://doi.org/10.1080/10408398.2022.2034735.

Hayat V. S., A. K. Sharma, and N. Arora, *Introduction to Industry 4.0*, pp. 29–59, Jan. 2023, https://doi.org/10.1007/978-981-19-8730-4_2.

Hussain S. H., T. B. Sivakumar, and A. Khang, "Cryptocurrency Methodologies and Techniques," *The Data-Driven Blockchain Ecosystem: Fundamentals, Applications, and Emerging Technologies* (1st ed.), 2(9), pp. 20–29. CRC Press (Eds.) 2022, https://doi.org/10.1201/9781003269281-2.

Khan H. and M. Javaid, "Role of Internet of Things (IoT) in Adoption of Industry 4.0," *Journal of Industrial Integration and Management*, p. 2150006, Mar. 2021, https://doi.org/10.1142/s2424862221500068.

Khang, A. "Material4Studies," *Material of Computer Science, Artificial Intelligence, Data Science, IoT, Blockchain, Cloud, Metaverse, Cybersecurity for Studies*, 2021. Retrieved from https://www.researchgate.net/publication/370156102_Material4Studies.

Khang A., S. Chowdhury, and S. Sharma, *The Data-Driven Blockchain Ecosystem: Fundamentals, Applications, and Emerging Technologies* (1st ed.). CRC Press, 2022a, https://doi.org/10.1201/9781003269281.

Khang A., *Advanced Technologies and AI-Equipped IoT Applications in High-Tech Agriculture* (1st ed.). IGI Global Press, 2023a, https://doi.org/10.4018/978-1-6684-9231-4.

Khang A., G. Rana, R. K. Tailor, and V. A. Hajimahmud, *Data-Centric AI Solutions and Emerging Technologies in the Healthcare Ecosystem*. CRC Press (Eds.) 2023b, https://doi.org/10.1201/9781003356189.

Khang A., V. Abdullayev, V. Hahanov, and V. Shah, *Advanced IoT Technologies and Applications in the Industry 4.0 Digital Economy* (1st ed.). CRC Press, 2024, https://doi.org/10.1201/9781003434269.

Khang A., V. Hahanov, G. L. Abbas, and V. A. Hajimahmud, "Cyber-Physical-Social System and Incident Management," *AI-Centric Smart City Ecosystems: Technologies, Design and Implementation* (1st ed.), 2(15). CRC Press, 2022b, https://doi.org/10.1201/9781003252542-2.

Khanh H. H. and A. Khang, "The Role of Artificial Intelligence in Blockchain Applications," *Reinventing Manufacturing and Business Processes through Artificial Intelligence*, 2, pp. 20–40. CRC Press, 2021, https://doi.org/10.1201/9781003145011-2.

Kumar K., D. Zindani, and J. P. Davim, *Industry 4.0*. Springer Singapore, 2019, https://doi.org /10.1007/978-981-13-8165-2.

Sadeeq Mohammed M., N. M. Abdulkareem, S. R. M. Zeebaree, D. Mikaeel Ahmed, A. Saifullah Sami, and R. R. Zebari, "IoT and Cloud Computing Issues, Challenges and Opportunities: A Review," *Qubahan Academic Journal*, vol. 1, no. 2, pp. 1–7, Mar. 2021, https://doi.org/10.48161/qaj.v1n2a36.

Nassereddine M. and A. Khang, "Applications of Internet of Things (IoT) in Smart Cities," *Advanced IoT Technologies and Applications in the Industry 4.0 Digital Economy* (1st ed.). CRC Press, 2024, https://doi.org/10.1201/9781003434269-6.

Paul S. et al., "Industry 4.0 Applications for Medical/Healthcare Services," *Journal of Sensor and Actuator Networks*, vol. 10, no. 3, p. 43, Jun. 2021, https://doi.org/10.3390 /jsan10030043.

Peralta G., M. Iglesias-Urkia, M. Barcelo, R. Gomez, A. Moran, and J. Bilbao, "Fog Computing Based Efficient IoT Scheme for the Industry 4.0," *IEEE Xplore*, May 01, 2017, https://ieeexplore.ieee.org/document/7945879 (accessed Dec. 07, 2020).

Rana G., A. Khang, R. Sharma, A. K. Goel, and A. K. Dubey, *Reinventing Manufacturing and Business Processes through Artificial Intelligence*. CRC Press (Eds.) 2021, https:// doi.org/10.1201/9781003145011.

Rath K. C., A. Khang, and D. Roy, "The Role of Internet of Things (IoT) Technology in Industry 4.0," *Advanced IoT Technologies and Applications in the Industry 4.0 Digital Economy* (1st ed.). CRC Press, 2024, https://doi.org/10.1201/9781003434269-1.

Suleiman Z., S. Shaikholla, D. Dikhanbayeva, E. Shehab, and A. Turkyilmaz, "Industry 4.0: Clustering of Concepts and Characteristics," *Cogent Engineering*, vol. 9, no. 1, Feb. 2022, https://doi.org/10.1080/23311916.2022.2034264.

Ungurean I., N. Gaitan, and V. G. Gaitan, "An IoT Architecture for Things from Industrial Environment," *IEEE Xplore*, May 01, 2014, https://ieeexplore.ieee.org/abstract/docu-ment/6866713/ (accessed Dec. 26, 2020).

Vrushank S., T. Vidhi, and A. Khang, "Electronic Health Records Security and Privacy Enhancement Using Blockchain Technology," *Data-Centric AI Solutions and Emerging Technologies in the Healthcare Ecosystem* (P 1) (1st ed.). CRC Press, 2023, https://doi .org/10.1201/9781003356189-1.

Xu D., E. L. Xu, and L. Li, "Industry 4.0: State of the Art and Future Trends," *International Journal of Production Research*, vol. 56, no. 8, pp. 2941–2962, Mar. 2018, https://doi .org/10.1080/00207543.2018.1444806.

Yu W. et al., "A Survey on the Edge Computing for the Internet of Things," *IEEE Access*, vol. 6, pp. 6900–6919, 2018, https://doi.org/10.1109/access.2017.2778504.

Zawra L. M., H. A. Mansour, A. T. Eldin, and N. W. Messiha, "Utilizing the Internet of Things (IoT) Technologies in the Implementation of Industry 4.0," *Proceedings of the International Conference on Advanced Intelligent Systems and Informatics 2017*, pp. 798–808, Aug. 2017, https://doi.org/10.1007/978-3-319-64861-3_75.

Zheng X., Z. Cai, and Y. Li, "Data Linkage in Smart Internet of Things Systems: A Consideration from a Privacy Perspective," *IEEE Communications Magazine*, vol. 56, no. 9, pp. 55–61, Sep. 2018, https://doi.org/10.1109/mcom.2018.1701245.

Zhi-hong Q. and W. Yi-jun, "IoT Technology and Application," *Acta Electonica Sinica*, vol. 40, no. 5, p. 1023, May 2012, https://doi.org/10.3969/j.issn.0372-2112.2012.05.026.

Zhou K., T. Liu, and L. Zhou, "Industry 4.0: Towards Future Industrial Opportunities and Challenges," *2015 12th International Conference on Fuzzy Systems and Knowledge Discovery (FSKD)*, Aug. 2015, https://doi.org/10.1109/fskd.2015.7382284.

5 Sustainable IT
A Comprehensive Guide to Green Computing and Green Internet of Things (IoT) Technologies

Neha Shrotriya, Parthivi Thakore, and Alex Khang

INTRODUCTION

The increasing growth of the tech sector has revolutionized the way businesses operate, communicate, and store data. However, this growth has come at a cost to the environment, with increasing energy consumption, carbon emissions, and e-waste. The result is that there is a growing need for businesses to adopt sustainable information technology (IT) practices to minimize their environmental impact. Green computing is an emerging field that focuses on designing, developing, and using computer resources in an eco-friendly way that minimizes energy consumption and reduces waste (Chiueh et al., 2005).

This chapter provides a comprehensive guide to green computing and its role in promoting sustainable IT practices. It explores the various strategies and technologies that will be useful to reduce and recover the environmental impact of IT, such as virtualization, cloud computing, green mobile computing, green big data analytics, and green internet of things (IoT) (Bhambri et al., 2022).

Virtualization involves creating more number of virtual servers or devices that run on a unique physical server. By doing this, businesses can reduce their energy consumption and hardware requirements, leading to significant cost savings. In addition, virtualization can increase the efficiency of servers, resulting in faster processing times and reduced downtime (Chiueh et al., 2005). Cloud computing involves storing and accessing data and applications over the internet, rather than on physical servers. This can significantly reduce energy consumption, as businesses do not need to operate and maintain their own physical servers. In addition, cloud computing providers often use renewable energy sources, further reducing their carbon footprint (Qian et al., 2009).

Green mobile computing involves designing mobile devices that use less energy and are more environmentally friendly. This can be achieved through the use of

DOI: 10.1201/9781003434269-5

energy-efficient hardware, longer battery life, and the ability to be easily recycled. In addition, green mobile computing can promote remote working, reducing the need for employees to travel and further reducing the carbon footprint of businesses (Ba et al., 2013).

Big data processing and analysis using green methods entail using environmentally friendly procedures. This can be done by reducing the amount of superfluous data, optimizing data storage, and using energy-efficient technology (Figure 5.1). Businesses can reduce their energy use and carbon impact while still gaining useful insights from their data by implementing green big data analytics (Liu et al., 2020).

Green internet of things involves designing and using internet-connected devices that are energy efficient and environmentally friendly. This can be achieved through the use of renewable energy sources, the optimization of device usage, and the reduction of unnecessary data. By adopting green internet of things, businesses can reduce their energy consumption and carbon footprint while still benefiting from the increased connectivity and data insights that internet of things (IoT) devices provide (Albreem et al., 2017).

Other topics discussed in this chapter include e-waste management, the benefits of green computing, and the metrics used to measure the success of sustainable IT practices. Proper e-waste management is essential for minimizing the environmental impact of technology, and the chapter explores various strategies for sustainably disposing of electronic waste.

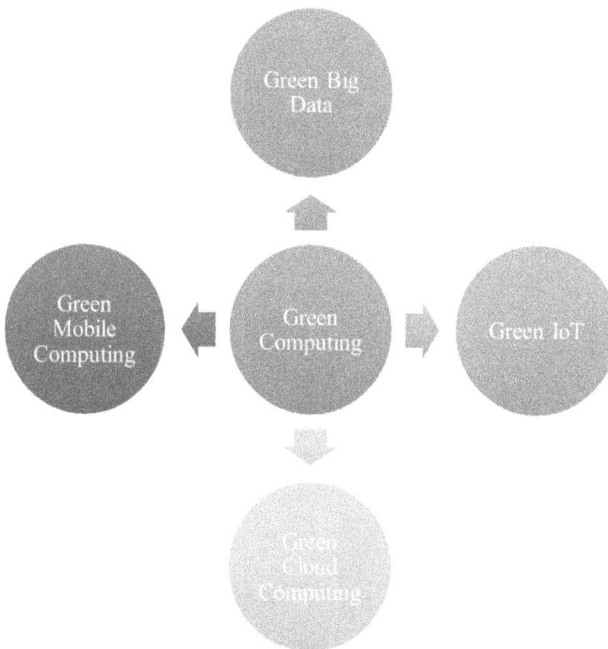

FIGURE 5.1   Parts of green computing

NEED FOR SUSTAINABLE IT AND GREEN COMPUTING

The technology industry has experienced rapid development, which has increased energy use, e-waste, and carbon emissions, making it a major driver of environmental deterioration. The information and communication technology (ICT) sector is thought to be responsible for 2% of worldwide carbon emissions, which is comparable to the carbon footprint of the aircraft sector (Figure 5.2). The energy usage of data centers, the foundation of the internet, has also grown significantly over the past ten years—by over 500%—and is expected to double by 2025 (Masanet et al., 2020).

These alarming statistics highlight the need for sustainable IT practices, and green computing is an emerging field that aims to minimize the environmental impact of technology. Green computing focuses on designing, developing, and using computer resources in an eco-friendly way that minimizes energy consumption and reduces waste.

There are several reasons why there is a need for green computing. First, the environmental impact of technology is significant and growing. The excessive energy consumption of data centers and the e-waste generated by the disposal of outdated hardware are just two examples of the impact technology has on the environment. Therefore, it is crucial to adopt sustainable IT practices to reduce the carbon footprint and minimize the waste generated by technology (Chiueh et al., 2005).

Second, green computing is essential to address the rising costs of energy consumption. The cost of energy is a significant expense for businesses, and adopting green computing practices can result in significant cost savings by reducing energy consumption.

Green computing is crucial for businesses looking to enhance their reputation and improve their corporate social responsibility (CSR) credentials. Consumers are becoming increasingly environmentally conscious, and they expect businesses to

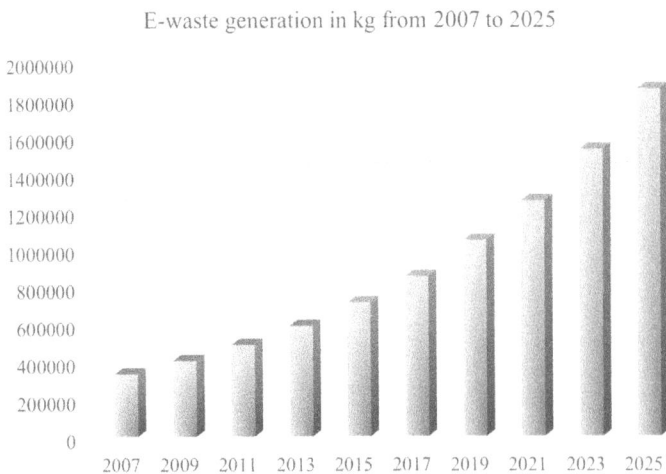

FIGURE 5.2   E-waste generation in India from 2007 to 2025

adopt sustainable practices. By adopting green computing practices, businesses can demonstrate their commitment to the environment and enhance their reputation (Ba et al., 2013).

In addition, green computing can lead to improved efficiency and productivity. For example, virtualization can increase the efficiency of servers, resulting in faster processing times and reduced downtime. Cloud computing can reduce the time and resources required to manage and maintain physical servers, allowing businesses to focus on their core operations. Green mobile computing can promote remote working, reducing the need for employees to travel and increasing productivity (Liu et al., 2020).

Overall, the need for green computing is clear. It is essential to minimize the environmental impact of technology, reduce costs, improve reputation, and enhance efficiency and productivity. Green computing practices can help businesses achieve these goals and promote a more sustainable future (Ba et al., 2013). Certain techniques and technologies can be adopted to certainly reduce the impact of IT on the environment like cloud computing, green big data, etc. Such technologies are further discussed in the chapter (Figure 5.3).

GREEN CLOUD COMPUTING

Users can utilize pooled computing resources, such as computers, storage, and apps, over the internet using the cloud computing processing paradigm. In recent years, cloud computing has grown in popularity due to its numerous advantages, including cost savings, freedom, scalability, and simplicity of administration (Yamini, 2012).

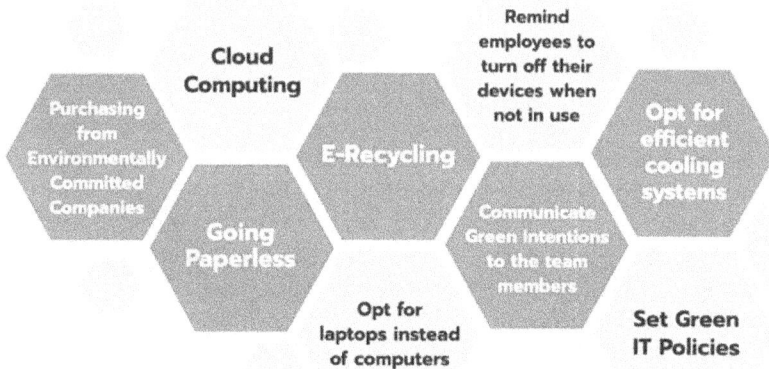

FIGURE 5.3 Steps to attain sustainability in IT

Cloud tech does have an effect on the world, though. Because they use a lot of electricity, cloud data centers increase the carbon impact of the IT sector. Due to this, green cloud computing has emerged with the intention of lessening the negative effects of cloud computing on the ecosystem (Qian et al., 2009).

The term "green cloud computing" describes the use of environmentally favorable procedures and tools to lessen cloud computing's energy usage and carbon impact. There are numerous methods that ecological cloud computing can be used (Chinnici and Quintiliani, 2013).

Hardware that uses less energy and generates less heat is available from cloud service providers, including servers and storage units. This can help save a lot of energy and lessen the need for refrigeration systems (Arthi and Shahul Hamead, 2013).

Virtualization plays a key role in managing and consolidating cloud data center resources, allowing for scalability and fault tolerance while reducing energy consumption. For example, virtual machines (VMs) can be migrated from low-utilization servers to idle servers to save energy. However, the network cost of VM migration must also be considered (Yashwant et al., 2015).

Renewable energy can be used to power cloud data centers and reduce their carbon footprint, but the cost and reliability of these resources remain concerns. Therefore, techniques have been developed such as balancing dynamic power workload, and capping of server power can help to solve these issues (Anan and Naseer, 2015).

Waste heat generated by cloud data centers can be reused in various ways, such as in cooling systems or district heating (Figure 5.4). However, the low quality of this heat can limit its usability (Usmin et al., 2014).

To optimize the allocation of resources in cloud data centers, multi-objective models are used with a focus on both task makes pan and energy consumption, and evolutionary algorithms are often employed to find near-optimal solutions to these complex problems (Yamini, 2012).

VIRTUALIZATION

In order to handle and consolidate cloud data center resources for better energy economy and scalability, virtualization is a method used in green cloud computing. In order to provide a higher-level interface for users and applications, it involves creating a virtual layer over the physical resources and abstracting the hardware layer interfaces. Through a variety of backup strategies, including resource migration and snapshots, this tier aids in the administration and aggregation of cloud data center resources (Kord and Haghighi, 2013).

In cloud data centers, flexible and fault-tolerant processes are the main goal of virtualization. By creating multiple VMs on a single physical server, virtualization allows for easy scaling of resources as the demand for computing power increases. If one VM fails, the other VMs on the same server can continue to operate without interruption (Figure 5.5). This ensures a high availability of computing resources and minimizes downtime (Fahimeh et al., 2014).

Virtualization is widely used in green cloud computing for resource consolidation and energy economy in addition to scalability and fault tolerance. The

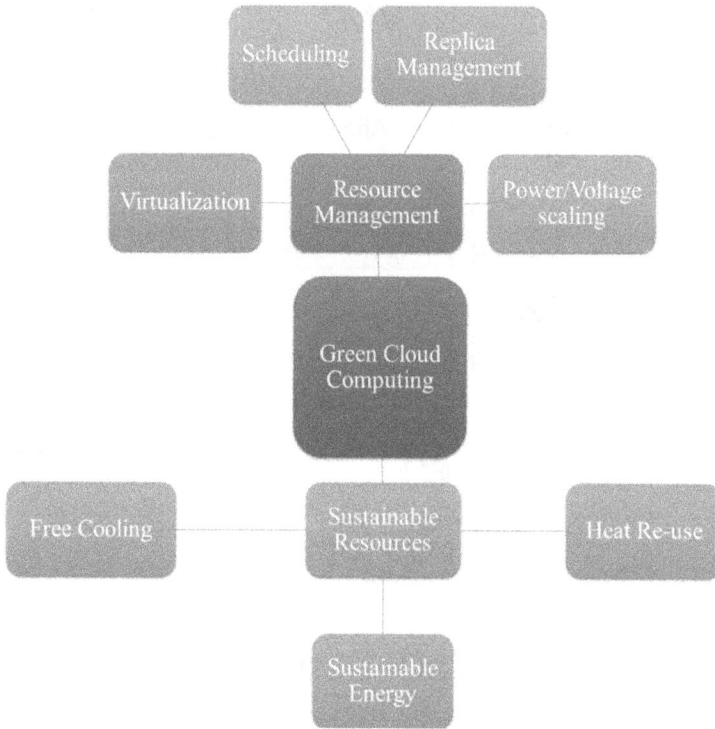

FIGURE 5.4 Green cloud computing structure

underutilization of resources, which results in unnecessary energy usage and elevated running costs, is one of the largest problems in data center operations. This problem is solved by virtualization, which enables the transfer of virtual resources from underutilized computers to servers that are used more frequently (Fahimeh et al., 2014). For instance, if a virtual resource is present on a server that is 40% used, it can be moved there while the first server is in low-power inactive state. With this method, the energy consumption of idle computers is reduced, which lowers the data center's total energy consumption (Jain et al., 2013).

Virtualization also enables the creation of dynamic power management policies that can adjust the consumption of power of physical servers based on the current demand for computing resources. For example, during periods of low demand, physical servers can be placed in low-power idle mode or even powered off, resulting in significant energy savings (Deng et al., 2016).

Both intra- and inter-data center setups can benefit from VM relocation while maintaining energy-efficient operations. To jointly optimize network and server resources, it is necessary to handle the network expenses associated with VM migration. By locating related and "talkative" VMs in close vicinity to each other on the best servers, a portion of the network is used for their interactions, which lowers the cost of the intra-data center VM migration network. Similarly, data deduplication

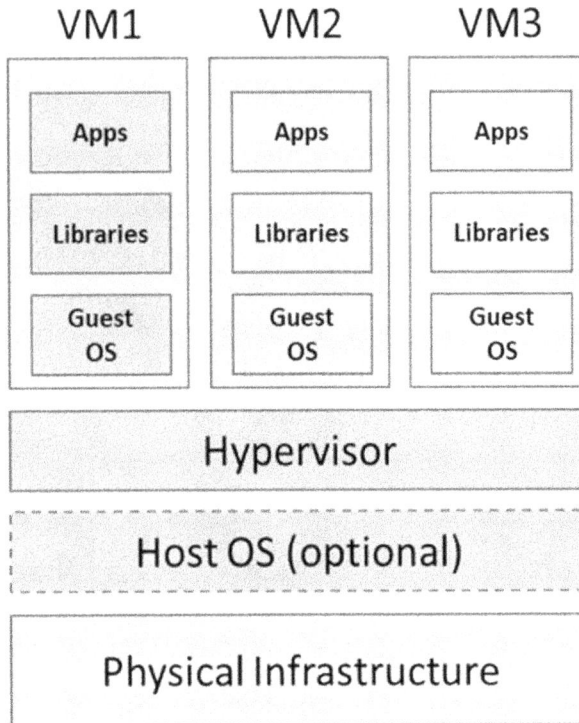

FIGURE 5.5   Virtualization in cloud computing

and compression methods over long-haul networks lower the cost of inter-data center VM movement (Beloglazov and Buyya, 2010).

The use of virtualization in green cloud computing also helps in reducing the carbon footprint of data centers. By consolidating resources and reducing energy consumption, virtualization contributes to the reduction of greenhouse gas emissions. Additionally, virtualization helps in resource allocation and management, leading to efficient resource usage and reduced waste. Table 5.1 showcases the various techniques proposed by various chapters to make cloud computing energy efficient and take steps toward green cloud computing (Rani et al., 2023).

ENERGY-EFFICIENT HARDWARE FOR GREEN CLOUD COMPUTING

Energy-efficient hardware is an essential component of green cloud computing. It plays a crucial role in reducing the power consumption and carbon footprint of cloud data centers (Beloglazov and Buyya, 2010). In this section, we will explain energy-efficient hardware components in detail.

Energy-efficient Processors: The processor is the heart of any computer system, including cloud data centers. Energy-efficient processors are designed to operate at lower voltages and frequencies than conventional processors, resulting in reduced

TABLE 5.1

Various Techniques Proposed to Make Cloud Computing Energy Efficient

S. no.	Proposed Techniques	Ref. no.	Virtualization	Objective	Advantages
1	Dynamic migration algorithm (DMA)	10	Yes	To reduce CO$_2$ emissions and electricity usage	Reduced electricity utilization and efficient resource use coexist
2	Automatic VM migration using trigger engine	8	Yes	Efficiency of energy	Pre-processed information about cloud service utilization
3	Replacement of elements with more energy-efficient ones	7	Yes	Analyze and verify the energy reductions achieved by DCs	The use of more energy-efficient components
4	Practical online bin packaging algorithm (EVISBP)	11	Yes	To reduce the amount of computers that are currently in use	Assists with traffic balancing and hotspot prevention
5	Decentralized architecture of energy-aware resource management system	18	Yes	To provide Quality of service (QoS) while lowering cloud DC's running costs	Takes into account diverse physical components
6	Green algorithm	12	Yes	To greatly lower energy use and dramatically reduce pollution	Direct effect on lowering operating expenses

power consumption (Deng et al., 2016). Advanced power management techniques, such as dynamic voltage and frequency scaling, are used to further reduce the power consumption of energy-efficient processors (Jain et al., 2013).

Solid-state Drives (SSDs): Solid-state drives consume less power than traditional hard disk drives (HDDs) due to their lack of moving parts (Figure 5.6). They are also faster and more reliable than HDDs. SSDs are particularly useful for cloud data centers that require fast and efficient storage solutions (Fahimeh et al., 2014).

Energy-efficient Memory: Energy-efficient memory modules consume less power than traditional memory modules. They use advanced power management techniques, such as reducing the operating voltage and frequency of the memory modules when they are not in use (Kord and Haghighi, 2013).

Power-efficient Networking: Power-efficient networking components, such as network interface cards (NICs) and switches, can significantly reduce the power

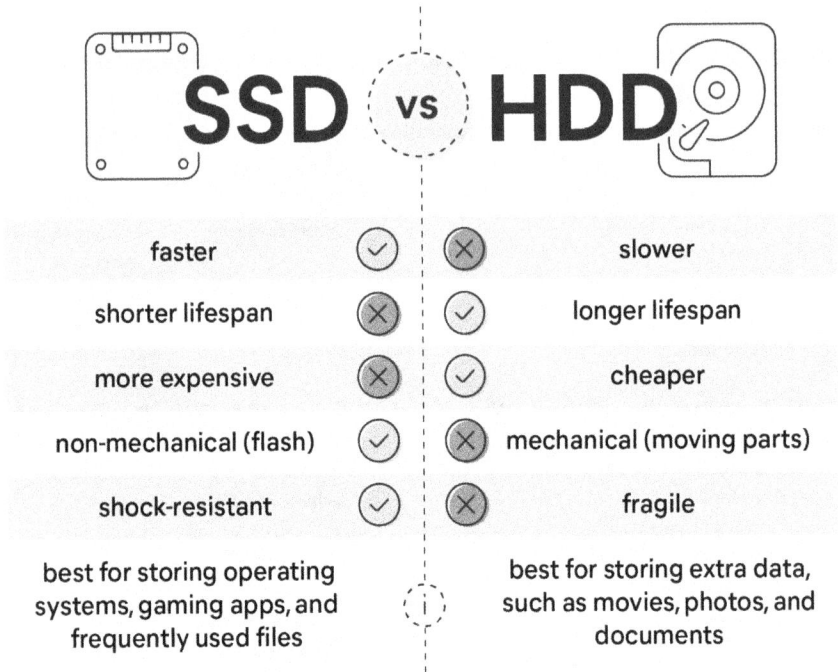

FIGURE 5.6 Comparative analysis of SSD and HDD

consumption of cloud data centers (Figure 5.7). These components are designed to operate at lower voltages and frequencies than conventional networking components (Yamini, 2012).

Energy-efficient Cooling Solutions: Cooling is a critical component of any data center, including cloud data centers. Energy-efficient cooling solutions, such as air-side economizers, liquid cooling, and containment systems, can significantly reduce the power consumption of data centers (Figure 5.8). These cooling solutions are designed to maintain the optimal temperature and humidity levels in data centers while consuming minimal power (Usmin et al., 2014).

Renewable Energy Sources: Cloud data centers can also use renewable energy sources, such as solar, wind, and hydropower, to power their operations. These energy sources are environmentally friendly and can significantly reduce the carbon footprint of cloud data centers (Arthi and Shahul Hamead, 2013).

GREEN MOBILE COMPUTING

Smartphones in recent generations are powerful devices with high storage capacity, capable of performing resource-intensive tasks. As a result, users are less reliant on desktop servers for computing tasks, which has led to an increase in the resource requirements of smartphone applications (Liu and Wu, 2015).

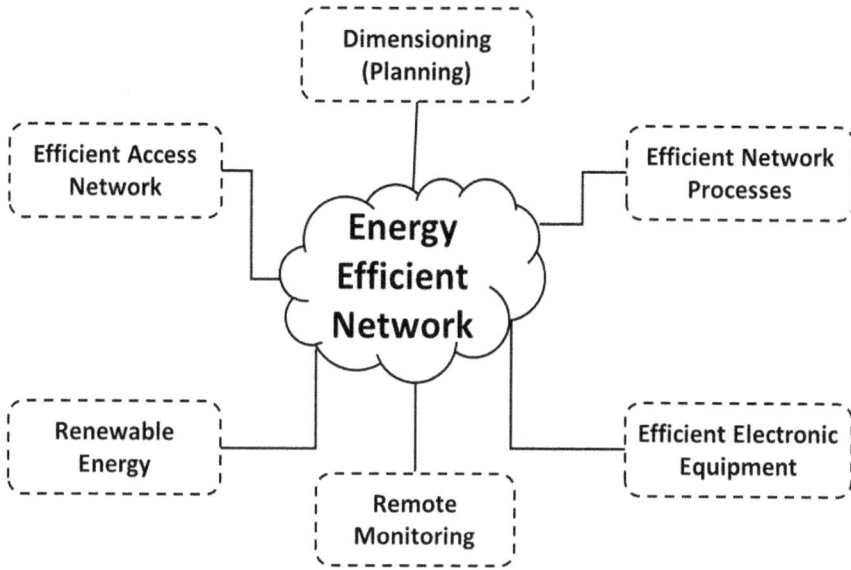

FIGURE 5.7   Energy-efficient network for green cloud computing

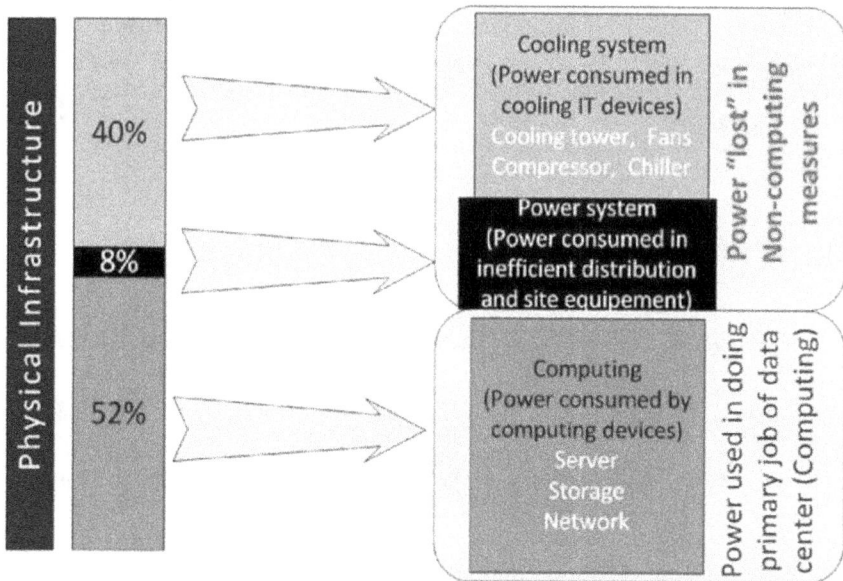

FIGURE 5.8   Energy-efficient cooling solution for green cloud computing

Modern smartphone applications often use sensors such as GPS, accelerometer, and wireless radios to provide context-aware services, which significantly increases computation, communication, and energy costs. To address the energy-performance trade-off, energy-efficient system designs are necessary for modern smartphones (Figure 5.9). Energy estimation techniques help identify energy-efficient design options for smartphone applications and system components, and can also identify rogue applications (Mishra and Ahmadi, 2016).

The overall energy budget of a smartphone device is greatly increased by efficient hardware administration. A smartphone's hardware components are built using CMOS technology, and both static and dynamic electricity are used in total (Figure 5.10). Dynamic frequency scaling and power limiting are two methods for lowering power usage (Kim and Shin, 2014).

Power-gating is a technique used to reduce static power consumption in a smartphone. This technique involves shutting off the power supply to a hardware module when it is not in use. By turning off the power supply, the component does not consume any static power, which can significantly improve battery life. Power-gating can be achieved using different techniques such as sleep transistors or power switches. Sleep transistors are used to disconnect the power supply from a component when it is not in use (Figure 5.11). Power switches are used to disconnect the power supply from a block of components when they are not in use (Biswas and Misra, 2015).

Dynamic frequency scaling is a technique used to reduce dynamic power consumption in a smartphone. This technique involves adjusting the frequency of the hardware module based on its workload. When a component is idle, its frequency is reduced to reduce the amount of dynamic power consumed. When the component is active, its frequency is increased to improve performance. Dynamic frequency scaling can be achieved using different techniques such as voltage scaling or clock scaling. Voltage scaling involves reducing the voltage supplied to a component when it is idle, which reduces the dynamic power consumption. Clock scaling involves

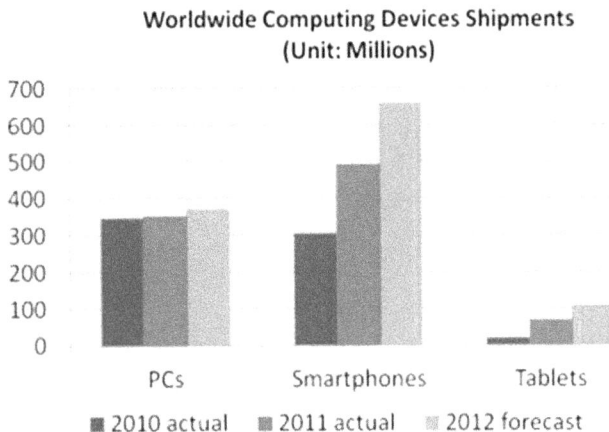

FIGURE 5.9   Shipment and purchase data of computing devices

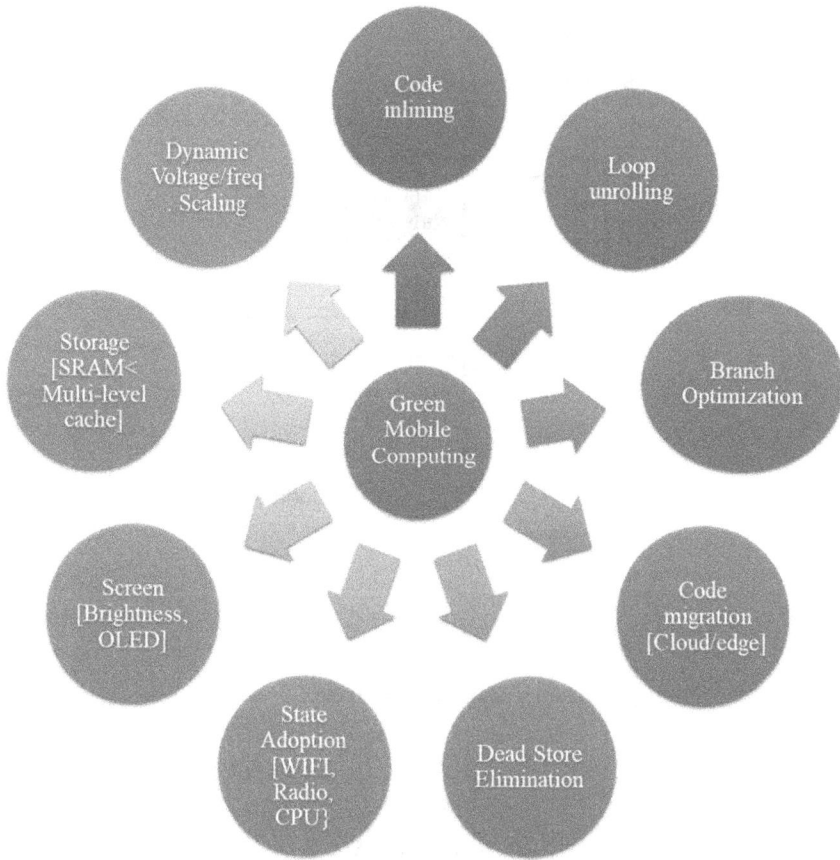

FIGURE 5.10 Techniques for green mobile computing

reducing the clock frequency of a component when it is idle, which reduces the dynamic power consumption (Yang and Song, 2016).

Tail power, which is the term for a smartphone component that is still operating at high power after finishing a job, can also greatly reduce battery life (Liu and Wu, 2015). At the component level, software tools like E-prof can assist in monitoring or estimating device energy consumption, but their profiling actions may affect device energy consumption (Mishra and Ahmadi, 2016).

Smartphone energy consumption can be greatly decreased by software-based green computing solutions like mobile cloud computing, energy bug manage-ment, and the creation of energy-efficient applications. Smartphones can offload energy-intensive tasks to distant cloud servers, thanks to mobile cloud computing (Figure 5.12). Before moving a job to resource-rich cloud servers, computational offloading choices take into consideration the overall processing time, resource usage, energy requirements, and privacy concerns of an application (Biswas and Misra, 2015). Energy bugs, which can be brought on by defective batteries, damaged

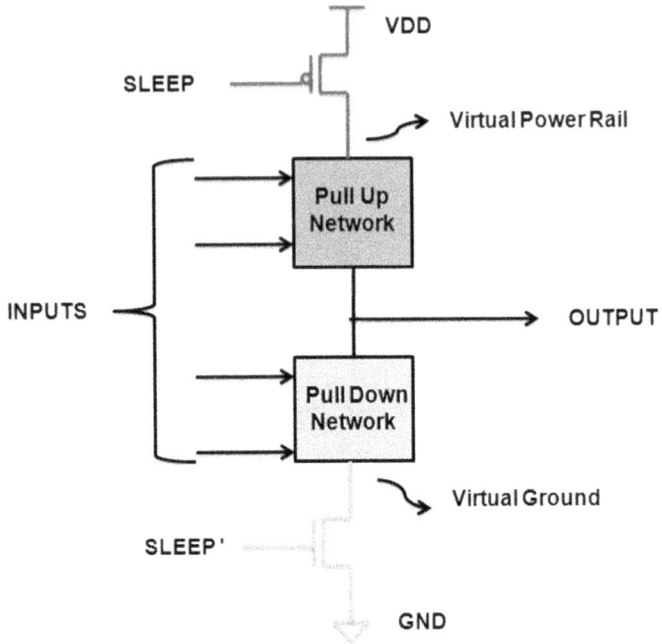

FIGURE 5.11   Power-gating for green mobile computing

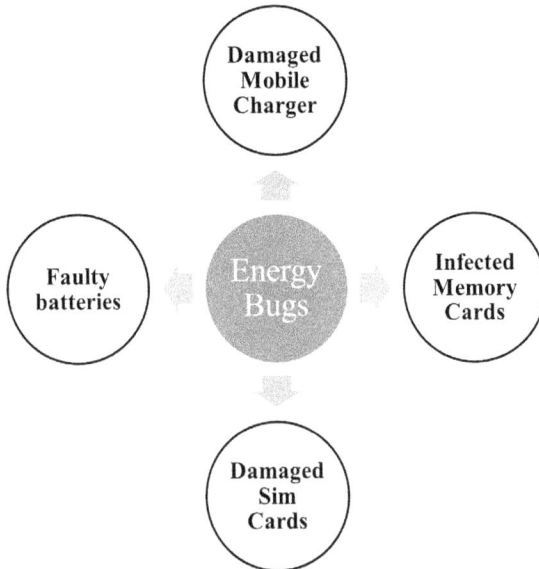

FIGURE 5.12   Energy bugs responsible for degrading mobile devices

mobile battery charges, infected memory cards, and damaged SIM cards, are another problem that may have an effect on how much energy is used by smartphones (Kim and Shin, 2014).

Component power models and code analysis-based estimation are two general groups for smartphone application energy estimation methods (Figure 5.13). While code analysis-based methods predict energy consumption-based cost energy of instructions within the source code of a program, component power model-based methods forecast the base energy consumption using SOC estimation techniques (Yang and Song, 2016).

The overall energy usage of a smartphone application is significantly impacted by inefficient code design, but the resource-optimal placement of classes and functions can cut down on energy use. The functioning of smartphones can also be enhanced by educating developers on energy-efficient application creation methods. Green mobile computing has been the subject of in-depth research that can be further analyzed for more in-depth observations (Kim and Shin, 2014).

There have been various types of research and projects on green mobile computing showcasing efficient utilization of mobile networks, which are showcased in

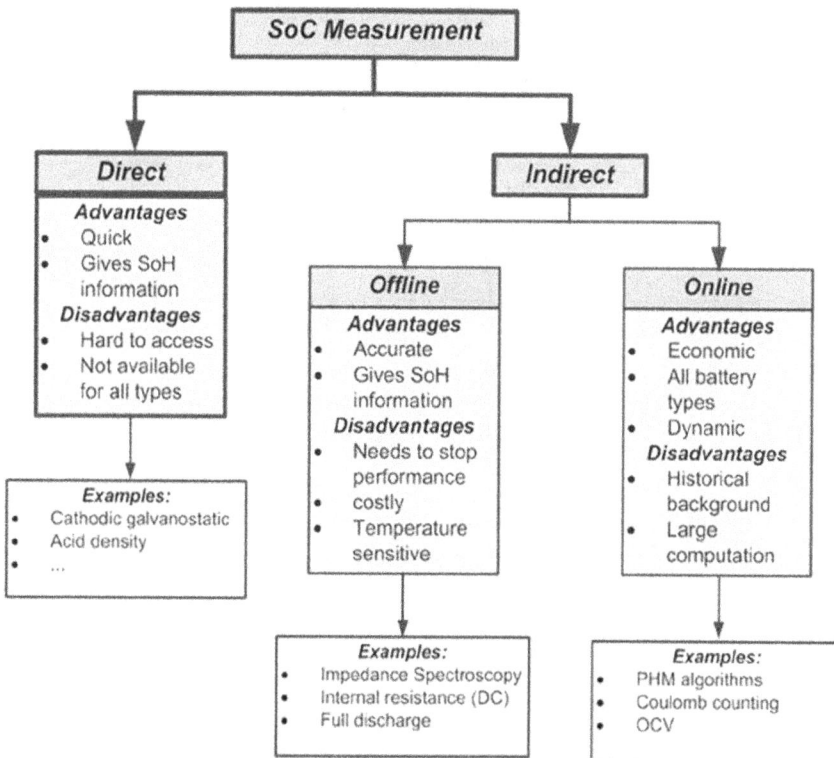

FIGURE 5.13   SoC measurement for green mobile computing

Table 5.2, and measuring metrics for the evaluation and computation of green mobile computing taxonomy, which is given in Table 5.3.

GREEN INTERNET OF THINGS

New technologies like IoT enable automatic communication between electronic devices without the assistance of humans or computers. A collection of practices known as "green IoT" aims to improve the energy economy and lessen the greenhouse impact in existing IoT services and apps (Yan et al., 2018). Green IoT also emphasizes topics like environmentally-friendly manufacturing, revision, and disposal. Collaborative efforts of facilitating technologies, communication strategies, and protocols result in actual IoT implementation. Facilitating technologies and greening strategies for IoT

TABLE 5.2
Summarization of Green Mobile Networks Projects

Project	Organizer	Participants	Targets	Working Emphasis
EARTH	European Commission FP7 IP (3 years/15 million)	Principal mobile carriers in Europe and study institutions	Networks of mobile	Energy-conscious distribution, design, and adaptive management techniques; wireless and network technology; and multi-cell collaboration
Green IT	METI & JEITA (Japan)	More than 100 businesses, institutions, and groups	Information technology sector	Energy efficiency at data centers, networks, and displays • policy and methods to promote green IT Business, academic, and governmental cooperation
GreenTouch	GreenTouch Consortium	Professionals from business and education	Networks of telecom and mobile	Optical, wireless, electrical, routing, design, and telco network innovation. Sustainable data networks
OPERA-Net	CELTIC/ EUREKA (3 years/5 million)	Principal cellphone carriers in Europe	Networks of mobile	A diverse internet radio network Network for mobile wireless connectivity Amplifier, test platform, and link-level power efficiency
GREEN-T	CELTIC (3 years/6 million)	Major cellphone carriers in Europe	Networks of mobile mainly 4G	Multiple cellular mobile device standards Cooperative tactics and cognitive radio QoS assurance

TABLE 5.3

Taxonomy for Green Mobile Computing

Metrics	Full Names	Creator	Targets	Calculation	Units	Remarks
PUE	Power usage effectiveness	Green grid	Data center	PUE = Total facility power/ IT equipment power	Ratio	Ranging from 1 to infinite
DCiE	Data center infrastructure efficiency	Green grid	Data center	DCiE = 1/PUE = IT equipment power/total facility power × 100%	Percentage	Ranging from 0 to 100%
DCP	Data center productivity	Green grid	Data center	DCP = Data center productivity work/total facility power	Ratio	Ranging from I to infinite
ECR	Energy consumption rating	ECR initiate (IXIA, Juniper)	ISP, ICT enterprises	ECR = Energy consumption/effective system capacity	Watt/Gbps	Energy normalized to capacity
ECRW	ECR-weighted	ECR initiate, (IXIA, Juniper)	ISP. ICT enterprises	ECRW = (0.35 × Ej + 0.4 × Ek + 0.25 × Ei)/Tj	Watt/Gbps	These are the energy consumption in full-load, half-load, and idle modes, respectively
TEER	Telecommunications equipment energy efficiency ratio	Verizon NEBS	Transport, switch, router, access, amplifier	$0.35 \times \text{TEEER} \log (0.35 * P_{max} + 0.4 * P50 + 0.25 * P_{min})/\text{throughput}$	log (Watt/ Gbps)	Referring to Energy Consumption Rating Weighted (ECRW) and Trans-Epithelial Electrical Resistance (TEER), P_{max}, P50, and P sleep are the power consumption at 100%, 50%, and 0% load utilization (formula may change based on different types of devices)
CCR	Consumer consumption rating	Juniper	Consumer network devices	$CCR = E/\sum A(j)$	Rad (dimension- less)	E is the power rating of a consumer network device; A is energy allowance per function; j is the set of all allowances claimed. Value 1 matches an average device

are emphasized in Table 5.4. Communication techniques and technologies that support green IoT are crucial in this situation (Al-Fuqaha et al., 2015).

Radio frequency identification (RFID) is a wireless technology that is widely used for tracking and identifying objects using radio waves. It consists of RFID tags and readers. RFID tags are microchips that can store data and are attached to objects, while readers are devices that send and receive signals to communicate with the tags (Ju et al., 2020).

Two types of RFID tags exist: these are active and passive. Active tags have onboard power batteries, which allow them to transmit signals to the reader without the need for external power (Figure 5.14). Passive tags, on the other hand, do not have their own power source and instead harvest energy from the reader's signal to power up and transmit their response (Ali et al., 2020).

Using energy-efficient communication protocols and algorithms, or "green RFID," can help you use less power and have a smaller environmental impact. Tag sizes should be decreased, and transmission algorithms and protocols that promote energy efficiency should be used, to accomplish green RFID (Rahman and Bhuiyan, 2020).

One important aspect of green RFID is optimizing the level of transmission power. Transmitting signals at high power levels consumes more energy and can lead to interference with other devices. Therefore, it is essential to use communication protocols that dynamically adjust the transmission power to the minimum required level for reliable communication (Ju et al., 2020).

Another key aspect is tag estimation. In a typical RFID system, multiple tags may be present in the reader's range, leading to tag collision and overhearing, which can result in inefficient use of the available energy. Communication protocols that support tag estimation can help mitigate these issues by allowing the reader to estimate the number of tags in its range and dynamically adjust its communication parameters accordingly (Al-Fuqaha et al., 2015).

Finally, green communication protocols also focus on avoiding tag collision and overhearing. Collision avoidance techniques ensure that only one tag is transmitting at any given time, while overhearing avoidance techniques reduce the amount of energy wasted by tags that are not being addressed by the reader. Wireless sensor network (WSN) comprises numerous sensor nodes with limited computing capability, storage capacity, and power (Ju et al., 2020) as shown in Figure 5.15.

The sensor nodes scan the environment using a variety of onboard sensors while being linked to a potent base station called a sink. GWSN makes use of methods like data gathering, radio optimization, wireless charging, radio mode activation, and energy-efficient navigation. GWSN optimizes energy use through periodic notification as opposed to ongoing tracking and synchronization without a date (Rahman and Bhuiyan, 2020).

Machine-to-machine (M2M) communication is a fundamental technology in the internet of things (IoT) ecosystem. It enables devices and machines to communicate and share data with each other without human intervention as shown in Figure 5.16. However, the massive number of nodes involved in M2M communication can lead to significant energy consumption, which can be a challenge in terms of sustainability and cost-effectiveness (Al-Fuqaha et al., 2015).

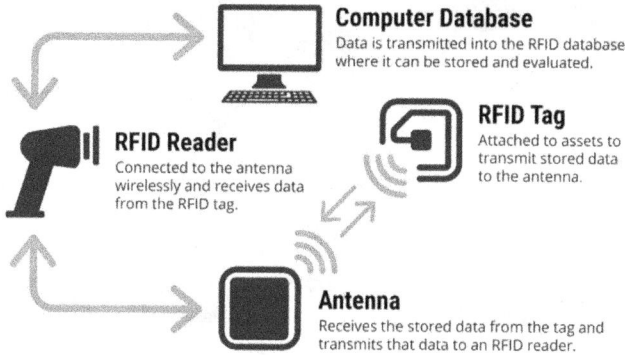

Basic RFID System

Computer Database
Data is transmitted into the RFID database where it can be stored and evaluated.

RFID Tag
Attached to assets to transmit stored data to the antenna.

RFID Reader
Connected to the antenna wirelessly and receives data from the RFID tag.

Antenna
Receives the stored data from the tag and transmits that data to an RFID reader.

FIGURE 5.14 RFID for green IoT

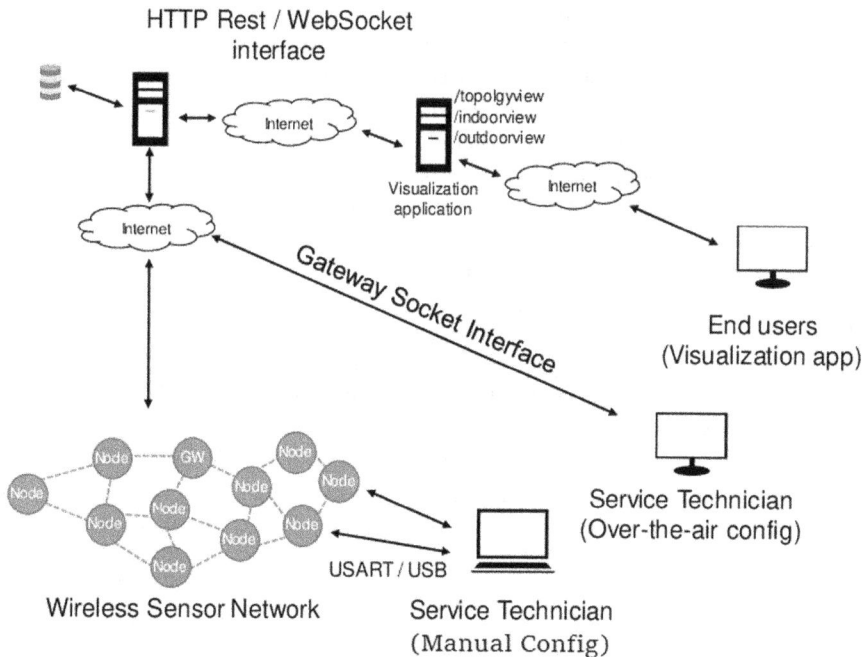

HTTP Rest / WebSocket interface

/topolgyview
/indoorview
/outdoorview

Visualization application

Internet

End users
(Visualization app)

Gateway Socket Interface

Service Technician
(Over-the-air config)

USART / USB

Wireless Sensor Network

Service Technician
(Manual Config)

FIGURE 5.15 Wireless sensor network for green IoT

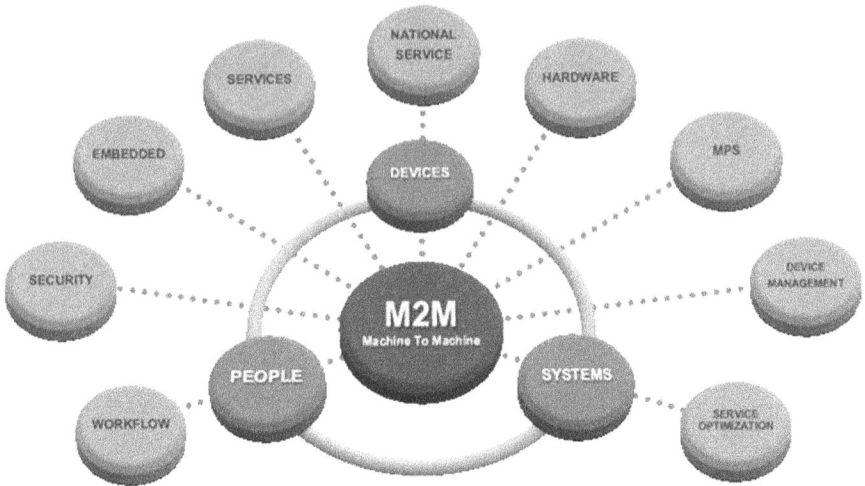

FIGURE 5.16   Machine to machine for green IoT

To address these challenges, green IoT technologies have emerged to improve energy efficiency in M2M communication. Ju et al. (2020) provide a comprehensive overview of these technologies, which include the following.

Intelligent transmission power control: One of the most effective ways to improve energy efficiency in M2M communication is to intelligently control the transmission power of the devices involved. By adjusting the transmission power dynamically based on the distance between the devices, the signal strength, and other factors, energy consumption can be reduced significantly (Ju et al., 2020).

Energy-efficient routing protocols: In M2M communication, routing protocols determine the path that data packets take to reach their destination (Figure 5.17). Energy-efficient routing protocols are designed to minimize the energy consumption of the devices involved in the communication by using the shortest path, reducing the number of hops, and balancing the energy consumption among the devices (Ali et al., 2020).

Activity scheduling: Devices in M2M communication may not need to be active all the time. Activity scheduling can be used to put the devices in sleep mode or turn them off when they are not in use. This can significantly reduce energy consumption while maintaining the required level of communication (Yan et al., 2018).

Energy harvesting: Energy harvesting techniques can be used to generate energy from ambient sources such as solar, thermal, or kinetic energy (Table 5.4). These techniques can be used to power low-power devices in M2M communication, such as sensors or actuators, without the need for external power sources (Al-Fuqaha et al., 2015).

GREEN BIG DATA

Big data creates new difficulties, such as handling vast quantities of heterogeneous, growing exponentially, organized, and unstructured data. For big data analytics to

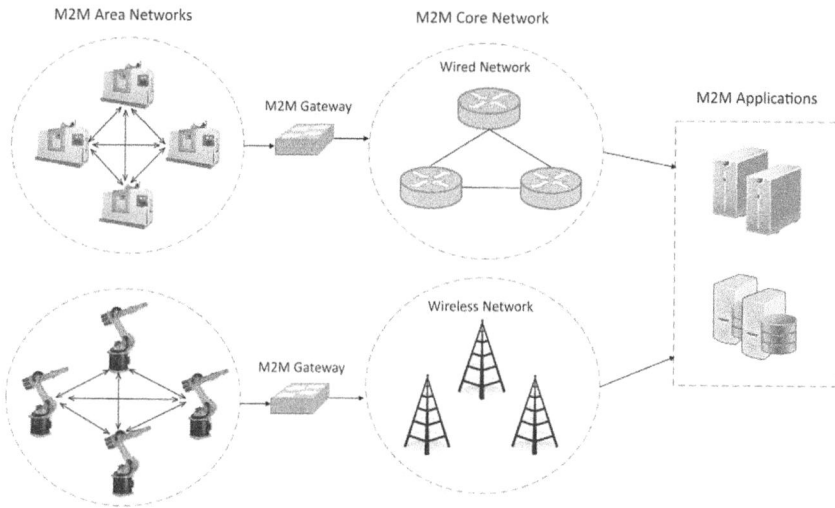

FIGURE 5.17 Machine-to-machine architecture

TABLE 5.4
IoT Enabling Technologies and Their Greening Strategies

IoT enabler	Type	Communication paradigm	Data transfer	Power source	Life-time	Greening strategies
RFID	Active tags	Two ways	Low	Battery	≤5	Energy-efficient
	Passive tags	One way	Very low	Harvested	∞	algorithms and protocols
Sensing network	Smart object	Central	Low	Battery	≤5	Sleep wake-up, data reduction
	Mobile sensing	P2P	High	Battery	≤2	mechanisms
Internet technologies	Cloud	Client/server	Very low	Grid	≤10 years	Turn off unrequired facilities, minimize
	Future internet	Distributed	Very low	Grid	∞	data path length

be as effective as possible for decision-making, it is essential to swiftly and correctly retrieve data from this enormous data set (Figure 5.18). The storing and analytics demands of big data cannot be handled by conventional data management systems; this is where NoSQL technology enters the picture, offering appropriate solutions for prompt data access and effective data processing (Li et al., 2015).

Big data analytics need highly available primary memory, flexible and effective storage, fast communication media, and always-on local physical or corporate cloud servers. Greening, or the process of optimizing energy consumption, resource utilization, and infrastructure scalability, is essential. Therefore, efficient resource

FIGURE 5.18   Green big data architecture

use, energy usage, and system scalability are necessary for green big data analytics (Gandomi and Haider, 2015).

Although high availability, dependability, and consistency are important for the growth of technological infrastructures, the green computing elements of analytics—energy conservation and resource optimization—are less frequently discussed in the literature. In order to prevent physical occupying, the big data analytics technology known as "cloud computing" provides resource outsourcing (Parastatidis and Joy, 2013). This enables multiple users with various analytics needs to use distantly available resources. The development of big data analytics through cloud computing is anticipated to lessen the reliance on personal computers in the new computing age (Tan et al., 2016).

Cloud computing offers lower energy consumption for carrying out complex computational operations on big data in addition to resource preservation. Big data analytics can be performed on a highly available platform called the cloud that minimizes the use of resources and energy. Examples of green computing strategies suggested for big data processing include GreenPlum and GreenHadoop (Huang et al., 2020).

Quick views on data at the petabyte scale are provided by the open-source data repository known as GreenPlum, along with efficient query processing through concurrent processing and optimization. Excellent insights on huge data sets with efficient use are guaranteed by GreenPlum's cost-based query improvement (Figure 5.19). On the other hand, GreenHadoop controls the supply and demand of energy sources linked to big data analytics through the use of green energy sources (Li et al., 2015).

Currently, social networks, healthcare, industries, trade, and business organizations are the main big data providers and users (Khang et al., 2023b). These sources of data are highly adaptable and provide essential statistics needs for quick decision-making. Data centers and processors residing on the cloud effectively manage this big data storage and processing burden, guaranteeing green analytics (Gandomi and Haider, 2015).

According to a research, cloud computing will reduce energy consumption by 38% by 2020. By 2040, it will be better to engage in the thought of recycling renewable and clean energy technology to discover new energy sources (Parastatidis and Joy, 2013). Renewable energy technology is advancing with reduced adaptation costs, effective green homes, and higher renewability requirements in order to reduce carbon pollution, achieve lower and steady energy costs, and have access to reliable energy sources (Tan et al., 2016).

Green big data analytics is crucial in maximizing energy consumption and the reusability of available sources in order to meet the extensive analytics requirements of big data. Big data analyses are combined with sustainable practices like recycling, reusing, and decreasing (Li et al., 2015). Despite the fact that the technology industry

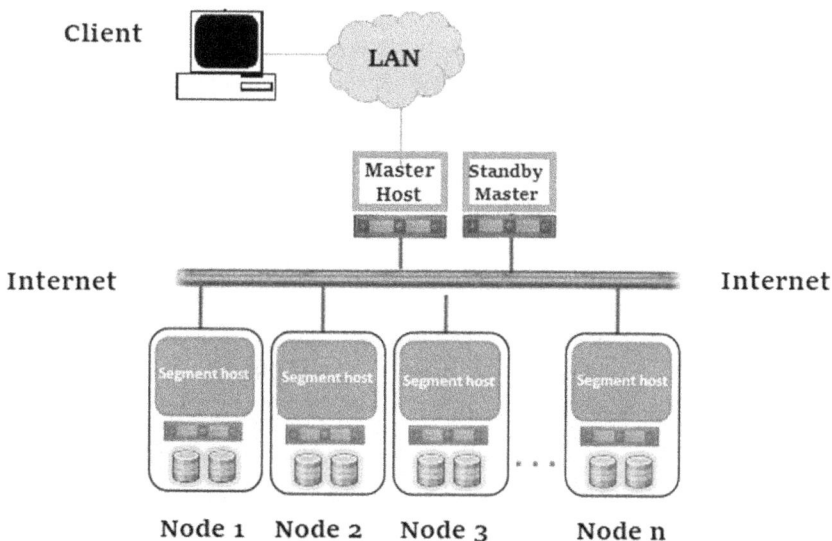

FIGURE 5.19   GreenPlum architecture

is more focused on analytics efficacy than environmental sustainability and computational intricacy, the application of green analytics to big data leads to decreased memory utilization and computational cost. Readers can review a comprehensive future perspective on green big data analytics if they are intrigued (Parastatidis and Joy, 2013).

ADVANTAGES OF GREEN COMPUTING

The practice of using, developing, and disposing of computing tools in an ecologically responsible way is known as "green computing." Green computing aims to lessen technology's negative effects on the ecosystem, consume less energy, and support sustainability. There are several advantages of green computing, including the following.

Reduced Energy Consumption: One of the primary advantages of green computing is reduced energy consumption. By using energy-efficient hardware, software, and infrastructure, organizations can significantly reduce their energy bills and carbon footprint. For example, replacing old servers with new, energy-efficient ones can reduce energy consumption by up to 80% (Masanet et al., 2020).

Lower Costs: Green computing can also help reduce costs associated with energy consumption, maintenance, and disposal. For instance, virtualization can help organizations save money on hardware, maintenance, and cooling costs by allowing multiple applications to run on a single server.

Improved Efficiency: Green computing can also lead to improved efficiency in IT operations. By optimizing computing resources and streamlining workflows, organizations can reduce downtime and improve productivity. For example, cloud computing can improve efficiency by allowing employees to access data and applications from anywhere, at any time (Li and Zhou, 2011).

Positive Public Image: Adopting green computing practices can also help organizations enhance their public image and reputation. By promoting sustainability and environmental responsibility, organizations can demonstrate their commitment to social responsibility, which can improve brand image and attract environmentally conscious customers.

Regulatory Compliance: Many countries and regions have strict environmental regulations that organizations must comply with. Green computing practices can help organizations comply with these regulations, avoid penalties, and reduce legal risks (Quesnel et al., 2013).

Environmental Benefits: Finally, green computing can help reduce the environmental impact of technology. By using energy-efficient hardware and promoting sustainability, organizations can help reduce greenhouse gas emissions, preserve natural resources, and protect the environment.

CONCLUSION

This chapter provides a detailed overview of the various technologies and practices that can be used to promote sustainability and reduce the environmental impact of information technology. The chapter explores the challenges and opportunities associated with green computing and provides recommendations for organizations looking to adopt sustainable IT practices (Rani et al., 2021).

This chapter concludes that sustainable IT practices are essential for organizations looking to reduce their environmental impact and improve their bottom line. The authors argue that green computing can help organizations reduce energy consumption, lower costs, improve efficiency, enhance their public image, comply with regulations, and promote environmental sustainability (Khang et al., 2024).

This chapter identifies several key trends and technologies that are driving the adoption of green computing practices, including green cloud computing that can help organizations reduce energy consumption and improve efficiency by allowing them to scale up or down their computing resources as needed. Big data analytics can help organizations reduce waste and improve efficiency by identifying areas where resources are being underutilized or wasted. The internet of things (IoT) can help organizations reduce energy consumption and improve efficiency by automating processes and optimizing resource use (Khang et al., 2023a).

Mobile computing can help organizations reduce energy consumption and improve efficiency by allowing employees to work remotely, reducing the need for commuting and office space. The chapter also highlights the importance of collaboration and knowledge sharing in promoting sustainable IT practices. The authors argue that by sharing best practices and collaborating on sustainability initiatives, organizations can achieve greater success in reducing their environmental impact and promoting sustainability (Kali et al., 2024).

In conclusion, this chapter on Sustainable IT provides a comprehensive overview of the benefits, challenges, and opportunities associated with green computing. The chapter highlights the importance of sustainable IT practices in promoting environmental sustainability, reducing costs, and improving efficiency. The authors provide recommendations for organizations looking to adopt green computing practices and emphasize the importance of collaboration and knowledge sharing in promoting sustainability (Nassereddine and Khang, 2024).

REFERENCES

Al-Fuqaha, A., Guizani, M., Mohammadi, M., Aledhari, M., & Ayyash, M. (2015). Internet of things: A survey on enabling technologies, protocols, and applications. *IEEE Communications Surveys and Tutorials*, 17(4), 2347–2376. https://doi.org/10.1109/COMST.2015.2444095

Albreem, M. A., El-Saleh, A. A., Isa, M., Salah, W., Jusoh, M., Azizan, M. M., & Ali, A. (2017, November). Green internet of things (IoT): An overview. In *IEEE 4th International Conference on Smart Instrumentation, Measurement and Application (ICSIMA)* (pp. 1–6). IEEE. https://ieeexplore.ieee.org/abstract/document/8312021/

Ali, S. A., Khan, M. N., & Lee, S. (2020). Green IoT: A survey on green wireless communication and networking techniques for internet of things. *IEEE Access*, 8, 24344–24363. https://doi.org/10.1109/ACCESS.2020.2965716

Anan, Muhammad, & Naseer, Nidal. (2015). *SLA- Based Optimization of Energy Efficiency for Green Cloud Computing*. IEEE. https://ieeexplore.ieee.org/abstract/document/7417712/

Arthi, T., & Shahul Hamead, H. (2013). Energy aware cloud service provisioning approach for green computing environment. IEEE. https://ieeexplore.ieee.org/abstract/document/6533371/

Ba, H., Heinzelman, W., Janssen, C. A., & Shi, J. (2013, April). Mobile computing-A green computing resource. In *2013 IEEE Wireless Communications and Networking Conference (WCNC)* (pp. 4451–4456). IEEE. https://ieeexplore.ieee.org/abstract/document/6555295/

Beloglazov, Anton, & Buyya, Rajkumar. (2010). Energy efficient resource management in virtualized cloud data centers. In *2010 10th IEEE/ACM International Conference on Cluster, Cloud and Grid Computing*. https://ieeexplore.ieee.org/abstract/document/5493373/

Bhambri, P., Rani, S., Gupta, G., & Khang, A. (2022). *Cloud and Fog Computing Platforms for Internet of Things*. CRC Press. https://doi.org/ 10.1201/9781003213888

Biswas, M., & Misra, S. (2015). Green mobile cloud computing: Challenges, architecture, and solutions. *Ad Hoc Networks*, 35, 96–108. Green mobile cloud computing: Challenges, architecture, and solutions. Ad Hoc Networks.

Chinnici, Marta, & Quintiliani, Andrea. (2013). An example of methodology to assess energy efficiency improvements in datacenters. In *2013 IEEE Third International Conference on Cloud and Green Computing*. https://ieeexplore.ieee.org/abstract/document/6686070/

Chiueh, S. N. T. C., & Brook, S. (2005). *A Survey on Virtualization Technologies*. Rpe Report, 142. http://132.248.181.216/MV/CursoMaquinasVirtuales/Bibliograf%C3%ADaMaqui nasVirtuales/VirtualizationSurveyTR179.pdf

Deng, Xiang, et al. Eco-aware online power management and load scheduling for green cloud datacenters. *IEEE Systems Journal*, 10(1), 78–87. https://ieeexplore.ieee.org/abstract/document/6877627/

Farahnakian, Fahimeh, Liljeberg, Pasi, Pahikkala, Tapio, Plosila, Juha, & Tenhunen, Hannu. (2014). Hierarchical VM management architecture for cloud data centers. In *2014 IEEE 6th International Conference on Cloud Computing Technology and Science*. https://ieeexplore.ieee.org/abstract/document/7037682/

Gandomi, A., & Haider, M. (2015). Beyond the hype: Big data concepts, methods, and analytics. *International Journal of Information Management*, 35(2), 137–144. https://doi.org/10.1016/j.ijinfomgt.2014.10.007

Huang, Z., Li, D., Wang, H., Huang, T., & Li, Z. (2020). Green big data analytics for internet of things: Opportunities and challenges. *Journal of Systems Architecture*, 105, 101727. https://doi.org/10.1016/j.sysarc.2020.101727

Jain, Anubha, et al. (2013). Energy efficient computing-green cloud computing. In *Energy Efficient Technologies for Sustainability (ICEETS), 2013 International Conference on*. IEEE. https://ieeexplore.ieee.org/abstract/document/6533519/

Ju, Z., Zhang, D., & Gu, F. (2020). Green internet of things: Challenges, perspectives, and opportunities. *IEEE Internet of Things Journal*, 7(2), 922–935. https://doi.org/10.1109/JIOT.2019.2942324

Khang, A., Abdullayev, Vugar, Hahanov, Vladimir, & Shah, Vrushank. (2024). *Advanced IoT Technologies and Applications in the Industry 4.0 Digital Economy* (1st Ed.). CRC Press. https://doi.org/10.1201/9781003434269

Khang, A., Gupta, S. K., Shah, V., & Misra, A. (Eds.) (2023a). *AI-aided IoT Technologies and Applications in the Smart Business and Production*. CRC Press. https://doi.org/10.1201/9781003392224

Khang, A., Rana, G., Tailor, R. K., & Hajimahmud, V. A. (Eds.) (2023b). *Data-Centric AI Solutions and Emerging Technologies in the Healthcare Ecosystem*. CRC Press. https://doi.org/10.1201/9781003356189

Kim, D. J., & Shin, K. G. (2014). Energy-efficient mobile computing. *IEEE Communications Magazine*, 52(6), 74–81. https://ieeexplore.ieee.org/abstract/document/5394036/

Kord, Negin, & Haghighi, Hassan. (2013). An energy-efficient approach for virtual machine placement in cloud based data centers. In *Information and Knowledge Technology (IKT), 5th Conference on*. IEEE. https://ieeexplore.ieee.org/abstract/document/6620036/

Li, M., Yu, S., Zheng, L., & Ruan, S. (2015). Green big data: A survey. *Frontiers of Computer Science*, 9(5), 685–697. https://doi.org/10.1007/s11704-015-4244-8

Li, Q., & Zhou, M. (2011, August). The survey and future evolution of green computing. In *2011 IEEE/ACM International Conference on Green Computing and Communications* (pp. 230–233). IEEE. https://ieeexplore.ieee.org/abstract/document/6061301/

Liu, J., Chen, M., & Liu, H. (2020). The role of big data analytics in enabling green supply chain management: A literature review. *Journal of Data, Information and Management*, 2(2), 75–83. https://link.springer.com/article/10.1007/s42488-019-00020-z

Liu, J., & Wu, H. (2015). A survey of green mobile networks: Opportunities and challenges. *Mobile Networks and Applications*, 20(1), 4–20. https://link.springer.com/article/10.1007/s11036-011-0316-4

Masanet, E., Shehabi, A., Lei, N., Smith, S., & Koomey, J. (2020). Recalibrating global data center energy-use estimates. *Science*, 367(6481), 984–986. https://www.science.org/doi/abs/10.1126/science.aba3758

Mishra, S., & Ahmadi, H. (2016). Green mobile cloud computing: A review. *Sustainable Computing: Informatics and Systems*, 10, 1–16. https://research.utwente.nl/files/6777200/1-s2.0-S2210537915000153-main.pdf

Nassereddine, M., & Khang, A. (2024). Applications of internet of things (IoT) in smart cities. *Advanced IoT Technologies and Applications in the Industry 4.0 Digital Economy* (1st Ed.). CRC Press. https://doi.org/10.1201/9781003434269-6

Parastatidis, S., & Joy, M. (2013). Making big data, green data. *IEEE Internet Computing*, 17(6), 76–80. https://doi.org/10.1109/MIC.2013.78

Patel, Yashwant Singh, Mehrotra, Neetesh, & Soner, Swapnil. (2015). Green cloud computing: A review on green IT areas for cloud computing environment. In *1st International Conference on Futuristic Trend in Computational Analysis and Knowledge Management (ABLAZE-2015)*. https://ieeexplore.ieee.org/abstract/document/7155006/

Qian, L., Luo, Z., Du, Y., & Guo, L. (2009). Cloud computing: An overview. In *Cloud Computing: First International Conference, CloudCom 2009, Beijing, China, and December 1–4, 2009. Proceedings 1* (pp. 626–631). Springer. https://link.springer.com/chapter/10.1007/978-3-642-10665-1_63

Quesnel, Flavien, Mehta, Hemant Kumar, & Menaud, Jean-Marc. (2013). Estimating the power consumption of an idle virtual machine. Green computing and communications (GreenCom) In *2013 IEEE and Internet of Things (iThings/CPSCom), IEEE International Conference on and IEEE Cyber, Physical and Social Computing*. IEEE. https://ieeexplore.ieee.org/abstract/document/6620036/

Rahman, M. A., & Bhuiyan, M. Z. A. (2020). Green IoT: A review on techniques and strategies for sustainable internet of things. *Journal of Network and Computer Applications*, 154, 102630. https://doi.org/10.1016/j.jnca.2020.102630

Rani, S., Bhambri, P., Kataria, A., Khang, A., & Sivaraman, A. K. (2023). *Big Data, Cloud Computing and IoT: Tools and Applications* (1st Ed.). Chapman and Hall/CRC. https://doi.org/10.1201/9781003298335

Rani, S., Chauhan, M., Kataria, A., & Khang, A. (Eds.) (2021). IoT equipped intelligent distributed framework for smart healthcare systems. *Networking and Internet Architecture*, 2, 30. https://doi.org/10.48550/arXiv.2110.04997

Rath, Kali Charan, Khang, A., & Roy, Debanik. (2024). The role of Internet of Things (IoT) technology in industry 4.0. *Advanced IoT Technologies and Applications in the Industry 4.0 Digital Economy* (1st Ed.). CRC Press. https://doi.org/10.1201/9781003434269-1

Tan, W., Wang, X., Zomaya, A., & Yang, L. T. (2016). A survey on green big data analytics. *The Journal of Supercomputing*, 72(8), 3057–3075. https://doi.org/10.1007/s11227-015 -1562-4

Usmin, S., Arockia Irudayaraja, M., & Muthaiah, U. (2014, June). *Dynamic Placement of Virtualized Resources for Data Centers in Cloud.* IEEE. https://ieeexplore.ieee.org/ abstract/document/7033745/

Yamini, R. (2012, March 30, 31). Power management in cloud computing using green algorithm. In *IEEE-International Conference on Advances in Engineering, Science and Management (ICAESM −2012)*. https://ieeexplore.ieee.org/abstract/document/6215585/

Yan, R., Zhang, Y., Li, Z., Xu, C., & Guo, S. (2018). Green internet of things for smart world. *IEEE Internet of Things Journal*, 5(5), 3577–3584. https://doi.org/10.1109/JIOT.2018 .2865313

Yang, Q., & Song, H. (2016). Energy-efficient mobile computing and communication systems: A tutorial review. *IEEE Communications Surveys & Tutorials*, 18(1), 503–525. https://ieeexplore.ieee.org/abstract/document/6861946/

6 Applications of Internet of Things (IoT) in Smart Cities

Mohamed Nassereddine and Alex Khang

INTRODUCTION

The massive growth of cities has recently become a significant issue that concerns governments, researchers, and companies. Nowadays, more than 56% of the world's population is concentrated in cities (*Times*, 2021). China and other developing countries in Asia and Africa have witnessed a rapid growth of urbanization. Indeed, China's urbanization exceeded 64% of its population in 2021 (Xinhua, 2022).

Urbanization enhances living standards by providing water supply, health and educational services, transportation, and sewerage systems. In cities, it is easy to find more job proposals. Indeed, cities are considered the regional economic centers that assist in improving regional economic prosperity. However, urbanization creates many challenges and problems; the increasing population and the enormous use of natural resources in cities may produce environmental and disorder problems (Bertinelli and Black, 2004). Recently, many cities in China have suffered from air and water pollution, environmental degradation, contagious diseases, and increasing crimes. Mexico City also faces a considerable increase in emissions of gases and particulate matter that significantly affect air quality (Calderón et al., 2015). Hence, these challenges force citizens, governments, and stakeholders to consider cities' environment and sustainable development and seek to find suitable technical solutions to reduce these problems (Yin et al., 2015).

The information and communication technology (ICT) revolution provides people with new techniques and tools that can face the urbanization challenges and solve its problems. The recent ICT integration with modern cities can envisage the concept of smart cities. A smart city represents a municipality where ICT improves operational efficiency, exchanges information with the public, and enhances citizens' benefits and government services. Recently, the improvement of smart cities has received more attention from researchers (Khang & Rani et al., 2022). Many new technologies and science, such as artificial intelligence (AI), machine learning (ML), the internet of things (IoT), and others, are used to enhance the efficacy of smart cities (Luckey et al., 2021; Mohapatra and Panda, 2019; Singh & Singh et al., 2020).

DOI: 10.1201/9781003434269-6

IoT revolutionizes the information technology (IT) sector. It consists of interconnecting computing devices communicating without needing human-to-human or human-to-computer interaction (Khang & Hajimahmud et al., 2024). The expression "internet of things" contains two main words: internet, which refers to a global network that connects computers worldwide using the standard internet protocol suite (TCP/IP), and things, which refers to any object or person that can be recognizable in the real world (Madakam et al., 2015). Therefore, IoT can be defined as an open and broad network of intelligent objects capable of exchanging information, arranging, and proceeding according to the current situations and status of the environment (Evangelos et al., 2011).

IoT has recently played a primary role in designing and improving intelligent cities. It allows users to manage, monitor, and control devices remotely in a real-time process. In addition, it provides the capability of collecting real-time data from connected devices. At the same time, the main characteristic of a smart city is the vast integration of information technology and the full-scale application of information resources. A smart city should contain innovative services, management, technology, and life. The integration of IoT in smart cities consists of installing sensors such as IR, lasers, scanners, and others. These sensors can communicate and exchange information through the internet via specific protocols. This process allows the smart city to perform intelligent recognition, location, tracking, monitoring, and management (Kim et al., 2017).

This chapter is organized into five sections. The new research and applications of smart cities are presented in the second section "Smart Cities." The IoT architecture is reviewed in the third section "Internet of Things." The role of IoT in improving the efficiency of intelligent cities is described in the fourth section "Internet of Things in Smart Cities." This chapter concludes the importance of integrating IoT in innovative cities in the final section "Conclusion."

SMART CITIES

Smart cities have significantly improved citizens' and urban environments' quality of life and services. They can work on physical goods in real time and provide important information to citizens concerning public safety, smart buildings, smart parking, smart agriculture (Khang et al., 2023), and transit (Al-Turjman et al., 2022).

This section seeks to present a summarized knowledge of a smart city: its different forms of definition in the literature, its history starting in the 1960s, its community of communication and information technology features, and elementary aspects (Khang & Gupta et al., 2023).

SMART CITIES DEFINITIONS

Smart city does not have any formal definition in theory that worldwide researchers and organizations carry. Many people have described the "smart city" vision and offered different types of definitions from different issues like industry, social, governmental, and others (Jebaraj & Khang et al., 2023). However, based on these definitions, smart cities are created, developed, driven, managed, and operated with new information communication and digital technologies. Researchers have defined intelligent cities in different ways (Laufs et al., 2020), for instance:

- An intelligent city uses ICT to enhance the quality and performance of urban services, including energy, transportation, and utilities, to save resource consumption and overall costs (Yeh, 2017).
- A smart city collects data from various electronic IoT sensors and uses that data to manage resources (Mitton et al., 2012).
- A smart city integrates digital technology into all city functions (Laufs et al., 2020).
- Smart cities are the most recent, implicitly more effective, and embodiment of the sustainable city (Lee et al., 2020).
- Smart cities use advanced ICT to enhance efficiency (competitiveness) and sustainability (energy saving) (Kourtit et al., 2017).
- A smart city's goal is to solve its inherent problems, reduce its outflow, and improve people's lifestyles (Yun & Lee, 2019).

Smart Cities History

Intelligent cities started in the 1960s when the Community Analysis Bureau in the USA initiated using databases, aerial photography, and cluster analysis to gather data and publish reports to run services, mitigate against catastrophes, and decrease deprivation. Therefore, the first generation of smart cities was created. In this generation, the smart city was equipped by technology providers to comprehend the essence of technology in daily human life (Yang, 2020). The second is New Urbanism, a school of urban planning developed in the 1980s that concentrates on comfort and reliability. One of the earliest cities that used computer-aided data analysis in the 1970s was the city of Los Angeles. It is a prime example of a smart city (Subhashini & Khang et al., 2023).

The term "smart city" appeared for the first time during the 1990s in different reports that present new technology tools to enhance services. Söderström et al. (2014) argued that our modern notion of the smart city is the main product of marketing language developed by information technology corporations such as Cisco, IBM, and Siemens. Figure 6.1 shows the evolution timeline of smart cities.

Above Figure 6.1 presents smart city evolution timeline, it is the first digital city practice was implemented in Amsterdam1 in 1994, aiming to facilitate dialogue between residents and politicians. It was a success story since the citizens increasingly accepted the effort, while it was followed by the implication of promoting Internet penetration in the city. After 1994, many projects were created to improve and practice the use of the smart city worldwide. Nowadays, intelligent cities cover more than smart devices and communications. It includes distributed generation and intelligent transportation systems that facilitate human life (Anthopoulos, 2017).

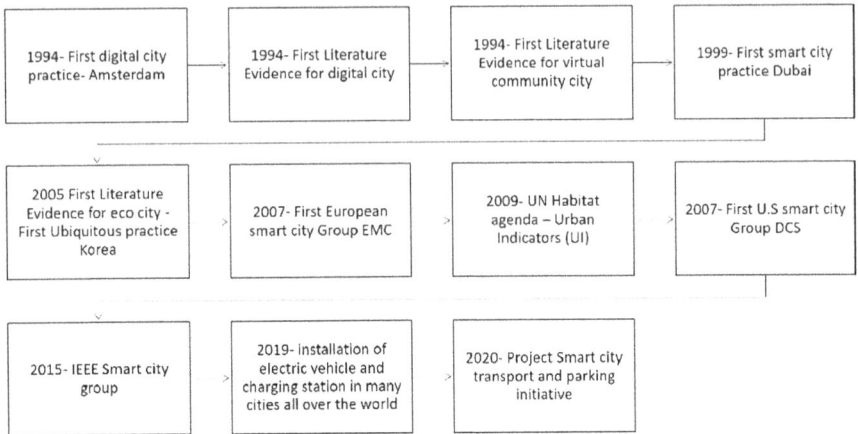

FIGURE 6.1 The timeline of an intelligent city between the years 1994 and 2020 (Das et al., 2019)

SMART CITIES MODEL

A smart city is an up-to-date town area that uses different communication technologies and several types of electronic sensors to gather specific data. These data are used later to efficiently manage resources, assets, and services. Therefore, the primary purpose of smart cities is to raise city functions and boost financial development (Wlodarczak, 2017). Likewise, these types of cities enhance the quality of life for residents by employing innovative technologies and data analysis. Figure 6.2 shows the conceptual model of a smart city (Shah & Suketu et al., 2023).

Figure 6.2 presents the main components of smart city. A smart city is a composition of the following:

- **Smart People**: In smart cities, all activities depend directly on recent technologies. Therefore, living in such cities requires people to maintain prior skills to use this technology and track the advances in day-to-day activities (Guelzim et al., 2016). Therefore, people in smart cities should be "smart" to have the education and the capability to profit from their smart city and enhance it.
- **An Intelligent Government**: Such government requires modern infrastructure for adequate communication with its residents. It facilitates citizens' daily government tasks like public transport through creative new business models like care sharing or social lending (Mellouli et al., 2014).
- **Innovative Economy**: It is based on technical innovation, resource efficiency, sustainability, and high social interest. An innovative economy adopts inventions and entrepreneurial endeavors and increases productivity and competitiveness to enhance residents' quality of life (Chen et al., 2020).

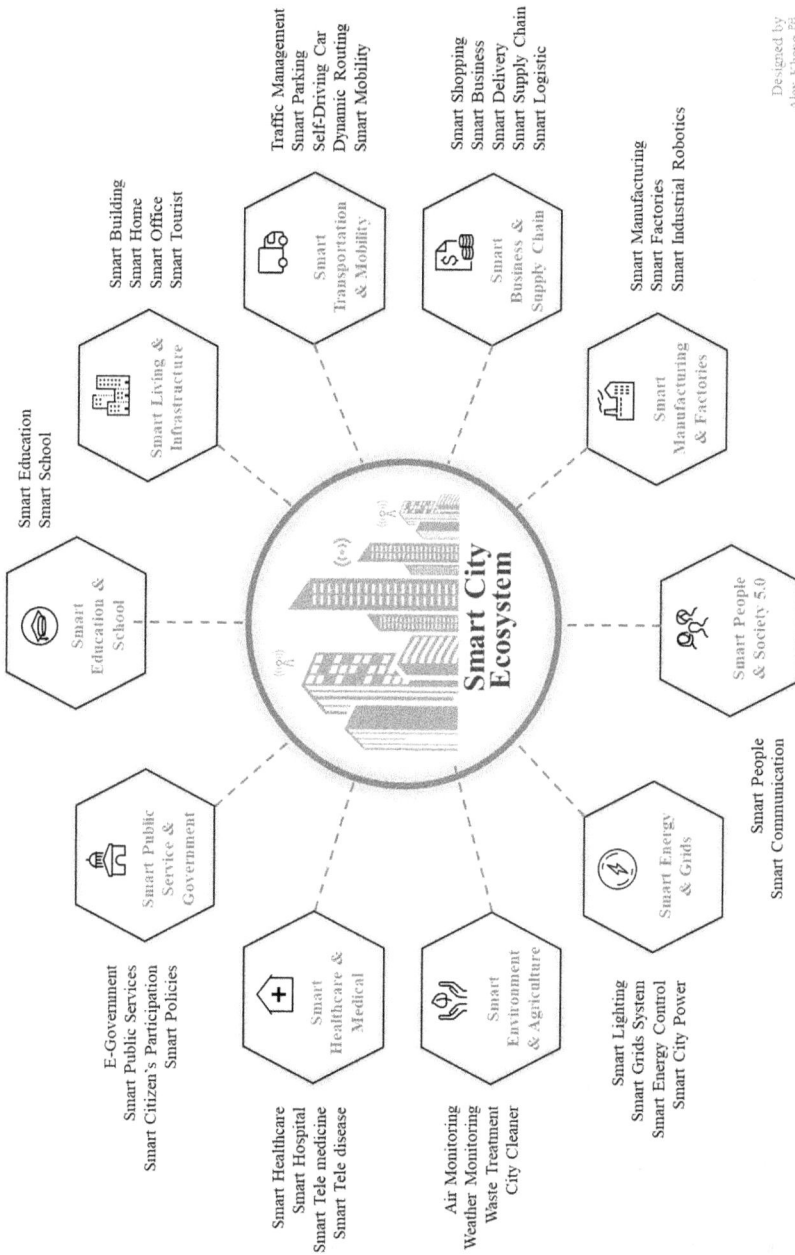

FIGURE 6.2 Smart city model Khang (2021)

- **Smart Living**: Surrounding improvements offer individuals the chance to use a new lifestyle. It involves natural and creative solutions desired to produce a more efficient, controllable, economical, productive, integrated, and sustainable life (Lewis, 2008).
- **Intelligent Traffic**: This is also called smart mobility. It is a crucial part of a smart city aiming to avoid congestion, improve transportation routes, decrease parking time, and enhance pedestrian safety. Thus, fuel consumption, gas emissions, and air pollution are reduced. Smart traffic enables sustainable mobility by developing intelligent public transport norms and engaging advanced innovative systems, including smart applications for ticketing, walking, and cycling (Kylili & Fokaides, 2015).
- **Innovative Environment**: A bright environment is defined as a small world where different intelligent devices constantly work to satisfy residents' lives (Hutton, 2005). It aims to pamper people's experience in every environment by replacing dangerous work, physical labor, and redundant tasks with automated agents. Three types of intelligent environments are defined in the literature:
  - Virtual computing environments enable intelligent devices to access relevant services anywhere and anytime.
  - Physical environments may be implanted with various smart devices, including tags, sensors, and controllers. These devices can have different forms like nano-, micro-, and macro-sized.
  - Human environments: Humans can be escorted by smart devices like mobile phones, use surface-mounted devices, and contain embedded devices.

SMART CITIES' BASIC ASPECTS

Figure 6.3 presents the primary aspects of a smart city. The fundamental three characteristics of a smart city are:

- **Infrastructure Development**: In the smart city, the main goal of the optimal development of infrastructure is to enhance economic, social, cultural, and metropolitan development. For this reason, communication technologies and channels should be improved. Therefore, services like lodging, amusement, telecommunications, industry, and many others can be linked to allow a city to grow and develop.
- **Competitive Environment**: Based on information and communication technologies (ICT), intelligent cities desire to create a competitive environment to develop urban sectors. Therefore, the development of new businesses can be enhanced, and the city's socio-economic performance can be improved.

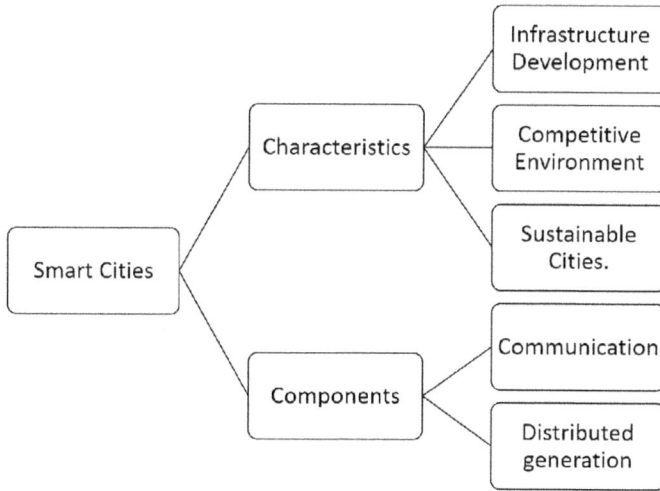

FIGURE 6.3   The primary aspects of a smart city

- **Sustainable Cities**: Sustainability is the primary strategic element of a smart city. Therefore, better consumption habits and more proper energy management should be adopted. In addition, renewable energies for preserving natural resources and environmental care become a priority.

Based on Figure 6.3, a smart city comprises two primary components: communication and distribution generation.

Communication in Smart Cities

Typically, smart cities consist of a vast population and infrastructures spread unequally in the municipality zone. Therefore, a large number of end-point applications will rise, and different kinds of communication technologies should be used in order to connect them. The most common communication technology used within smart cities is represented in Figure 6.4.

Figure 6.4 presents the technologies that play an important role in smart cities' communications. These technologies are classified into two main classes (Haidine et al., 2016):

- Indoor communication in which the sender and receiver are relatively closed (in the same room, home, or building)
- Outdoor communications in which the sender and receiver are relatively far from each other

Usually, wireless communications are the most used technology. The actual internet is nowadays mainly accessed via Wi-Fi or 3G/4G. The mobility and ease of

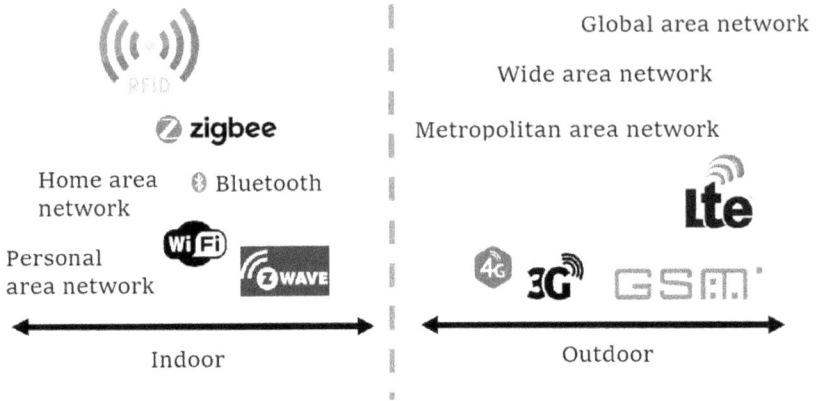

FIGURE 6.4   Main technologies used in smart city communication (Haidine et al., 2016)

deployment are the main advantages of wireless communication. Nevertheless, wired communications will also be used in the communications topography for intelligent cities, particularly when very high speeds are needed (Table 6.1).

Communication technologies enable integration between heterogeneous smart devices and objects. These technologies cover Wi-Fi, Bluetooth, Zigbee, Z-wave, LTE, and others:

- Wi-Fi, also called WLAN, allows the mobile user to link to a local network using a wireless connection. The group of IEEE 802.11 defines the technologies and presents their different types (IEEE 802.11 a/b/g/n/p/aka/ah). The 802.11 standards are based on CSMA/CA for the way sharing and Wired Equivalent Privacy (WEP) for the encryption algorithm.
- Bluetooth (802.15.1) and Zigbee (802.15.4) are low-range communication technologies. They are appropriate for personal or home-area network-based applications.

TABLE 6.1

A Comparison between the Different Technologies Used in Smart Cities

Technology	Protocol	Range	Data Rates
Bluetooth	IEEE 802.15.1	1–100 m	1 Mbps
Z-wave	—	100 m	9.6–100 Kbps
Zigbee	IEEE 802.15.4	10–20 m	250 Kbps
Wi-Fi	IEEE 802.11 (a/b/g/n)	100 m	1–54 Mbps
LTE	3 GPP	30 km	300 Mbps (Downlink), 75 Mbps (Uplink)

- Z-Wave is low-power RF communications technology designed for home automation.
- LTE (Long-Term Evolution) is a prototype for wireless broadband communication for mobile devices and data terminals, based on the GSM/EDGE and UMTS/HSPA standards. It enhances those standards' capabilities and rates using a new radio interface and core network improvements.

These communication technologies are present in most intelligent city applications like smart grids and metering, smart traffic, bright street lighting, smart homes, and intelligent health monitoring (Makani et al., 2022). However, each technology has pros and cons, as illustrated in Table 6.2.

Distribution Generation

Distributed generation (DG) represents the technologies used in smart cities to generate electricity like wind systems, solar panels, or combined heat and power. Usually, in smart cities, electricity is generated near or where it will be used. Distributed generation may decrease the cost of transmission and distribution. Additional benefits include mass production and flexibility, energy efficiency, and emission reduction (Khalil, 2021). According to Ackermann et al. (2001), DG is an electric power source linked to the distribution network from one side or on the customer side of the meter. In addition, according to Söderman and Pettersson (2006), the distributed energy system is considered a complex system including several energy suppliers and consumers, district heating pipelines, heat storage facilities, and power transmission lines in a region. DG should not only be twisted with renewable energy

TABLE 6.2
Advantages and Limitations of Communication Technologies within a Smart City (Yaqoob et al. 2017)

Technology	Advantages	Limitations
Bluetooth	Cheap and easy to install	Short-range communication secure flaws
Z-wave	Power saving, collision avoidance, and low cozy	A bit slower
Zigbee	A lot simpler than Zigbee	• Mobility management is complicated
Wi-Fi	Lack of wires and mobility	• Security flaws • High signal attenuation • Limited service radius • Less stable compared to wired connections
LTE	• Backwards compatibilityHigh-spectrum efficiency stations for data transmission • Reduce the problem of lagging in internet connection	Higher cost due to the usage of and future-proofing additional antennas at the network base

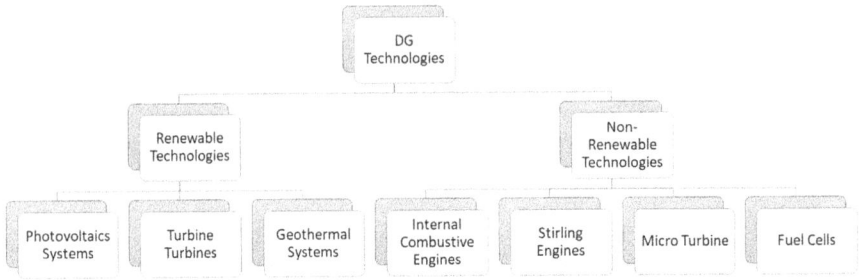

FIGURE 6.5 Distributed generation technologies for power generation. Source: Razavi et al. (2019)

generation. Renewables can be used in DG and are encouraged by certain lobbying groups, though non-renewable technologies could also be considered in DG systems. Some DG technologies, such as internal combustion engines and gas turbines, are ancient. However, with the changes in the utility industry, several recent technologies have been created or progressed regarding commercialization. Figure 6.5 shows recent DG technologies (Razavi et al., 2019).

INTERNET OF THINGS

The internet of things (IoT) describes real-world objects, also known as things that are connected to form a network. Computing devices, digital equipment, mechanical machines, animals, and people are all examples of objects. Each object has a unique identifier known as a UID. Objects are also outfitted with sensors, software, and other technologies that enable them to communicate information without needing human-to-human or human-to-computer interaction. In 1998, academics developed the concept of the internet of things. However, most of its implementations are still in the works (Khajenasiri et al., 2017). More than seven billion devices are now classified as IoT devices (Balaji et al., 2019).

Many organizations and industries are rapidly adopting IoT to improve operational efficiency, provide better customer service, and make better decisions. An IoT ecosystem comprises web-enabled intelligent devices that receive, send, and behave based on their ecological status. They utilize embedded systems such as central processing units (CPU), sensors, and communication hardware (Redelinghuys et al., 2020).

In this section, the IoT architecture is described based on the information presented and adopted in the literature. In addition, IoT technologies are illustrated due to their significant ability to improve people's lives and decrease human efforts. Moreover, protecting and securing interconnected devices in IoT networks are critical to preventing attacks and infiltration; thus, security and privacy are briefly discussed later in this section. Finally, some practical and trendy IoT applications are presented (Nassereddine & Khang et al., 2024).

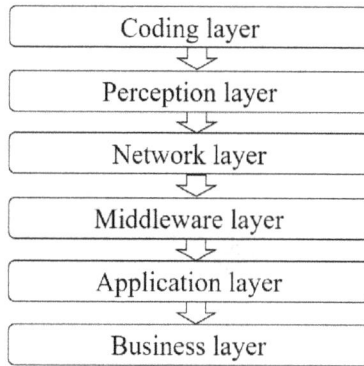

FIGURE 6.6   The internet of things architecture (Kumar and Mallick, 2018)

IoT Architecture

There is no universally accepted architecture for the internet of things. Various researchers have proposed various architectures. However, it should include at least two devices with sensor layers that can communicate and exchange data (see Figure 6.6). Thus, the six-layer IoT architecture is briefly described in this section.

The architecture of IoT consists of six layers are described as follows:

- **Code Layer (Zhang et al., 2012)**: The foundation of IoT is the coding layer, which identifies the objects of interest. Different IDs are assigned to objects in this layer to distinguish between them.
- **Perception Layer (Bandyopadhyay & Sen, 2011)**: It offers a physical sense of connected elements. It shows data sensors in different forms, such as RFID tags, IR sensors, or other sensors that collect values like temperature, humidity, speed, and location of the objects. This layer gathers valuable information using the sensor devices connected to objects. After that, the perception layer converts available information into digital signals that are passed later to the network.
- **Network Layer (Zhang, 2011)**: This layer receives the information in digital form per the perception layer and transfers it to the Middleware layer. The transfer is performed using transmission mediums such as 3G, 4G, Wi-Fi, Zigbee, and others. This layer usually uses protocols like IPv4, IPv6, DDS, and others.
- **Middleware Layer (Muccini & Moghaddam, 2018)**: This layer aims to process information from the network layer using an intelligent processing unit. Based on the result of the processing step, a fully automated action can be taken. This layer should include new technologies such as cloud computing to guarantee direct access to the database to store helpful information for later use.
- **Application Layer (Burhan et al., 2018)**: This layer performs applications of IoT based on processed data received from the middleware layer. IoT applications include smart homes, farms, renewable systems, and cities.

- **Business Layer (Kumar & Mallick, 2018)**: This layer's responsibilities are to manage the IoT application and services.

IoT Technologies

IoT development requires integrating computing systems in which objects are uniquely identified. These objects should be able to collect available data and interact with each other. This process requires the use of new technologies. In this section, the most relevant technologies that are integrated into IoT development are presented, which are as follows:

- **Cloud Computing**: Uses IoT to form networks of sensors, actuators, computers, and other devices with limited processing capability and storage (Stergiou et al., 2018). Usually, the massive communication between interconnected devices produces an enormous amount of data generated. Cloud computing is used in IoT systems to provide virtually unlimited storage and processing power. Cloud systems are developing to deliver outstanding support to the internet of things (IoT), as shown in Figure 6.7.
- **Wireless Sensor Network (WSN)**: Represents the technology that allows intelligent devices to communicate and access the internet wirelessly (Figure 6.8). Recently, it has been one of the most significant technologies (Abdul-Qawy et al., 2020). Indeed, WSN is an aggregation of several low-power, low-cost, and multipurpose sensor nodes transmitting wireless over a short distance. Sensor network has donated to several applications and has developed to implement the technology into the innovative city environment.

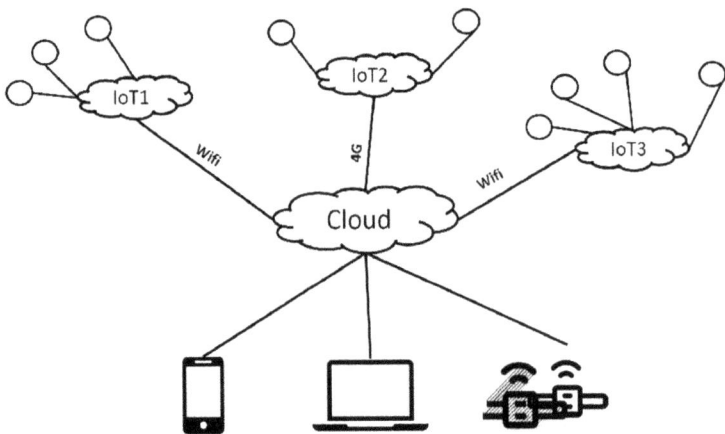

FIGURE 6.7   Cloud computing in the internet of things (Dang et al., 2019)

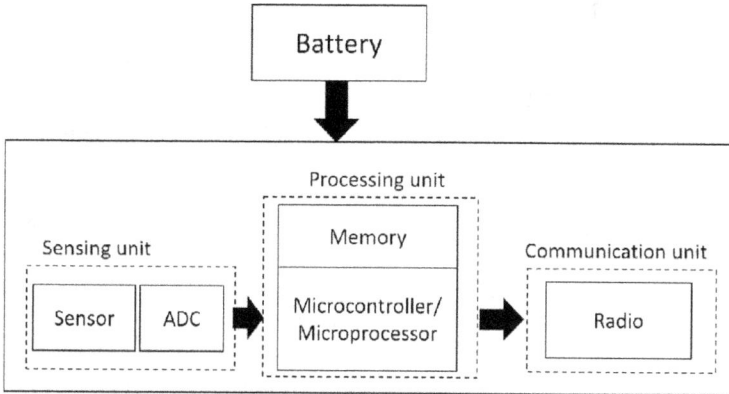

FIGURE 6.8 The essential components of wireless sensor network (Farsi et al., 2019)

The sensor gathers analog data from the physical environment. These data are converted to a digital signal using analog-to-digital circuit (ADC). A microprocessor or a microcontroller is the central part of the processing unit. It achieves intelligent data processing and manipulation. A radio system, usually a short-range radio, is the central part of a communication unit that performs data transmission and reception. In addition, a tiny battery is used to power the entire system. Using all these components, WSN can perform many tasks like data collection, network analysis, correlation, and fusion of its data with data received from other sensors.

Radio frequency identification (RFID) (Cui et al., 2019) is considered a new standard of sensing and communication for information systems. It integrates the characteristic of wireless information and power transfer (WIPT), energy-efficient sensing capabilities, and object identification. RFID sensor labels featuring contactless sensing, wireless information transfer, wireless powered, lightweight, non-line-of-sight transmission, flexible, and pasteable are a vital enabling technology for future internet of things (IoT) applications, such as manufacturing, logistics, healthcare, and smart cities.

Nanotechnologies (Mezaall et al., 2018) recognize smaller and improved versions of connected objects. These objects can be sensors, actuators, or any other smart device. Such nano-devices made of nano-objects can be connected to construct a new version of IoT called the internet of nano-things. The goal of using nano-devices is to decrease energy consumption overall system.

Optical technologies (Miladić-Tešić et al., 2020) have been developed to produce many new technologies such Li-Fi and Cisco's bidirectional optical that could be a significant step forward in the development of IoT. Li-Fi, an epoch-making visible light communication (VLC) technology, can supply outstanding connectivity on a higher bandwidth for the interconnected objects in IoT systems. Besides, bidirectional technology gives a 40G Ethernet for big data from various IoT devices.

SECURITY AND PRIVACY IN IoT

Due to the shortage of appropriate infrastructures and servers to exchange messages among objects, authentication becomes highly challenging in IoT systems. Indeed, objects have lower resources compared to PCs, PDAs, and cell phones that hold complex computing operations. Recently, some solutions to authentication problems have been suggested.

However, proposed solutions are still weak and suffer from serious problems. They cannot help in detecting and solving the man-in-the-middle attack problem. Data integrity solutions instruct that a hostile cannot make a transaction without being detected by the system. In IoT, when RFID systems and sensor networks are integrated using the internet, a new problem arises. Sensor nodes or RFID tags are often applied in broad areas where they should be isolated. Therefore, data can be easily changed by hostiles. Usually, memory should be protected in tag technologies, and other solutions have been proposed for WSNs. In addition, transferred messages may be covered using the Keyed-Hash Message Authentication Code (HMAC) scheme. Recently, several cryptographic methodologies have been offered to reinforce security. Such solutions cannot be devoted to the IoT. Indeed, some IoT components like RFID tags and sensor nodes have limited communication and computations capabilities and small energy storage (Govinda and Saravanaguru, 2016).

IoT APPLICATIONS

IoT has emerged as a critical technology in our lives. It will keep evolving as long as humans benefit from it. Many homes and offices around the world now use IoT technologies. Numerous electronic devices, such as lights, ovens, fans, heaters, and others, have recently been integrated with sensors and actuators that help with energy efficiency, monitor, and control the amount of heating and level of light, lowering costs and increasing energy savings (Whitmore et al., 2015).

In addition, IoT is used to monitor air quality. Indeed, embedded sensors collect data on air components such as carbon monoxide (CO), nitrogen dioxide (NO_2), temperature, and humidity. This information can determine the air pollution level and take precautions if it exceeds the standard level (Xiaojun et al., 2015).

Smart farming has recently emerged as one of the IoT applications. IoT systems may collect and process information such as temperature, solar intensity, plant wetness, and soil status to send notifications to farmers about water and plant treatment (Dagar et al., 2018; Ratnaparkhi et al., 2020).

IoT allows healthcare providers, such as physicians, to extend their capabilities beyond the confines of their clinical settings. Indeed, home monitoring systems have recently been used to allow patients and doctors to track their health records and automatically recognize situations requiring a physical presence (Kodali et al., 2015).

Furthermore, interconnected sensors, lights, and intelligent devices are used in smart cities to collect, process, and analyze data. Recently, the smart city has used this data to improve infrastructure, public utilities, and benefits. The following section will provide a detailed description of IoT integration in smart cities (Ahmad & Muthmainnah et al., 2023).

INTERNET OF THINGS IN SMART CITIES

The IoT uses the internet to connect various heterogeneous objects. Therefore, all existing things must be connected to the internet to access these objects quickly. Indeed, in smart cities, a network with internet connection is crucial to remotely survey intelligence objects such as power usage monitoring, light management, and air conditioner management. For this purpose, sensors can be spread to different locations to collect and interpret data for utilization refinement. Figure 6.9 shows the main application of the IoT in a smart city (Khang & Rani et al., 2022).

This section represents four application domains: smart homes, intelligent parking, healthcare systems, and traffic management in public transportation.

SMART HOME

Typically, innovative home products use IoT technology in order to communicate and exchange information. IoT provides all home devices with internet access, which enlarges the possibilities of home networks. Using this network, people can monitor what happens in their house, watching through any smart device like a smartphone or computer. IoT applications allow devices to connect and share information

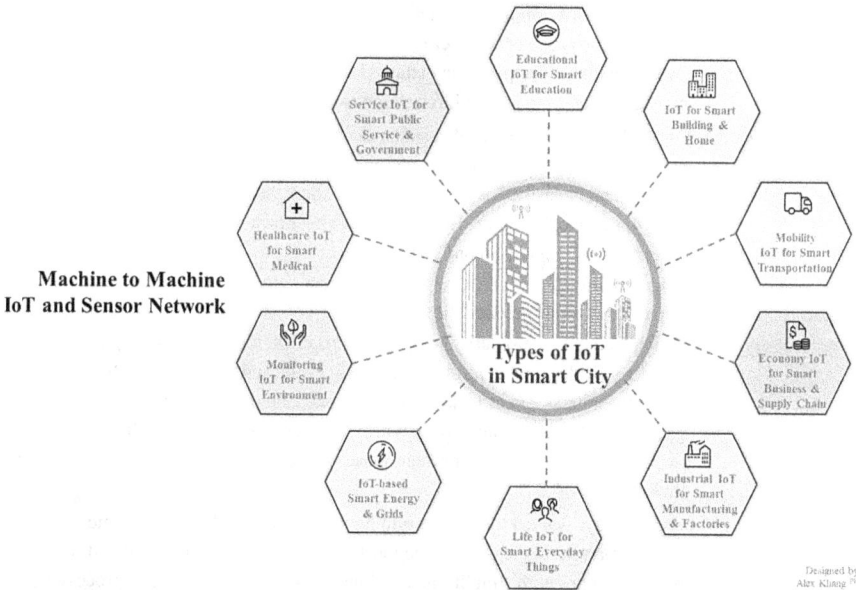

FIGURE 6.9 IoT applications in smart cities (Khang, 2021)

without human intervention. Some examples of innovative home products with their functions are listed as follows (Lobaccaro et al., 2016):

- Cameras that track the outer part of houses even during nights.
- A video door phone delivers more than a doorbell by receiving facial information about people at the house's door.
- Motion sensors send an alert when there is motion around houses, and they can even detect the difference between pets and thieves.
- Door handles can unlock with scanned fingerprints or a code, eliminating the necessity of using house keys.
- Audio systems broadcast the music from stereos to any room with connected speakers.
- Remote controls, keypads, and tabletop controllers are the means of triggering intelligent home applications. Devices also have built-in web servers that permit users to access their information online.

Consequently, some benefits of smart homes are summarized in Table 6.3.

TABLE 6.3
Benefits with a Description of Intelligent Homes

Benefit	Description
Control and monitoring	The IoT greatly enhances how people can manage and survey all the home processes. For example, the fridge can inform about whether the yogurt becomes terrible in two days or whether it is necessary to add milk to the shopping list. Every day the IoT application gathers data concerning how the household works, manages it, and sends vital information to the residential
Cost and energy savings	A smart home provides detailed statistics about each smart device's work and energy consumption. Therefore, optimization of energy usage and cost-saving becomes more easily. Also, residential can allow the intelligent home system to do everything independently. The system can shut down unused devices to reduce power consumption if no one is home
Environment impact	Besides cutting down bills, energy savings help reduce a negative environmental impact and live a "greener" life
Better security	The innovative home security system is an excellent tool for protecting properties since it provides total control of everything inside and outside of the house. Security cameras, smart locks, motion, smoke, and other sensors—all these gadgets work together and notify whether something goes wrong. In addition, the owners can quickly check for their houses while sitting in their offices or far from their houses
Comfort	People always wanted to have everything more comfortably and conveniently, and IoT in smart homes is perhaps the most significant achievement in this direction. It perfectly copes with simplifying and improving people's life: many processes in their home run autonomously, yet the overall control remains in their hands

SMART PARKING

Car parking facilities are one of the significant issues in smart cities. Nowadays, finding an available parking spot within a city is very difficult for drivers, and it may become harder with the increasing number of personal cars. This situation pushes researchers on smart cities to launch actions to enhance the efficiency of parking resources, hence reducing parking search times (Khanna and Anand, 2016). A parking problem can be resolved if the driver has prior information about the availability of parking spaces near his destination. The international parking institute introduced a valuable survey in 2018 to reflect the growth in the number of creative ideas in parking systems (Al-Turjman and Malekloo, 2019).

Nowadays, some parking systems help drivers by delivering real-time information about available parking spaces. These systems are based on IoT. The parking areas should be equipped with specific sensors to monitor occupancy. In addition, a quick processing unit should be used to gain valuable insights from data collected from different sources (Sakthivel & Shashi et al., 2022). An intelligent parking system is a complex system that is composed of two main parts (Figure 6.10):

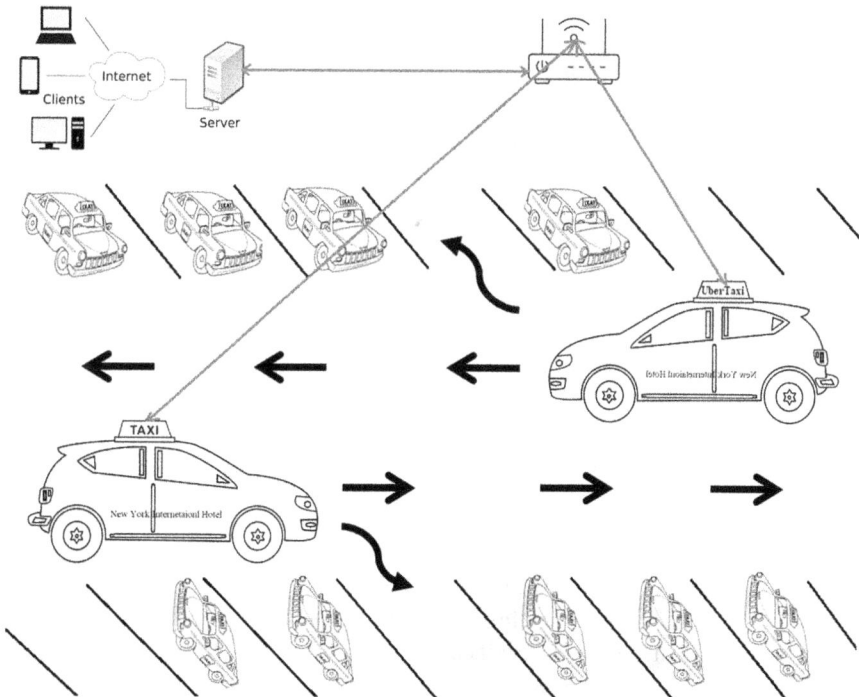

FIGURE 6.10 IoT technology used in the general architecture of intelligent parking (Khang, 2021)

- Hardware devices detect and display the occupancy level of parking spaces, such as a router, server, processing unit, and parking display sign.
- Software components manage the allocation of available parking spaces and redirect the cars accordingly, such as mobile and web applications.

HEALTHCARE

Before the internet of things, patients could interact with their doctors only during a visit or via telephone or message communication tools. Doctors or hospitals could not constantly survey patients' health and give appropriate recommendations. Recently, due to the internet of things (IoT), doctors can remotely monitor their patients. This monitoring process could allow the healthcare sector to keep patients safe and healthy and authorize physicians to provide excellent care. In addition, IoT increases patient concentration and fulfillment as communications with doctors become more comfortable and efficient. Likewise, IoT helps decrease the length of hospital stay of the patient by allowing the doctor to monitor the patient's health remotely. It also can avoid re-admissions, reduce healthcare costs enormously, and improve treatment results (Lodhi et al., 2017).

The healthcare industry used IoT to enhance the connection among this sector's individuals (physicians, patients, and workers) and strengthen their relationships by reducing space and time in providing healthcare solutions. This section discusses IoT applications in healthcare for patients, physicians, hospitals, and insurance companies (Khang & Rana et al., 2023).

IoT for Patients

In past years, the integration of IoT in the healthcare sector has changed people's lives, especially aged patients. It allows a continuous pursuit of health conditions due to IoT smart devices. It has a significant influence on people who live alone. Moreover, a mechanism sends signals to the family members or health providers when changes in the routine activities of a patient occur (Akkaş et al., 2020). Furthermore, as the use of wirelessly connected sensors and smartphones has grown, so has the remote healthcare monitoring application. IoT health monitoring can help prevent sickness.

Furthermore, it can aid in the diagnosis of a person's health without the requirement for a doctor's presence. Valsalan et al. (2020) presented a portable physiological monitoring application that can continually display patient health data such as heart rate and body temperature. It may also monitor the critical parameters of the room. This program uses communication channels such as the Wi-Fi module to continue the check, display, and control device to display the patient's health condition and store this information. Using this IoT health application, authorized individuals can have access to data saved on any IoT platform, and ailments are diagnosed remotely by doctors based on the values obtained.

Debauche et al. (2019) suggested the "Fog IoT Cloud-Based Health Monitoring System" as a novel IoT healthcare monitor system. This technology used physiological and environmental cues to transmit vital information such as patients' everyday actions and workouts. This technology enables healthcare personnel to monitor the patient's health status as well as behavioral changes in elderly or independent people. Furthermore, it examines patients' restoration and recovery methods. This Fog-IoT system is built on a wireless sensor network, gateway, and Lambda cloud architecture, which can store and process data. It allows healthcare workers to quickly access data and automatically validate identified irregularities (Khang & Hahanov et al., 2023).

Srinivasan et al. (2023) considered the case of patients with paralysis. Indeed, these individuals are frequently unable to express their requirements because they are unable to speak correctly or use sign language. The authors devised a technique that allows the crippled patient to use any part of his body to print text on the LCD. Patients can use this system to communicate with doctors, therapists, or loved ones at home or at work over the internet.

IoT for Physicians

Due to home monitoring equipment embedded with IoT, physicians can monitor patients' health effectively. They can detect patients' dedication to treatment programs and immediate medical attention demands. Indeed, IoT allows healthcare professionals to be more attentive and communicate easily with patients. In addition, data collected from IoT devices is instrumental. It may help doctors identify the best treatment process for patients and achieve the expected outcomes (Alhasan et al., 2022).

Haghi et al. (2020) offered a novel platform for extended control parameters. This platform derives the most important parameters from an ambient domain. The IoT gateway serves as a conduit between the wearable sensors and the IoT server. It allows for bidirectional communication between the user and medics in real time. The physician can observe in real time and set up the necessary parameters for measurement via the IoT gateway utilizing this platform. It may also activate and disable the sensors on the wearables. As a result, the suggested platform is not limited to certain populations, but can be used in daily practice or medical research.

A new IoT-based system for collecting patient data was proposed (Gera et al., 2021). This system is called patient data collection. It can assist in the creation of patient primary reports, clinic patient care, and diagnostics. This system aims to offer exceptional patient help in remote areas. It has the intelligence to examine the collected data via IoT sensors and suggests a proposal for a check-up. It offers also an interface between the patient, the physician, the pharmacist, and the diagnostician.

IoT for Hospitals

IoT devices labeled with sensors track medical equipment locations like wheelchairs, oxygen pumps, and other hospital equipment. In addition, IoT is used in hygiene monitoring in order to decrease infection rates. IoT devices also help in supporting pharmacy inventory management, monitoring the environment, checking refrigerator temperature, and many other tasks. It is a very useful technology in hospital sectors (Thangaraj et al., 2015).

Valanarasu (2019) suggested an intelligent and safe framework for the hospital environment based on IoT and AI. This solution addresses the shortcomings of the present hospital information system. The proposed application virtually aids in the treatment, diagnosis, patient monitoring, and maintenance of hospital information in electronic format.

54Leng et al. (2020) presented an IoT-based system that aids in the integration of new technologies such as LoRa, Wi-Fi, and other data collecting devices in the hospital. These devices will be utilized for data collection, transferring the data to the cloud platform for further analysis over a secure connection, and displaying the results to users in real time via the user interface.

IoT for Health Insurance Companies

Health insurers can widely benefit from the use of IoT-connected intelligent devices. Indeed, insurance companies can collect data captured via health monitoring devices to use later in their underwriting and claims operations. This data helps them to detect fraud claims and identify underwriting opportunities. Recently, insurers suggested that their clients use and share health data generated by IoT devices to keep track of their everyday activities and fidelity to treatment plans and preventive health measures. It will help them in seriously decreasing claims. In addition, IoT devices can also allow insurance companies to validate claims via the data caught by these devices (Singh & Sharma et al., 2020; Table 6.4). This table presents the benefits of IoT in the healthcare sector, with their corresponding descriptions (Joyia et al., 2017).

TABLE 6.4
Some Benefits of IoT in Healthcare and IoT Integration in the Healthcare Sector

Benefit	Description
Cost reduction	IoT enables real-time patient monitoring, reducing doctor visits and hospital stays
Improved treatment	It enables physicians to make evidence-based informed decisions and brings absolute transparency
Faster disease diagnosis	Continuous patient monitoring and real-time data help diagnose diseases at their early stages. Sometimes, it is beneficial in detecting diseases before symptoms occur
Proactive treatment	Continuous health monitoring opens the doors for providing proactive medical treatment
Error reduction	Data generated through IoT devices help in effective decision-making and ensure smooth healthcare operations with reduced errors, waste, and system costs (Anh & Vladimir et al., 2024)
Drugs and equipment management	Drugs and medical equipment are managed and used efficiently through connected devices, leading to cost reduction (Khang & Abdullayev et al., 2024)

TRANSPORTATION TRAFFIC

With the vast population increase in cities worldwide and the enormous production of all kinds of vehicles by manufacturers, the number of vehicles on the roads will only continue to grow. Therefore, traffic congestion increases naturally, particularly in significant urban areas and during peak rush hour. This situation pushed the researcher, government, and urban planners to enhance the traffic management systems to make both transportation systems and fuel consumption safe. Recently, new approaches, called intelligent transportation systems, have been proposed to resolve these problems. They can help resolve the problems by integrating existing technology like IoT with the road and city infrastructure (Avatefipour and Sadry, 2018).

Many applications of IoT integration in ITS have been recently developed, such as theft vehicle detecting techniques, prioritized vehicle scheduling, non-stop toll-collecting system, and traffic violation monitoring. In the future, the interconnections of automobiles—people and roads—vehicles are envisioned as extensive systems to reduce traffic congestion, decrease air pollution, and provide secure transportation (Abishu et al., 2021). In this section, only clustering and controlling methods of the ITS system are presented (Figure 6.11) (Sakthivel & Shashi et al., 2022) (Goel & Khang et al., 2023).

V2V communication, the central part of ITS, is a communication technology that reduces vehicle crash numbers. It is based on vehicular ad hoc networks (VANETs), consisting of a wireless network that allows vehicles to communicate and share information about their driving behaviors. This information may include acceleration, current position, and travel direction. This technology's importance is improving the

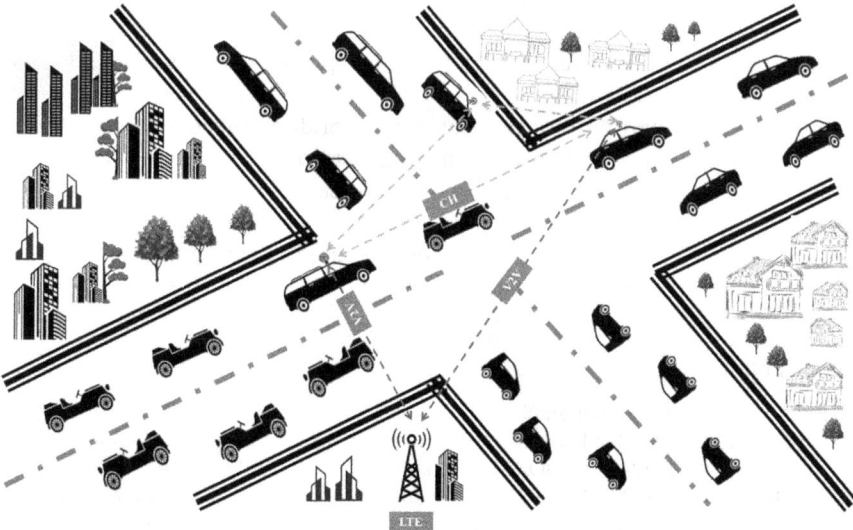

FIGURE 6.11 Clustering strategy in ITS. V2V represents vehicle-to-vehicle communication. CH represents cluster head (Khang, 2021)

roads' safety by providing incident alerts before a driver sees or detects them. In the clustering method, V2V communication allows CH to send and receive data over a communication channel to collect traffic conditions within a specific zone (oval parts represented in Figure 6.11). Therefore, the driver has a good time preventing crashes or other problems.

A roadside unit (RSU) collects traffic data from a fixed sensing area along a road and transmits it to traffic control devices and a central traffic management center. These devices also serve as a data source for intelligent vehicles to gather future traffic data (Kouonchie et al., 2022).

CONCLUSION

Smart cities have been a dream for several societies for a long time. Recently, remarkable advancements have been made to present innovative cities in the real world. The growth development of the internet of things and cloud technologies has enormously advanced intelligent cities. A smart city is a city that combines digital technologies into its infrastructure to secure human life and make it more comfortable. It allows residents to benefit from all services while maintaining and conserving natural resources. Smart cities offer better use of the area, less traffic, less air pollution, and more efficient civic services (Vicky & Amar et al., 2023).

In this chapter, a review of the concept of smart cities and IoT has been presented. A small history of smart cities has also been presented. Then, the innovative city model has been described briefly. After that, the six layers' architecture of IoT is presented. In addition, some IoT technologies like cloud computing, wireless sensor networks, nanotechnologies, and optical technologies have been introduced briefly. Finally, the integration of IoT in smart cities' application has been described. The authors focus on four main applications of IoT in smart cities: smart homes, healthcare, smart parking, and transportation traffic (Khang & Ragimova et al., 2022).

In the future, smart cities may include mega bridges, super-connected streets, more renewable energy systems, and more underground spaces. In addition, big data, the internet of things, and artificial intelligence may be integrated to make the healthcare sector, smart parking, and intelligent transportation systems more efficient. Furthermore, these technologies should be used to reduce pollution and fuel consumption (Rani & Bhambri et al., 2022).

REFERENCES

Abdul-Qawy, A. S. H., Almurisi, N. M. S., & Tadisetty, S. (2020). Classification of energy saving techniques for IoT-based heterogeneous wireless nodes. *Procedia Computer Science, 171*, 2590–2599. DOI: 10.1016/j.procs.2020.04.281

Abishu, H. N., Seid, A. M., Yacob, Y. H., Ayall, T., Sun, G., & Liu, G. (2021). Consensus mechanism for blockchain-enabled vehicle-to-vehicle energy trading in the internet of electric vehicles. *IEEE Transactions on Vehicular Technology, 71*(1), 946–960. DOI: 10.1109/TVT.2021.3129828

Ackermann, T., Andersson, G., & Söder, L. (2001). Distributed generation: A definition. *Electric Power Systems Research, 57*(3), 195–204.DOI: 10.1016/S0378-7796(01)00101-8

Al Yakin, A., Muthmainnah, K. A., Mukit, A., & Zuber, M. (2023). Personalized social-collaborative IoT-symbiotic platforms in smart education ecosystem. *Smart Cities: IoT Technologies, Big Data Solutions, Cloud Platforms, and Cybersecurity Techniques* (1st Ed.). CRC Press. https://doi.org/10.1201/9781003376064-15

Akkaş, M. A., Sokullu, R., & Çetin, H. E. (2020). Healthcare and patient monitoring using IoT. *Internet of Things, 11*, 100173. DOI: 10.1016/j.iot.2020.100173

Alhasan, A., Audah, L., Ibrahim, I., Al-Sharaa, A., Al-Ogaili, A. S., & M. Mohammed, J. (2022). A case-study to examine doctors' intentions to use IoT healthcare devices in Iraq during COVID-19 pandemic. *International Journal of Pervasive Computing and Communications, 18*(5), 527–547. DOI: 10.1108/IJPCC-10-2020-0175

Al-Turjman, F., & Malekloo, A. (2019). Smart parking in IoT-enabled cities: A survey. *Sustainable Cities and Society, 49*, 101608. DOI: 10.1016/j.scs.2019.101608

Al-Turjman, F., Zahmatkesh, H., & Shahroze, R. (2022). An overview of security and privacy in smart cities' IoT communications. *Transactions on Emerging Telecommunications Technologies, 33*(3), e3677. DOI: 10.1002/ett.3677

Anh, P. T. N., Hahanov, V., Triwiyanto, R. N. A., İsmibeyli, R., Abdullayev, V., & Alyar, A. V. (2024). AI models for disease diagnosis and prediction of heart disease with artificial neural networks. *Computer Vision and AI-integrated IoT Technologies in Medical Ecosystem* (1st Ed.). CRC Press. https://doi.org/10.1201/9781003429609-9

Anthopoulos, L. G., & Anthopoulos, L. G. (2017). The rise of the smart city. *Understanding Smart Cities: A Tool for Smart Government or an Industrial Trick?* 5–45. DOI: 10.1007/978-3-319-57015-0_2

Avatefipour, O., & Sadry, F. (2018, May). Traffic management system using IoT technology-A comparative review. In *2018 IEEE International Conference on Electro/Information Technology (EIT)* (pp. 1041–1047). IEEE. DOI: 10.1109/EIT.2018.8500246

Balaji, S., Nathani, K., & Santhakumar, R. (2019). IoT technology, applications and challenges: A contemporary survey. *Wireless Personal Communications, 108*, 363–388. DOI: 10.1007/s11277-019-06407-w

Bandyopadhyay, D., & Sen, J. (2011). Internet of things: Applications and challenges in technology and standardization. *Wireless Personal Communications, 58*, 49–69. DOI: 10.1007/s11277-011-0288-5

Bertinelli, L., & Black, D. (2004). Urbanization and growth. *Journal of Urban Economics, 56*(1), 80–96. DOI: 10.1016/j.jue.2004.03.003

Burhan, M., Rehman, R. A., Khan, B., & Kim, B. S. (2018). IoT elements, layered architectures and security issues: A comprehensive survey. *Sensors, 18*(9), 2796. DOI: 10.3390/s18092796

Calderón-Garcidueñas, L., Kulesza, R. J., Doty, R. L., D'Angiulli, A., & Torres-Jardón, R. (2015). Megacities air pollution problems: Mexico City Metropolitan Area critical issues on the central nervous system pediatric impact. *Environmental Research, 137*, 157–169. DOI: 10.1016/j.envres.2014.12.012

Chen, T. L., Kim, H., Pan, S. Y., Tseng, P. C., Lin, Y. P., & Chiang, P. C. (2020). Implementation of green chemistry principles in circular economy system towards sustainable development goals: Challenges and perspectives. *Science of the Total Environment, 716*, 36998. DOI: 10.1016/j.scitotenv.2020.136998

Cui, L., Zhang, Z., Gao, N., Meng, Z., & Li, Z. (2019). Radio frequency identification and sensing techniques and their applications—A review of the state-of-the-art. *Sensors, 19*(18), 4012. DOI: 10.3390/s19184012

Dagar, R., Som, S., & Khatri, S. K. (2018, July). Smart farming–IoT in agriculture. In *2018 International Conference on Inventive Research in Computing Applications (ICIRCA)* (pp. 1052–1056). IEEE. DOI: 10.1109/ICIRCA.2018.8597264

Dang, L. M., Piran, M. J., Han, D., Min, K., & Moon, H. (2019). A survey on internet of things and cloud computing for healthcare. *Electronics, 8*(7), 768. DOI: 10.3390/electronics8070768.

Das, A., Sharma, S. C. M., & Ratha, B. K. (2019). The new era of smart cities, from the perspective of the internet of things. In *Smart Cities Cybersecurity and Privacy* (pp. 1–9). Elsevier. DOI: 10.1016/B978-0-12-815032-0.00001-9.

Debauche, O., Mahmoudi, S., Manneback, P., & Assila, A. (2019). Fog IoT for health: A new Architecture for patients and elderly monitoring. *Procedia Computer Science, 160,* 289–297. DOI: 10.1016/j.procs.2019.11.087

Evangelos, A. K., Nikolaos, D.T., & Anthony, C. B. (2011). Integrating RFIDs and smart objects into a unifiedinternet of things architecture. *Advances in Internet of Things.* DOI: 10.4236/ait.2011.11002

Farsi, M., Elhosseini, M. A., Badawy, M., Ali, H. A., & Eldin, H. Z. (2019). Deployment techniques in wireless sensor networks, coverage and connectivity: A survey. *IEEE Access, 7,* 28940–28954. DOI: 10.1109/ACCESS.2019.2902072

Gera, S., Mridul, M., & Sharma, S. (2021, April). IoT-based automated health care monitoring system for smart city. In *2021 5th International Conference on Computing Methodologies and Communication (ICCMC)* (pp. 364–368). IEEE. DOI: 10.1109/ICCMC51019.2021.9418487

Goel, A. K., & Khang, A., AI-aided IoT technologies and applications in the smart business and production. *Implementation of Smart Vehicle Parking System Using Internet of Things (IoT).* P (8), (2023). CRC Press. https://doi.org/10.1201/9781003392224-8

Govinda, K., & Saravanaguru, R. A. K. (2016). Review on IOT technologies. *International Journal of Applied Engineering Research, 11*(4), 2848–2853. DOI: 10.1016/j.future.2023.01.016

Guelzim, T., Obaidat, M. S., & Sadoun, B. (2016). Introduction and overview of key enabling technologies for smart cities and homes. *Smart Cities and Homes* (pp. 1–16). Morgan Kaufmann. DOI: 10.1016/B978-0-12-803454-5.00001-8

Haidine, A., El Hassani, S., Aqqal, A., & El Hannani, A. (2016). The role of communication technologies in building future smart cities. *Smart Cities Technologies, 1,* 1–24. DOI: 10.5772/64611

Haghi, M., Neubert, S., Geissler, A., Fleischer, H., Stoll, N., Stoll, R., & Thurow, K. (2020). A flexible and pervasive IoT-based healthcare platform for physiological and environmental parameters monitoring. *IEEE Internet of Things Journal, 7*(6), 5628–5647. DOI: 10.1109/JIOT.2020.2980432

Hutton, D. M. (2005). Smart environments: Technology, protocols and applications. *Kybernetes, 34*(6), 903–904. DOI: 10.1108/03684920510595580

Joyia, G. J., Liaqat, R. M., Farooq, A., & Rehman, S. (2017). Internet of medical things (IoMT): Applications, benefits and future challenges in healthcare domain. *Journal of Communication, 12*(4), 240–247. DOI: 10.12720/jcm.12.4.240-247

Jebaraj, L., Khang, A., Chandrasekar, V., Pravin, A. R., & Sriram, K. (2023). Smart city concepts, models, technologies and applications. *Smart Cities: IoT Technologies, Big Data Solutions, Cloud Platforms, and Cybersecurity Techniques* (Eds.). CRC Press. https://doi.org/10.1201/9781003376064-1

Khajenasiri, I., Estebsari, A., Verhelst, M., & Gielen, G. (2017). A review on Internet of Things solutions for intelligent energy control in buildings for smart city applications. *Energy Procedia, 111,* 770–779. DOI: 10.1016/j.egypro.2017.03.239

Khalil, E. E. (2021). Distributed energy in smart cities and the infrastructure. In *Solving Urban Infrastructure Problems Using Smart City Technologies* (pp. 249–268). Elsevier. DOI: 10.1016/B978-0-12-816816-5.00012-7

Khang, A. (2021). Material4Studies. *Material of Computer Science, Artificial Intelligence, Data Science, IoT, Blockchain, Cloud, Metaverse, Cybersecurity for Studies.* https://www.researchgate.net/publication/370156102_Material4Studies

Khang, A. (2023). *Advanced Technologies and AI-Equipped IoT Applications in High-Tech Agriculture* (1st Ed.). IGI Global Press. https://doi.org/10.4018/978-1-6684-9231-4

Khang, A., Abdullayev, V. A., Hahanov, V., & Shah, V. (2024). *Advanced IoT Technologies and Applications in the Industry 4.0 Digital Economy* (1st Ed.). CRC Press. https://doi.org/10.1201/9781003434269

Khang, A., Abdullayev, V. A., Hrybiuk, O., & Shukla, A. K. (2024). *Computer Vision and AI-integrated IoT Technologies in Medical Ecosystem* (1st Ed.). CRC Press. https://doi.org/10.1201/9781003429609

Khang, A., Gupta, S. K., Rani, S., Karras, D. A. (2023). *Smart Cities: IoT Technologies, Big Data Solutions, Cloud Platforms, and Cybersecurity Techniques* (Eds.). CRC Press. https://doi.org/10.1201/9781003376064

Khang, A., Hahanov, V., Litvinova, E., Chumachenko, S., Triwiyanto, H. V. A., Ali, R. N., Alyar, A. V., & Anh, P. T. N. (2023). The analytics of hospitality of hospitals in healthcare ecosystem. *Data-Centric AI Solutions and Emerging Technologies in the Healthcare Ecosystem.* P (4). CRC Press. https://doi.org/10.1201/9781003356189-4

Khang, A., Ragimova, N. A., Hajimahmud, V. A., Alyar, V. A. (2022). Advanced technologies and data management in the smart healthcare system. In *AI-Centric Smart City Ecosystems: Technologies, Design and Implementation* (1st Ed.), 16 (10), CRC Press. https://doi.org/10.1201/9781003252542-16

Khang, A., Rana, G., Tailor, R. K., & Hajimahmud, V. A. (2023). *Data-Centric AI Solutions and Emerging Technologies in the Healthcare Ecosystem* (Eds.). CRC Press. https://doi.org/10.1201/9781003356189

Khang, A., Rani, S., & Sivaraman, A. K. (2022). *AI-Centric Smart City Ecosystems: Technologies, Design and Implementation* (1st Ed.). CRC Press. https://doi.org/10.1201/9781003252542

Khanna, A., & Anand, R. (2016, January). IoT based smart parking system. In *2016 International Conference on Internet of Things and Applications (IOTA)* (pp. 266–270). IEEE. DOI: 10.1109/IOTA.2016.7562735

Kim, T. H., Ramos, C., & Mohammed, S. (2017). Smart city and IoT. *Future Generation Computer Systems, 76,* 159–162. DOI: 10.1016/j.future.2017.03.034

Kodali, R. K., Swamy, G., & Lakshmi, B. (2015, December). An implementation of IoT for healthcare. In *2015 IEEE Recent Advances in Intelligent Computational Systems (RAICS)* (pp. 411–416). IEEE. DOI: 10.1109/RAICS.2015.7488451

Kouonchie, P. K. N., Oduol, V., & Nyakoe, G. N. (2022, April). Roadside units for vehicle-to-infrastructure communication: An overview. In *Proceedings of the Sustainable Research and Innovation Conference* (pp. 69–72). DOI: 10.1016/j.trpro.2017.05.163

Kourtit, K., Nijkamp, P., & Steenbruggen, J. (2017). The significance of digital data systems for smart city policy. *Socio-Economic Planning Sciences, 58,* 13–21. DOI: 10.1016/j.seps.2016.10.001

Kumar, N. M., & Mallick, P. K. (2018). The Internet of Things: Insights into the building blocks, component interactions, and architecture layers. *Procedia Computer Science, 132,* 109–117. DOI: 10.1016/j.procs.2018.05.170.

Kylili, A., & Fokaides, P. A. (2015). European smart cities: The role of zero energy buildings. *Sustainable Cities and Society, 15,* 86–95. DOI: 10.1016/j.scs.2014.12.003

Laufs, J., Borrion, H., & Bradford, B. (2020). Security and the smart city: A systematic review. *Sustainable Cities and Society*, 55, 102023. https://doi.org/10.1016/j.scs.2020.102023.

Lee, J. Y., Woods, O., & Kong, L. (2020). Towards more inclusive smart cities: Reconciling the divergent realities of data and discourse at the margins. *Geography Compass*, *14*(9), e12504. DOI: 10.1111/gec3.12504

Leng, J., Lin, Z., & Wang, P. (2020, April). An implementation of an internet of things system for smart hospitals. In *2020 IEEE/ACM Fifth International Conference on Internet-of-Things Design and Implementation (IoTDI)* (pp. 254–255). IEEE. DOI: 10.1109/IoTDI49375.2020.00034

Lewis, T. (2008). *Smart Living: Lifestyle Media and Popular Expertise* (Vol. 15). Peter Lang. DOI: 10.1080/10304310903085756

Lobaccaro, G., Carlucci, S., & Löfström, E. (2016). A review of systems and technologies for smart homes and smart grids. *Energies*, *9*(5), 348. DOI: 10.3390/en9050348

Lodhi, M. K., Ansari, R., Yao, Y., Keenan, G. M., Wilkie, D., & Khokhar, A. A. (2017). Predicting hospital re-admissions from nursing care data of hospitalized patients. In *Advances in Data Mining. Applications and Theoretical Aspects: 17th Industrial Conference*, ICDM, New York, July 12–13, 2017, Proceedings 17 (pp. 181–193). Springer International Publishing. DOI: 10.1007/978-3-319-62701-4_14

Luckey, D., Fritz, H., Legatiuk, D., Dragos, K., & Smarsly, K. (2021). Artificial intelligence techniques for smart city applications. In *Proceedings of the 18th International Conference on Computing in Civil and Building Engineering: ICCBE 2020* (pp. 3–15). Springer International Publishing. DOI: 10.1007/978-3-030-51295-8_1

Madakam, S., Lake, V., Lake, V., & Lake, V. (2015). Internet of Things (IoT): A literature review. *Journal of Computer and Communications*, *3*(05), 164. DOI: 10.4236/jcc.2015.35021

Makani, S., Pittala, R., Alsayed, E., Aloqaily, M., & Jararweh, Y. (2022). A survey of blockchain applications in sustainable and smart cities. *Cluster Computing*, *25*(6), 3915–3936. DOI: 10.1007/s10586-022-03625-z

Mellouli, S., Luna-Reyes, L. F., & Zhang, J. (2014). Smart government, citizen participation and open data. *Information Polity*, *19*(1–2), 1–4. DOI: 10.3233/IP-140334

Mezaall, Y. S., Yousif, L. N., Abdulkareem, Z. J., Hussein, H. A., & Khaleel, S. K. (2018). Review about effects of IOT and Nano-technology techniques in the development of IONT in wireless systems. *International Journal of Engineering & Technology*, *7*(4), 3602–3606. DOI: 10.14419/ijet.v7i4.19615

Miladić-Tešić, S. D., Marković, G. Z., & Nonković, N. P. (2020). Optical technologies in support of the smart city concept. *Tehnika*, *75*(2), 209–215. DOI: 10.5937/tehnika2002209M

Mitton, N., Papavassiliou, S., Puliafito, A., & Trivedi, K. S. (2012). Combining Cloud and sensors in a smart city environment. *EURASIP Journal on Wireless Communications and Networking*, *2012*(1), 1–10. DOI: 10.1186/1687-1499-2012-247

Mohapatra, B. N., & Panda, P. P. (2019). Machine learning applications to smart city. *ACCENTS Transactions on Image Processing and Computer Vision*, *5*(14), 1. DOI: 0.19101/TIPCV.2018.412004

Muccini, H., & Moghaddam, M. T. (2018). IoT architectural styles: A systematic mapping study. In *Software Architecture: 12th European Conference on Software Architecture, ECSA 2018*, Madrid, September 24–28, 2018, Proceedings 12 (pp. 68–85). Springer International Publishing. DOI: 10.1007/978-3-030-00761-4_5

Nassereddine, M., & Khang, A. (2024). Applications of Internet of Things (IoT) in smart cities. *Advanced IoT Technologies and Applications in the Industry 4.0 Digital Economy* (1st Ed.). CRC Press. https://doi.org/10.1201/9781003434269-6

Rani, S., Bhambri, P., Kataria, A., & Khang, A. (2022). Smart city ecosystem: Concept, sustainability, Design principles and technologies. *AI-Centric Smart City Ecosystems: Technologies, Design and Implementation* (1st Ed.), 1 (20). CRC Press. https://doi.org/10.1201/9781003252542-1

Ratnaparkhi, S., Khan, S., Arya, C., Khapre, S., Singh, P., Diwakar, M., & Shankar, A. (2020). Smart agriculture sensors in IOT: A review. *Materials Today: Proceedings*. DOI: 10.1016/j.matpr.2020.11.138

Razavi, S. E., Rahimi, E., Javadi, M. S., Nezhad, A. E., Lotfi, M., Shafie-khah, M., & Catalão, J. P. (2019). Impact of distributed generation on protection and voltage regulation of distribution systems: A review. *Renewable and Sustainable Energy Reviews, 105*, 157–167. DOI: 10.1016/j.rser.2019.01.050

Redelinghuys, A. J. H., Kruger, K., & Basson, A. (2020). A six-layer architecture for digital twins with aggregation. In *Service Oriented, Holonic and Multi-agent Manufacturing Systems for Industry of the Future*. Proceedings of SOHOMA 2019 9 (pp. 171–182). Springer International Publishing. DOI: 10.1007/978-3-030-27477-1_13

Sakthivel, M., Gupta, S. K., Karras, D. A., Khang, A., Dixit, C. K., & Haralayya, B. (2022). Solving vehicle routing problem for intelligent systems using Delaunay triangulation. In *2022 International Conference on Knowledge Engineering and Communication Systems (ICKES)*. https://ieeexplore.ieee.org/abstract/document/10060807/

Singh, S., Sharma, P. K., Yoon, B., Shojafar, M., Cho, G. H., & Ra, I. H. (2020). Convergence of blockchain and artificial intelligence in IoT network for the sustainable smart city. *Sustainable Cities and Society, 63*, 102364. DOI: 10.1016/j.scs.2020.102364

Singh, S. K., Singh, R. S., Pandey, A. K., Udmale, S. S., & Chaudhary, A. (Eds.). (2020). *IoT-Based Data Analytics for the Healthcare Industry: Techniques and Applications*. Academic Press. DOI: 10.1016/b978-0-12-821472-5.09992-0

Söderman, J., & Pettersson, F. (2006). Structural and operational optimisation of distributed energy systems. *Applied Thermal Engineering, 26*(13), 1400–1408. DOI: 10.1016/j.applthermaleng.2005.05.034

Söderström, O., Paasche, T., & Klauser, F. (2014). Smart cities as corporate storytelling. *City, 18*(3), 307–320. DOI: 10.1080/13604813.2014.906716

Srinivasan, L., Selvaraj, D., Dhinakaran, D., & Anish, T. P. (2023). IoT-Based solution for paraplegic sufferer to send signals to physician via Internet. *arXiv preprint arXiv:2304.10840*. DOI:10.48550/arXiv.2304.10840

Stergiou, C., Psannis, K. E., Kim, B. G., & Gupta, B. (2018). Secure integration of IoT and cloud computing. *Future Generation Computer Systems, 78*, 964–975. DOI: 10.1016/j.future.2016.11.031

Subhashini, R., & Khang, A. (2023). The role of Internet of Things (IoT) in smart city framework. *Smart Cities: IoT Technologies, Big Data Solutions, Cloud Platforms, and Cybersecurity Techniques* (Eds.). CRC Press. https://doi.org/10.1201/9781003376064-3

Thangaraj, M., Ponmalar, P. P., & Anuradha, S. (2015, December). Internet Of Things (IOT) enabled smart autonomous hospital management system-A real world health care use case with the technology drivers. In *2015 IEEE International Conference on Computational Intelligence and Computing Research (ICCIC)* (pp. 1–8). IEEE. DOI: 10.1109/ICCIC.2015.7435678

Times, Statistics. (2021, September 27). *World Urban Population*. https://statisticstimes.com/demographics/world-urban-population.php (accessed August 7, 2022).

Tyagi, V., Saraswat, A., Bansal, S., & Khang, A. (2023). An analysis of securing IoT Devices from Man-in-the-Middle (MIMA) and Denial of Service (DoS). *Smart Cities: IoT Technologies, Big Data Solutions, Cloud Platforms, and Cybersecurity Techniques* (Eds.). CRC Press. https://doi.org/10.1201/9781003376064-20

Valanarasu, M. R. (2019). Smart and secure IoT and AI integration framework for hospital environment. *Journal of ISMAC*, *1*(03), 172–179. DOI: 10.36548/jismac.2019.3.004

Valsalan, P., Baomar, T. A. B., & Baabood, A. H. O. (2020). IoT-based health monitoring system. *Journal of Critical Reviews*, *7*(4), 739–743. DOI: 10.31838/jcr.07.04.137

Shah, V., Jani, S., & Khang, A. (2023). Automotive IoT: Accelerating the automobile industry's long-term sustainability in smart city development strategy. *Smart Cities: IoT Technologies, Big Data Solutions, Cloud Platforms, and Cybersecurity Techniques* (Eds.). CRC Press. https://doi.org/10.1201/9781003376064-9

Whitmore, A., Agarwal, A., & Da Xu, L. (2015). The Internet of Things—A survey of topics and trends. *Information Systems Frontiers*, *17*, 261–274. DOI: 10.1007/s10796-014-9489-2

Wlodarczak, P. (2017). Smart cities–enabling technologies for future living. *City Networks: Collaboration and Planning for Health and Sustainability*, 1–16. DOI: 10.1007/978-3-319-65338-9_1

Xiaojun, C., Xianpeng, L., & Peng, X. (2015, January). IOT-based air pollution monitoring and forecasting system. In *2015 International Conference on Computer and Computational Sciences (ICCCS)* (pp. 257–260). IEEE. DOI: 10.1109/ICCACS.2015.7361361

Xinhua. (2022, February 22). *China's urbanization rate hits 64.72% in 2021*. http://english.www.gov.cn/archive/statistics/202202/22/content_WS62149dc7c6d09c94e48a5517.html (accessed August 7, 2022).

Yang, C. (2020). Historicizing the smart cities: Genealogy as a method of critique for smart urbanism. *Telematics and Informatics*, *55*, 101438. DOI: 10.1016/j.tele.2020.101438

Yaqoob, I., Hashem, I. A. T., Mehmood, Y., Gani, A., Mokhtar, S., & Guizani, S. (2017). Enabling communication technologies for smart cities. *IEEE Communications Magazine*, *55*(1), 112–120. DOI: 10.1109/MCOM.2017.1600232CM

Yeh, H. (2017). The effects of successful ICT-based smart city services: From citizens' perspectives. *Government Information Quarterly*, *34*(3), 556–565. DOI: 10.1016/j.giq.2017.05.001

Yin, C., Xiong, Z., Chen, H., Wang, J., Cooper, D., & David, B. (2015). A literature survey on smart cities. *Science China Information Sciences*, *58*(10), 1–18. DOI: 10.1007/s11432-015-5397-4

Yun, Y., & Lee, M. (2019). Smart city 4.0 from the perspective of open innovation. *Journal of Open Innovation: Technology, Market, and Complexity*, *5*(4), 92. DOI: 10.3390/joitmc5040092

Zhang, M., Sun, F., & Cheng, X. (2012, October). Architecture of internet of things and its key technology integration based-on RFID. In *2012 Fifth international symposium on computational intelligence and design* (Vol. 1, pp. 294–297). IEEE. DOI: 10.1109/ISCID.2012.81

Zhang, Y. (2011, September). Technology framework of the Internet of Things and its application. In *2011 International Conference on Electrical and Control Engineering* (pp. 4109–4112). IEEE. DOI: 10.1109/ICECENG.2011.6057290

7 Internet of Things (IoT)-Powered Communication in Smart Energy Systems

Ankit Jain, Anita Shukla, Puspraj Singh Chauhan, and Raghvendra Singh

INTRODUCTION

Energy is a crucial element for a country's economic development as well as for expanding human civilization and the growth of the world's population. Because of increased economic growth and changing consumption habits, there is an ever-increasing demand for energy. Initially, the primary source of energy is fossil fuels. But increasing demand leads to excessive usage of fossil fuels, which in turn, results threat of an energy crisis as these resources are running out. Burning fossil fuels has caused the atmospheric concentration of carbon dioxide to rise, which has resulted in extreme weather patterns (Annual Energy Outlook, 2020; Mani et al., 2017).

As a result of global industrialization, the climate has become abnormal and humanity is struggling to survive. To meet the challenges of energy crises, alternative energy sources are developed with little to no emissions of greenhouse gases like carbon dioxide (CO_2) because they do not come from fossil fuels. As a result, the greenhouse effect that causes climate change is not exacerbated by the energy produced from non-conventional sources. Solar energy, geothermal energy, hydroelectric energy, biomass energy, wind energy, and nuclear energy are some examples of alternative energy sources (Efe et al., 2022).

The demand for energy is rising in many countries throughout the world as populations and income levels rise. Between 2017 and 2018, the world's energy demand climbed by 2.3%, and between 2018 and 2019, it increased by 3.1%. After decreasing by 4.5% in 2020, the world's energy consumption returned with a 5% growth in 2021 (Karam et al., 2022; Mohassel et al., 2014). Unquestionably, this shows that the current energy demand is likely to soar in the near future. Yet, we have found various other energy sources; still dependencies on fossil fuels persist to fulfill the required demands and they will be depleted in next 20 years.

The immense challenges brought on by rapid digitization have greatly increased the demand for energy. Global energy consumption is projected to increase by 56% by 2040. One example of a global attempt to reduce energy consumption in buildings and cities is the EU's 2050 roadmap, which seeks to cut energy and petrol emissions

DOI: 10.1201/9781003434269-7

by around 40% (Amin and Wollenberg, 2005) and the reference therein (Karthick et al., 2021; Mani et al., 2017). To cut down on energy use and lessen the dramatic environmental impacts, it is necessary to create an efficient and effective energy-saving technology. In the next section, we thoroughly discussed the necessary steps taken by various researchers in this direction.

LITERATURE REVIEW

The increasing population also raised energy usage, specifically electric energy. Extreme usage of electric energy has become one of the most important issues that most nations are currently facing. In this regard, the research community has contributed a lot in developing low-power devices that consume very low energy as compared to traditional systems. Also, they have contributed to developing an efficient methodology that keeps track of this energy consumption that helps in identifying the actual usage of energy by different systems (Saleem et al., 2017; Chelloug and El-Zawawy, 2017).

The rapid advancement of information and communication technologies (ICT) has significantly changed both the technological landscape and industrial automation processes (Tailor et al., 2022). As a result of the blending of these technical breakthroughs with energy systems, the internet of energy (IoE), a cutting-edge idea that enables an energy system to be smart, has advanced (Mishra et al., 2018). The internet of energy (IoE), which includes both thermal and electrical energy sources, is viewed as a single entity and a subclass of the internet of things (IoT).

Using the internet and advanced communication and technology infrastructures, the IoT's primary objective is to connect distant objects and systems, including devices, people, processes, and data (Govindarajan et al., 2018). A promising future for decreasing operational and environmental issues with energy systems is indicated by the application of IoT in power and energy systems. There is optimism about the advancement of existing technology and the creation of new ones. However, this has drawbacks due to high investment costs (Karam et al., 2022).

There are reviews and research findings related to integrating IoT into smart energy system (SES) that have been published in the literature. A Commercial Building Energy Management System (CBEMS) designed to track and manage energy use and power problems has been proposed by Karthick et al. (2021). Advanced reduced instruction set computing machines are used by the system to manage all operations. Using data on humidity and light intensity to wirelessly regulate lighting and fans, Vignesh Mani et al. built an IoT-based smart energy management system (Mani et al. 2017). These sensor data are used to regulate the appliances intelligently rather than simply turning them on or off. Additionally, the system continuously calculates how much energy is consumed daily by the appliances and displays that information to the user (Hahanov et al., 2022).

In Haque et al. (2019), Mishra et al. designed IoT-based smart energy meter using the Raspberry Pi devices. They have modified the old meter by adding the circuitry to monitor it through web pages and mobile applications. Govindarajan et al. (Muliadi et al., 2020) designed a low-cost energy monitoring system with the aid of Arduino and used a Zigbee communication module to wirelessly monitor the system health (Rani et al., 2021). A smart city is one where all the devices are managed smartly including the main power grid. With this motivation, in Hariharan et al. (2021) the authors have developed a widely used chip-based IoT electrical meter that will record all meter data and wirelessly transmit the necessary data to the main server. The system is featured with intimating the consumer about the pending fees and is capable of disconnecting the connection if needed (Khang et al., 2022).

Real-time web-based electrical parameters such as current, voltage, and power monitoring systems for households are developed in Hasan et al. (2021). The ESP8266 controller, PZEM004T current sensor, relays are used to construct a sensing peripheral node for the measurement system. A home energy monitoring system for appliances has been developed (Macheso and Thotho, 2022). It can be used to estimate the energy consumption of a household and inform the user about the electricity consumption through an Android app that enables him to view the units of electricity used and an estimate of the bill at the end of the month.

An efficient power monitoring and switching system that reduces the current error of 0.6% from the pre-existing system that provides a current error of 7.2% is proposed by Pimplea et al. (2021). Khang et al. (2024a) suggested an IoT-based smart energy meter to track energy usage of residential loads, power billing, and smart grid power management automatically. The system is built around a low-cost ESP32 microcontroller that receives data from sensor nodes through non-intrusive current transformer (CT) sensors (SCT013 and HLW8012) and a voltage sensor. The system sends the consumer's consumption data via a mobile app or the Blynk server.

FEATURES OF THE PROPOSED SYSTEM

This system now has many features, which significantly improve its functionality and reliability. Using the ESP32, we built an internet of things-based electricity energy meter and we used the Blynk app to track data. We still have to manually record measurements in the meter reading room with the present technology. Thus, keeping track of and monitoring your electricity consumption is a laborious effort. We can utilize the internet of things (IoT) to automate this, which saves time and money by automating distant data collection. Recently, the smart energy meter has won much praise from all across the world (Kali et al., 2024). Below is a list of the qualities:

- We must choose both the current and voltage sensors in order to measure the current and voltage and determine the amount of power being utilized.
- SCT-013 is the top current sensor currently on the market. This SCT-013 noninvasive clamp meter sensor, which measures AC current up to 100A, is of the split core variety.
- The AC voltage sensor module ZMPT101B is the best voltage sensor. When using a voltage transformer to detect an accurate AC voltage, the ZMPT101B AC voltage sensor is the ideal option.
- We can measure all the necessary parameters needed for an electricity energy meter using the SCT-013 current sensor and ZMPT101B voltage

sensor. The SCT-013 current sensor and ZMPT101B voltage sensor will be connected to the ESP32 Wi-Fi module, and the data will be sent to the Blynk application.

- The energy meter data is stored in the Electrically Erasable Programmable Read-Only Memory (EEPROM) of the ESP32, ensuring that readings continue even when there is no power. On the Blynk application dashboard, the voltage, current, power, and total unit consumed in kWh are displayed.

ALGORITHM AND IMPLEMENTATION

In this work, we suggest creating a model through the use of control engineering in order to solve the stated issue and also to include additional beneficial characteristics. The main goal of the methodical approach is taken to create the system's hardware in stages and built a broad system to gather input from outside sources (sensors) and alter it to produce a certain output. The block diagram of proposed AC energy meter is shown in Figure 7.1.

ESP32 Pin Out

The ESP WROOM32 Wi-Fi + BLE module is the basis for the ESP32 development board depicted in Figure 7.2. It is a tiny, simple system development board that fits neatly onto a solderless breadboard and is powered by the most recent ESP-WROOM-32 module. It includes all of the ESP-WROOM-32's essential support components, including the USB-UART Bridge, reset and boot-mode buttons, a Low-dropout regulator (LDO) regulator, and a micro-USB connector.

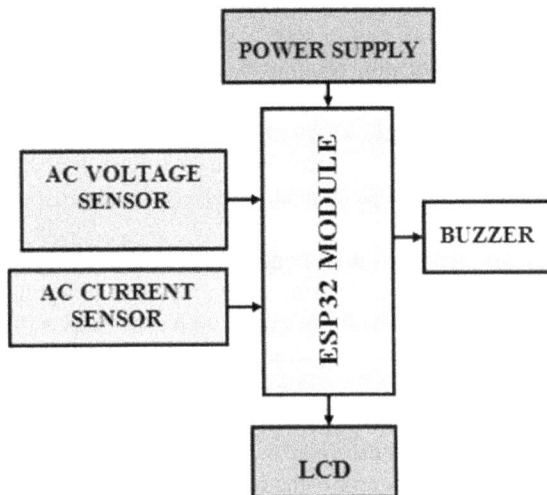

FIGURE 7.1   Block diagram of proposed AC energy meter

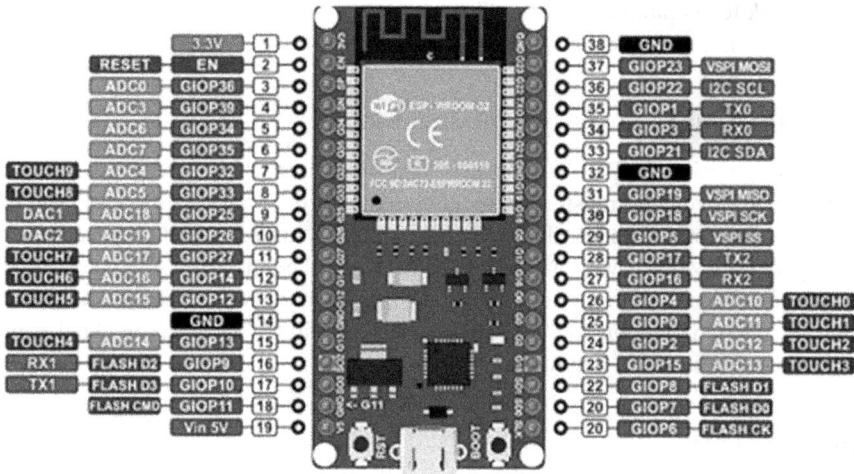

FIGURE 7.2   Pin mapping of ESP32 development board

SCT-013 Current Sensor

The noninvasive AC current sensor SCT-013 in Figure 7.3 is a split core type clamp meter sensor that can measure AC current up to 100A. Alternating current can be measured with current transformers (CTs), which are sensors. They are very helpful for calculating the total electricity usage of buildings. There is no need to perform any high-voltage electrical work in order to clip the SCT-013 current sensors directly to the live or neutral wire.

Specifications of SCT-013 AC current sensor are as follows:

- Input current: 0–30A AC
- Output signal: DC 0–1V
- Nonlinearity: 2–3%
- Build-in sampling resistance (RL): 62Ω
- Turn ratio: 1800:1
- Resistance grade: Grade B

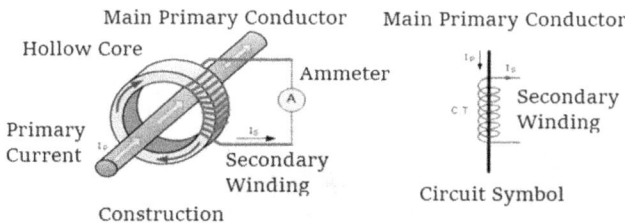

FIGURE 7.3   SCT-013 AC current sensor

- Work temperature: −25°C to +70°C
- Dielectric strength (between shell and output): 1000 V AC/1 min 5Ma

ZMPT101B AC Single-Phase Voltage Sensor

The high-precision ZMPT101B voltage transformer used to measure the precise AC voltage with a voltage transformer is the foundation of the ZMPT101B AC single-phase voltage sensor module depicted in Figure 7.4. This is the best option for measuring the AC voltage with an ESP32 or Arduino. The module's analogue output may be modified, and they can measure voltage up to 250V AC. The module has a multi-turn trim potentiometer that can be used to calibrate and tune the Analog-to-digital converter (ADC) output.

16 × 2 LCD Display

A liquid crystal display (LCD) screen, such as the one in Figure 7.5, is a type of electronic display module that has several uses. A very fundamental module that is frequently utilized in many different devices and circuits is the 16 × 2 LCD display. These modules are preferable over multi-segment LEDs with seven segments and additional segments. The benefits include LCD's low cost, ease of programming, lack of restrictions on the display of special and even customized characters, animations, and many other helpful features (unlike in seven segments). With a 16 × 2 LCD, there are two lines that can each display 16 characters. Each character on this LCD is presented using a 5 × 7 pixel matrix. Command and data registers are present

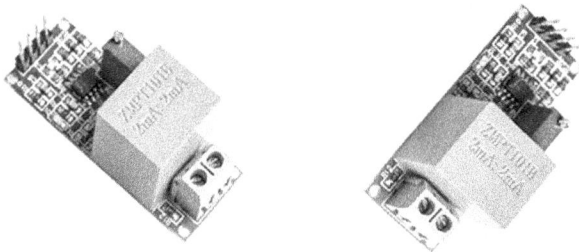

FIGURE 7.4   ZMPT101B AC single-phase voltage sensor

FIGURE 7.5   16 × 2 LCD

on this LCD. The command instructions sent to the LCD are stored in the command register. An instruction to the LCD to do a preset action, such as initializing it, clearing its screen, establishing the cursor location, controlling the display, etc., is known as a command.

DEVELOPMENT OF CIRCUIT

The system is built with an ESP32 module, chosen for its modest and compact size, as shown in the circuit design in Figure 7.6. The Fritzing software was used to construct the design, and the connection diagram is straightforward. The Vin and GND of the ESP32, which is a 5V supply, are linked to the VCC and GND pins of the SCT-013 current sensor and ZMPT101B voltage sensor, respectively. The GPIO34 of ESP32 is connected to the output analogue pin of the SCT-013 current sensor, while the GPIO35 of ESP32 is connected to the output analogue pin of the ZMPT101B voltage sensor. The circuit also needs a 10uF capacitor, two 10K resistors, one 100ohm resistor, and two 10K resistors.

The AC wires must be connected to the input AC terminal of the voltage sensor in order to measure current and voltage. One live or neutral wire only needs to be placed into the clip portion of the device to use it as a current sensor. You may also use a 16 × 2 I2C LCD panel; however, it requires extra connections. Connect the ESP32's 5V, GND, GPIO21, and GPIO22 to the LCD display's Voltage common collector (VCC), Ground-voltages in the circuit (GND), Serial data (SDA), and Serial clock (SCL), respectively.

IMPLEMENTATION OF PCB

Use of Dip Trace was used to create this Printed circuit boards (PCB) (Figure 7.7 (a, b)).

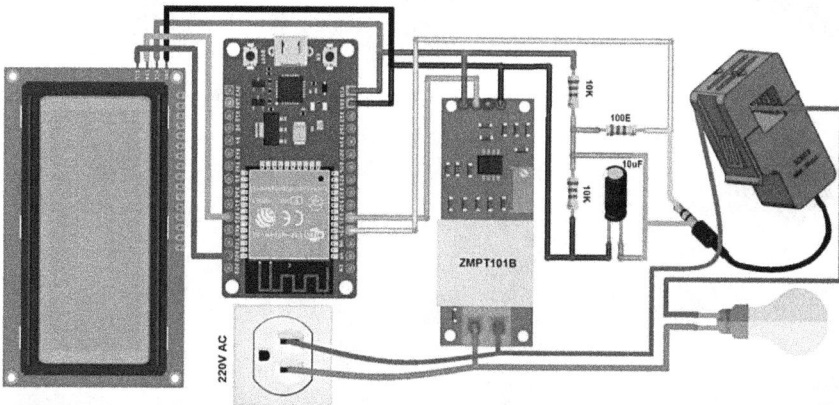

FIGURE 7.6 Circuit diagram of AC energy meter

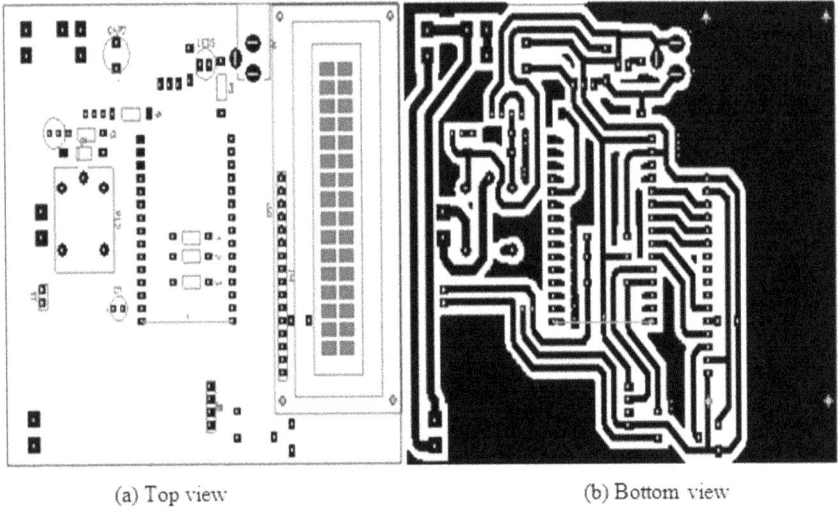

(a) Top view   (b) Bottom view

FIGURE 7.7   PCB prepared using Dip Trace

SETTING UP BLYNK APPLICATION

Just go to https://blynk.io and click on start free as shown in Figure 7.8. After that, register with Gmail ID. Enter your email address and password to join up for the Blynk IoT cloud server (Figure 7.9), and then hit the icon of the new project to assign the project name.

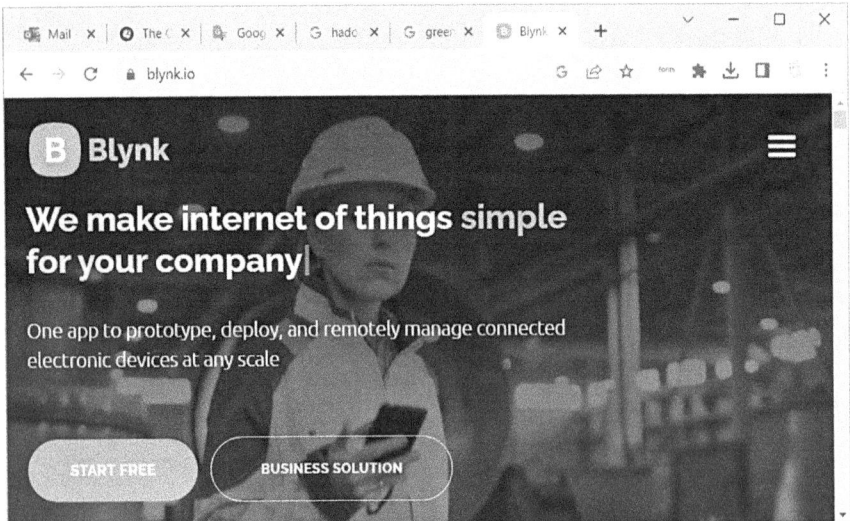

FIGURE 7.8   Blynk cloud user interface

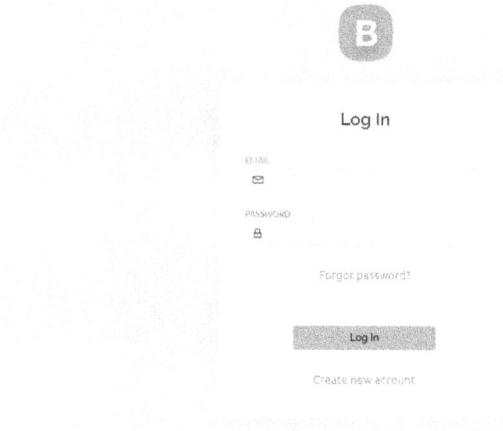

FIGURE 7.9   Log in with registered mail ID

Choose the ESP32 board and then pick Wi-Fi as the connection type. Lastly, press the Create button as displayed in Figure 7.10.

OVERVIEW OF CENTRAL LIBRARY

Since the establishment of the National Institute of Technical Teachers Training and Research (NITTTR), Sector-26, Chandigarh, a library has been kept up. In November 2000, it moved to the Sir J. C. Bose academic and administrative block, where it is now located on the first floor. The library is equipped with a sizable collection of books, periodicals, and magazines to benefit the students' development of knowledge and skills. Additionally, it offers an e-library service with a fast internet connection. The use of numerous electrical and electronic components throughout

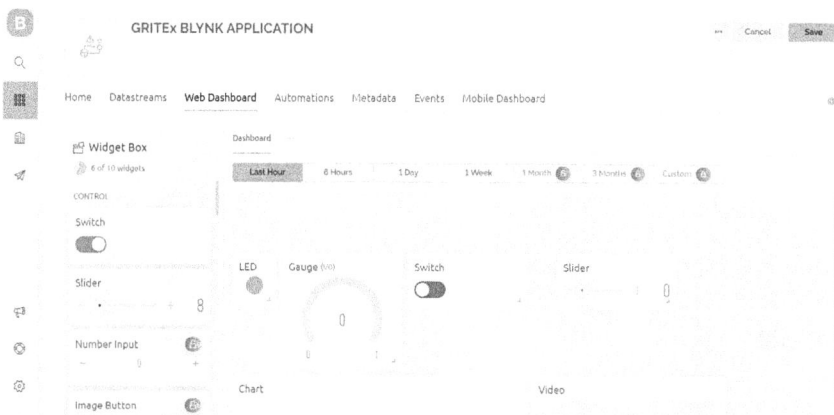

FIGURE 7.10   Blynk server dashboard

the library space results in significant energy consumption. We performed an energy audit on this central library. The energy audit report includes detailed information about the total electrical energy used by the utilities in NITTTR's Sector-26 central library and makes necessary recommendations for energy conservation opportunities (ECOs). In order to ascertain the advantages and cost-effectiveness of the ECOs, they are discovered and analyzed. The usage of the proposed energy meter is our final recommendation for reducing electrical energy use.

Major Energy Use and Areas

The major areas in the library for energy consumption are the book storage section, reading section, and e-library center (Figure 7.11).

Layout of Central Library: Existing Layout Plan

Figure 7.12 shows the existing layout plan of central library (NITTTR, Sector-26, Chandigarh).

Energy and Utility System Description

Electronics and electrical utilities are used in various parts of the library such as in reception, e-library, chief librarian chamber, magazine section, reading section, employee cabins, server room, store room, classroom, and book storage section. All these utilities are listed in Table 7.1.

Brief Description of Utility Consumptions

For the audit specification, all the utilities are highly essential. Table 7.2 shows a brief description of the utilities and describes the amount of energy consumed in watt/hour by each individual.

Now the calculation has been done for each type of utility per day, respectively, and this leads to total energy consumption of the entire library in terms of kW. Considering per unit cost in Chandigarh is Rs. 4.50/- (approx.).

Tube Light

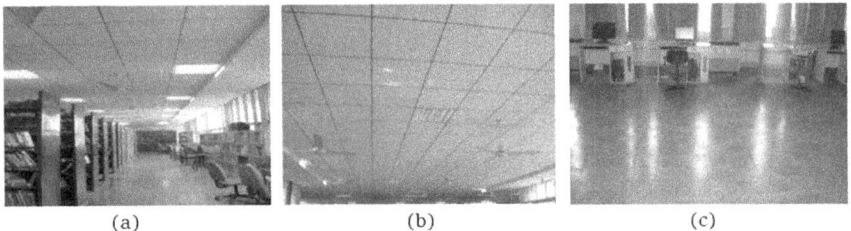

(a)   (b)   (c)

FIGURE 7.11   (a) Book storage section. (b) Reading section. (c) E-library center

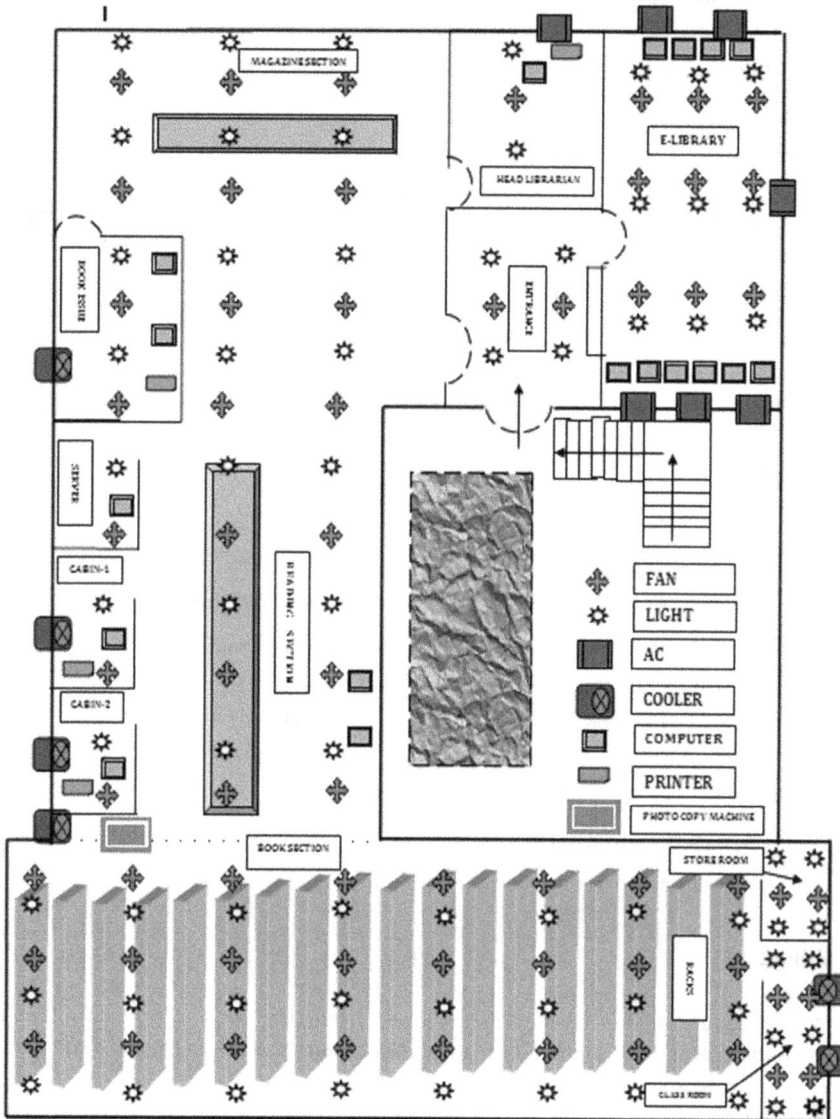

FIGURE 7.12 Existing layout of central library (NITTTR, Sector-26, Chandigarh)

- No. of light = 78 × 4=312 (14 watt each) (as shown in Figure 7.3)
- Total power consumption during one day (08 hours) = 312 × 14 × 08 = 34944 watt = 34.94 kW
- Per unit cost in Chandigarh is Rs. 4.50/- (approx.)
- Total cost per day (08 hour) = 34.94 × 4.50 = Rs. 157.24/-

TABLE 7.1
List of Electronics and Electrical Utilities Used in Various Parts of the Library

Location	Names of Electrical Utilities
Reception	Tube lights and fans
E-library	Computers, air conditioners, tube lights, and fans
Chief librarian chamber	Computer, air conditioner, printer, tube light, and fan
Magazine section	Tube lights and fans
Reading section	Tube lights, fans, and computers
Employee cabins	Tube lights, fans, coolers, computers, and printer
Server room	Tube light, fan, and computers
Book storage section	Photo copy machine, tube lights, and fans
Storeroom	Tube lights and fans
Classroom	Tube lights, fans, and cooler

TABLE 7.2
Amount of Energy Consumed in Watt/Hour

Name of the Utility	Watt/Hour Consumption
Ceiling fan	60
Tube light	14
Air cooler	200
Air conditioner	2000
Computer	80
Photocopy machine	250

Ceiling Fan

- No. of fans = 63 (60 watt each) (as shown in Figure 7.3)
- Total power consumption during one day (08 hours) = $63 \times 60 \times 08 = 30240$ watt = 30.24 kW
- Per unit cost in Chandigarh is Rs. 4.50/- (approx.)
- Total cost per day (08 hour) = 30.24×4.50 = Rs. 136.08/-

Air Cooler

- No. of cooler = 06 (200 watt each)
- Total power consumption during one day (08 hours) = $06 \times 200 \times 08 = 9600$ watt = 9.6 kW
- Per unit cost in Chandigarh is Rs. 4.50/- (approx.)
- Total cost per day (08 hour) = 9.6×4.50 = Rs. 43.2/-

Air Conditioner

- No. of AC = 7 (2000 watt each)
- Total power consumption during one day (08 hours) = 07 × 2000 × 08 = 112,000 watt = 112 kW
- Per unit cost in Chandigarh is Rs. 4.50/- (approx.)
- Total cost per day (08 hour) = 112 × 4.50 = Rs. 504/-

Personal Computer

- No. of PC = 18 (80 watt each)
- Total power consumption during one day (08 hours) = 18 × 80 × 08 = 11520 watt = 11.52 kW
- Per unit cost in Chandigarh is Rs. 4.50/- (approx.)
- Total cost per day (08 hour) = 11.52 × 4.50 = Rs. 51.84/-

Photo Copy Machine

- No. of photocopy machine = 01 (250 watt each)
- Total power consumption during one day (08 hours) = 01 × 250 × 08 = 2000 watt = 2 kW
- Per unit cost in Chandigarh is Rs 4.50/- (approx.)
- Total cost per day (08 hour) = 2 × 4.5 = Rs. 9/-

ENERGY CONSERVATION RECOMMENDATIONS

The following are the energy conservation recommendations for the existing layout of library, and the necessary replacements have been designed in the preceding proposed layout plan.

- Reduction in the number of lights, fans, and air conditioners in the various parts of the library.
- Replacement of the switchboards in a few places.
- Replacement of old type fan regulator with electronic fan regulators.
- Use natural light source (sunlight) efficiently.
- Use dark at night controller circuit in the main circuitry.
- Use 15A socket instead of 05A socket for computers, ACs, and coolers.
- LEDs light should prefer.
- Use tube lights with electronic chock.
- Proper placement of racks so that natural light may come.

The development of energy-efficient lighting equipment such as low-loss chokes, high-lumen tubes, low-wattage bulbs for the same Lux levels, and energy-efficient luminaries is a result of technological breakthroughs in lighting technology.

Layout of Central Library: Proposed Layout Plan

Figure 7.13 shows the proposed layout plan of the central library (NITTTR, Sector-26, Chandigarh).

Description of Proposed System

As per the energy conservation recommendations, certain reductions and replacements in the utilities have been done in the various sections of the library.

Tube Light: We observed that four Compact fluorescent lamp (CFL) ceiling light tubes (14 watt each) are arranged in one combination of light, and we recommended that two CFL (14 watt each) tubes used should be arranged in one combination of light.

- No. of light = 78 × 2=156 (14 watt each)
- Total power consumption during one day (08 hours) = 156 × 14 × 08 = 17472 watt = 17.472 kW
- Per unit cost in Chandigarh is Rs. 4.50/- (approx.)
- Total cost per day (08 hour) = 17.472 × 4.50 = Rs. 78.62/-

Ceiling Fans: We observed that in some places extra ceiling fans are placed.

- In e-library: Three unnecessary fans
- In entrance hall: One unnecessary fan
- In central section: Seven unnecessary fans
- In book section: Eight unnecessary fans
- Total unnecessary fans: Nineteen fans
- Recommended no. of fans: 63 − 19 = 44 fans (60 watt each)
- Total power consumption during one day (08 hours) = 44 × 60 × 08 = 21120 watt = 21.12 kW
- Per unit cost in Chandigarh is Rs. 4.50/- (approx.)
- Total cost per day (08 hour) = 21.12 × 4.50 = Rs. 95.04/-

Air conditioners: We observed that in e-library six ACs (2000 watt each) are available but we recommended that four ACs are sufficient.

- So recommended no. of ACs: 06 − 02=04 ACs
- So no. of ACs = 04 (2000 watt each)
- Total power consumption during one day (08 hours) = 04 × 2000 × 08 = 64000 watt= 64 kW
- Per unit cost in Chandigarh is Rs. 4.50/- (approx.)
- Total cost per day (08 hour) = 64 × 4.50 = Rs. 288/-

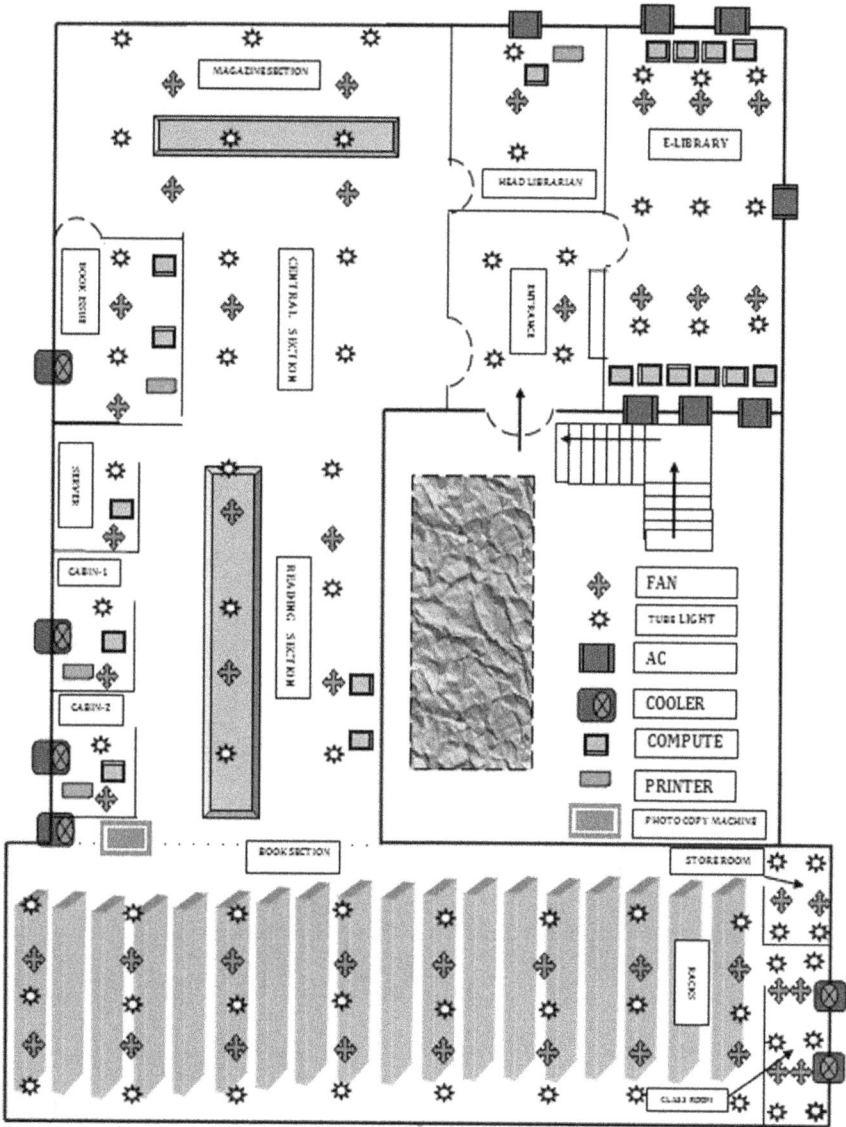

FIGURE 7.13 Proposed plan of central library (NITTTR, Sector-26, Chandigarh)

ENERGY-SAVING CALCULATIONS AND COST BENEFITS

Tables 7.3 and 7.4 depict the existing plan and proposed plan for energy-saving calculation.

Now the energy consumption calculation for certain electrical utilities is required to compare the existing layout with the proposed layout, and the following description explains the proposed system.

TABLE 7.3
Energy-Saving Calculation for Existing Plan

				Existing Plan	
Sl. No.	Equipment	Wattage	Quantity (Existing)	Total Wattage Consumed (8 hour/day) in kW	Cost/Day (8 Hour/Day)
1	Ceiling fan	60	63	30.24	136.08
2	Tube light	$14 \times 4 = 56$	78	34.94	157.248
3	Air cooler	200	6	9.6	43.2
4	Air conditioner	2000	7	112	504
5	Computer	80	18	11.528	51.84
6	Photocopy machine	250	01	2	9
			Total	200.308	901.36

TABLE 7.4
Energy-Saving Calculation for Proposed Plan

				Proposed Plan	
Sl. No.	Equipment	Wattage	Quantity (Proposed)	Total Wattage Consumed (8 Hour/Day) in kW	Cost/day (8 Hour/Day)
1	Ceiling fan	60	44	21.12	95.04
2	Tube light	$14 \times 2 = 28$	78	17.472	78.62
3	Air cooler	200	6	9.6	43.2
4	Air conditioner	2000	4	64	288
5	Computer	80	18	11.528	51.84
6	Photocopy machine	250	01	2	9
			Total	125.72	556.7

- Total power saving in kW (per day) = 200.308 − 125.72 = 74.588 kW
- Total power saving in kW (per year) = 74.588 × 30 × 12 = 26,851.68 kW
- Total saving in rupees (per day) = 901.36 − 556.7 = Rs. 335.66/-
- Total saving in rupees (per year) = 335.66 × 30 × 12 = Rs. 120,837.6/-

AREA NEEDS TO BE IMPROVED

We found some faulty area in the library where we observed some utilities such as regulators, switch board along with switches, and extension sockets need to be improved as shown in Figure 7.14. So regulators can be replaced with electronic regulators, replace the old switches with new ones, and heavy types of equipment should be driven using 15A socket instead of 5A socket.

OFF AT DARK CIRCUIT

A light-activated turn-off circuit is shown in Figure 7.15. We can install this circuit in different places of the library to reduce the electrical energy consumption. In the

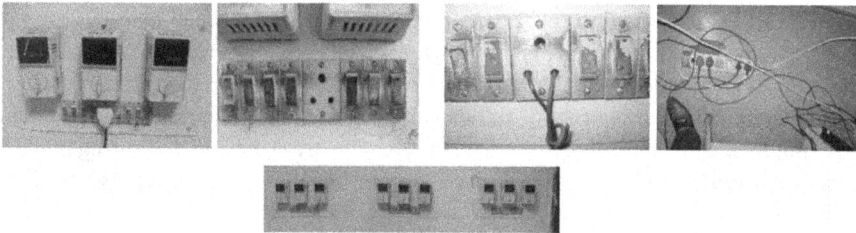

FIGURE 7.14 Displays some utilities such as regulators, switch board along with switches, and extension sockets

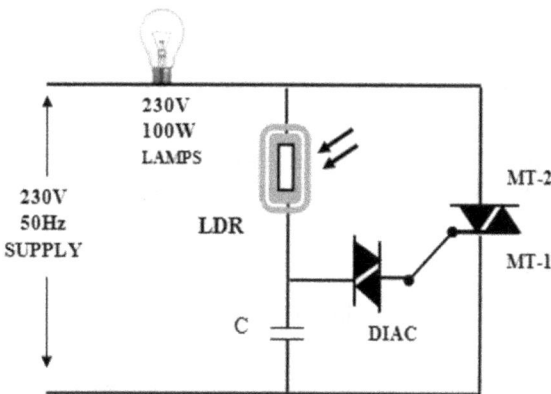

FIGURE 7.15 Off at dark circuit

presence of light, the resistance of the Light dependent resistor (LDR) is very low (approx. 100 Ω) and allows the flow of current through it, simultaneously charging capacitor C. The voltage across the capacitor is higher than the breakover voltage of Diode for alternating current (DIAC), and it enables DIAC to fire. So as soon as DIAC starts charging, it sends the required fire pulse at the Three-terminal electronic component (TRIAC) to trigger it. As TRIAC conducts, lamps glow.

RESULTS AND DISCUSSION

The internet of things (IoT) has the ability to efficiently send data on wireless energy consumption while also detecting electricity use. We have to go to the meter reading room to record readings with the current technology. So it takes a lot of work to measure and manage our energy usage (Khanh and Khang, 2021).

A revolution in the tracking and measuring of electricity use will result from the creation of an IoT-based smart electricity energy meter that uses the ESP32 and Blynk platforms. By eliminating manual meter readings, the IoT-based solution saves time and money. Accurate readings of voltage, current, power, and total energy spent can be acquired by using the SCT-013 current sensor and ZMPT101B voltage sensor. Through the Blynk dashboard, the data is accessible from anywhere. The ESP32's EEPROM stores the energy meter data, ensuring continual readings even in the event of a power outage. This project offers a chance to streamline and automate the process of monitoring electricity consumption.

The creation of an internet of things (IoT)-based smart electricity energy meter utilizing the ESP32 and Blynk will revolutionize the tracking and measuring of electricity consumption. With the IoT-based system, manual meter readings are eliminated, saving both time and money. Accurate measurements of voltage, current, power, and total energy spent can be made with the help of the ZMPT101B voltage sensor and SCT-013 current sensor. The Blynk dashboard allows access to the data from anywhere. The energy meter data is saved in the ESP32's EEPROM, ensuring continual measurements even in the event of power interruptions. With the help of this project, it will be possible to streamline and automate the process of monitoring electricity use (Hajimahmud et al., 2022).

Energy audit provides specific details regarding the total electrical energy used by the utilities in the central library at NITTTR, Sector-26, Chandigarh, and makes required energy conservation opportunities (ECOs) recommendations. So, the ECOs are found and assessed to ascertain their advantages and cost-effectiveness. To reduce electrical energy use, we have also recommended using the planned energy meter.

CONCLUSION

To replace the traditional electricity meter, a cutting-edge digital device called a smart energy meter was developed. Installing a smart meter will help us avoid estimated costs by enabling us to manage and reduce our energy use. It is equipped with the capacity to gather data remotely and send meter readings to energy suppliers. The current project's objective is to measure the amount of electricity consumed by household appliances. Our daily lives and social interactions play a significant role in the worldwide problem of energy conservation. The audit report promotes energy efficiency at the Chandigarh, NITTTR Central Library. Only after careful observation and evaluation of the library's whole building process, including the installation of its electrical utilities, was this made possible (Khang et al., 2024b).

However, our team has discovered the most effective technique to save energy by decreasing a few electrical utilities, replacing certain utilities with energy-saving equipment, and also by introducing an automatic controller circuit like "Off at dark circuit." Following advice on energy conservation, the total cost benefit has also been computed since the overall energy consumption was cut by 38.3%. Again, this can be lessened by utilizing natural light sources (such as sunlight) effectively and by enhancing airflow throughout the library. The proposed "Off at dark circuit" will provide the best results for the most energy savings because the library only operates for 8 hours per day.

REFERENCES

Abhilasha Mani V., L. Gunasekhar, and S. Sankaranarayanan, "IoT Based Smart Energy Management System," *International Journal of Applied Engineering Research*, vol. 12, no. (6), pp. 5455–5462, 2017. https://education.nationalgeographic.org/resource/alternative-energy-use/, accessed on May 2023

Al-Obaidi K. M., M. Hossain, et al., "A Review of Using IoT for Energy Efficient Buildings and Cities: A Built Environment Perspective," *Energies*, vol. 15, no. (16), pp. 1–30, 2022. https://yearbook.enerdata.net/total-energy/world-consumption-statistics.html, accessed on May 2023

Amin S. M. and B. F. Wollenberg, "Toward a Smart Grid: Power Delivery for the 21st Century," *IEEE Power and Energy Magazine*, vol. 3, no. (5), pp. 34–41, 2005. https://ieeexplore.ieee.org/abstract/document/1507024/

Annual Energy Outlook, Energy Information Administration. Department of Energy, pp. 1–15 (2020). https://www.academia.edu/download/43247432/EIA_AEO_2010.pdf

Chelloug S. A. and M. A. El-Zawawy, "Middleware for Internet of Things: Survey and Challenges," *Intelligent Automation and Soft Computing*, vol. 3, pp. 70–95, 2017. https://www.tandfonline.com/doi/abs/10.1080/10798587.2017.1290328

Fahrezi M. Y., I. S. Areni, E. Palantei, and A. Achmad, "A Smart Home Energy Consumption Monitoring System Integrated with Internet Connection," *2020 IEEE International Conference on Communication, Networks and Satellite (Comnetsat)*, Batam, 2020, pp. 75–80. https://ieeexplore.ieee.org/abstract/document/9328960/

Govindarajan R., S. Meikandasivam, and D. Vijayakumar, "Energy Monitoring System Using Zigbee and Arduino," *International Journal of Engineering and Technology*, vol. 7, pp. 608–611, 2018. https://www.researchgate.net/profile/Govindarajan-Ramalingam/publication/334626461_Energy_Monitoring_System_Using_Zigbee_and_Arduino/links/5d3
6df30299bf1995b42bd15/Energy-Monitoring-System-Using-Zigbee-and-Arduino.pdf

Haque M. M., Z. H. Choudhury, and F. M. Alamgir, "IoT Based Smart Energy Metering System for Power Consumers," *2019 2nd International Conference on Innovation in Engineering and Technology (ICIET)*, Dhaka, 2019, pp. 1–6. https://ieeexplore.ieee.org/abstract/document/9290661/

Hahanov V., A. Khang, E. Litvinova, S. Chumachenko, V. A. Hajimahmud, and V. A. Alyar, "The Key Assistant of Smart City - Sensors and Tools," *AI-Centric Smart City Ecosystems: Technologies, Design and Implementation* (1st ed.), 17(10), (2022). CRC Press. https://doi.org/10.1201/9781003252542-17

Hajimahmud V. A., A. Khang, V. Hahanov, E. Litvinova, S. Chumachenko, and V. A. Alyar, "Autonomous Robots for Smart City: Closer to Augmented Humanity," *AI-Centric Smart City Ecosystems: Technologies, Design and Implementation* (1st ed.), 7(12), (2022). CRC Press. https://doi.org/10.1201/9781003252542-7

Hariharan R. S., R. Agarwal, M. Kandamuru, and H. A. Gaffa, "Energy Consumption Monitoring in Smart Home System," *IOP Conference Series, Materials Science and Engineering*, vol. 1085, 2021. Phys. Rev. 47, pp. 777–780. https://iopscience.iop.org/article/10.1088/1757-899X/1085/1/012026/meta

Hasan M. K., M. M. Ahmed, B. Pandey, H. Gohel, S. Islam, and I. F. Khalid, "Internet of Things-Based Smart Electricity Monitoring and Control System Using Usage Data," *Wireless Communications and Mobile Computing*, 2021. Phys. Rev., 47, pp. 777–780. https://www.hindawi.com/journals/wcmc/2021/6544649/

Karthick T., R. S. Charles, D. N. J. Jeslin, and K. Chandrasekaran, "Design of IoT Based Smart Compact Energy Meter for Monitoring and Controlling the Usage of Energy and Power Quality Issues with Demand Side Management for a Commercial Building," *Sustainable Energy, Grids and Networks*, vol. 26, p. 100454, 2021. https://www.sciencedirect.com/science/article/pii/S2352467721000254

Khang A., N. A. Ragimova, V. A. Hajimahmud, and V. A. Alyar, "Advanced Technologies and Data Management in the Smart Healthcare System," *AI-Centric Smart City Ecosystems: Technologies, Design and Implementation* (1st ed.), 16(10), (2022). CRC Press. https://doi.org/10.1201/9781003252542-16

Khang A., V. Abdullayev, V. Hahanov, and V. Shah, *Advanced IoT Technologies and Applications in the Industry 4.0 Digital Economy* (1st ed.), (2024a). CRC Press. https://doi.org/10.1201/9781003434269

Khang A., R. Gujrati, H. Uygun, R. K. Tailor, and S. S. Gaur, "Data-Driven Modelling and Predictive Analytics in Business and Finance," (2024b). CRC Press. https://doi.org/10.1201/9781032600628

Khanh, H. H., Khang, A., "The Role of Artificial Intelligence in Blockchain Applications," *Reinventing Manufacturing and Business Processes through Artificial Intelligence*, vol. 2, no. (20–40), 2021. CRC Press. https://doi.org/10.1201/9781003145011-2

Macheso P. S. and D. Thotho, "ESP32 Based Electric Energy Consumption Meter," *International Journal of Computer Communication and Informatics*, vol. 4, no. (1), pp. 23–35, 2022. https://www.researchgate.net/profile/Paul-Macheso/publication/362733776_ESP32_Based_Electric_Energy_Consumption_Meter/links/6305c20d61e4553b95340a35/ESP32-Based-Electric-Energy-Consumption-Meter.pdf

Mishra J. K., S. Goyal, V. A. Tikkiwal, and A. Kumar, "An IoT Based Smart Energy Management System," *2018 4th International Conference on Computing Communication and Automation (ICCCA)*, Greater Noida, 2018, pp. 1–3. https://ieeexplore.ieee.org/abstract/document/8777547/

Mohassel R. R., A. Fung, F. Mohammadi, and K. Raahemifar, "A Survey on Advanced Metering Infrastructure," *International Journal of Electrical Power and Energy Systems*, vol. 63, pp. 473–484, 2014. https://www.sciencedirect.com/science/article/pii/S0142061514003743

Orumwense, E. F. and K. Abo-Al-Ez, "Internet of Things for Smart Energy Systems: A Review on Its Applications, Challenges and Future Trends," *AIMS Electronics and Electrical Engineering*, vol. 7, no. (1), pp. 50–74, 2022. https://www.researchgate.net/profile/Efe-Orumwense/publication/366507689_Internet_of_Things_for_smart_energy_systems_A_review_on_its_applications_challenges_and_future_trends/links/63be62fa097c7832caa72cd9/Internet-of-Things-for-smart-energy-systems-A-review-on-its-applications-challenges-and-future-trends.pdf

Pimplea M., S. Thopateb, A. Nikamc, and S. Gadekard, "IOT Based Smart Energy Meter Using ESP 32," *3rd International Conference on Communication and Information Processing (ICCIP-2021)*, 2021. https://papers.ssrn.com/sol3/papers.cfm?abstract_id =3921974

Rani S., M. Chauhan, A. Kataria, and A. Khang, "IoT Equipped Intelligent Distributed Framework for Smart Healthcare Systems," *Networking and Internet Architecture* (Eds.), vol. 2, p. 30, 2021. https://doi.org/10.48550/arXiv.2110.04997

Rath, K. C., A. Khang, and D. Roy, "The Role of Internet of Things (IoT) Technology in Industry 4.0," *Advanced IoT Technologies and Applications in the Industry 4.0 Digital Economy* (1st ed.), (2024). CRC Press. https://doi.org/10.1201/9781003434269-1

Saleem Y., N. Crespi, M. H. Rehmani, et al., "Internet of Things-Aided Smart Grid: Technologies, Architectures, Applications, Prototypes, and Future Research Directions," *IEEE Access*, vol. 7, pp. 62962–63003, 2017. https://ieeexplore.ieee.org/ abstract/document/8701687/

Tailor R. K., R. Pareek, and A. Khang, "Robot Process Automation in Blockchain," Khang A., S. Chowdhury, and S. Sharma (Eds.), *The Data-Driven Blockchain Ecosystem: Fundamentals, Applications, and Emerging Technologies* (1st ed.), 8(13), 149–164. (2022). CRC Press. https://doi.org/10.1201/9781003269281-8

8 Application of Internet of Things (IoT) and Sensors Technologies for Agriculture

Ushaa Eswaran and Alex Khang

INTRODUCTION

Over the past two decades, the internet's broad use has delivered countless advantages to businesses and people everywhere. The ability to produce and consume services in real time was the main advantage of this breakthrough. The internet of things (IoT) is a recent development that promises to deliver the same benefit through its cutting-edge technology and give a way to improve the user's perception and ability by changing the working environment. IoT provides a variety of solutions across a range of industries, including healthcare, retail, traffic, security, smart homes, smart cities, and agriculture. Considering the requirement for ongoing monitoring and management in agriculture, IoT adoption is viewed as the best option.

IoT is applied in agriculture at several points along the chain of industrial output (Medela et al., 2013). Precision farming, livestock, and greenhouses are the three main IoT applications in agriculture, which are organized into various monitoring domains. Wireless sensor networks (WSNs), which assist farmers in gathering pertinent data through sensing devices, are used to monitor all these applications with the assistance of various IoT-based sensors and devices. Some IoT-based setups use cloud services to analyze and interpret remote data, which enables academics and agriculturalists to make better decisions. With the development of modern technology, environment monitoring solutions now provide more management and decision-making options (Rani et al., 2021).

A specially designed system for landslip risk monitoring has been created that enables speedy deployments in dangerous environments without human input (Giorgetti et al., 2016). The created system's ability to handle node failures and reorganize the network's subpar communication lines on its own is what makes it more intriguing. A large-scale IoT management system is suggested in Zheng et al. (2016) that keeps track of environmental factors such as wind, soil, atmosphere, and water. The sub-domains to which they belong have also been used to identify IoT-based

DOI: 10.1201/9781003434269-8

agricultural monitoring systems. The selected sub-domains include monitoring of the soil, air, water, temperature, water quality, diseases, locations, environmental conditions, pests, and fertilizationt (Anh et al., 2024).

The IoT paradigm also enhances human connection in the real world by using inexpensive electronic gadgets and communication protocols. IoT also keeps an eye on many environmental factors to produce detailed, real-time maps of temperature, radiation levels, noise levels, and air and water pollution (Torres et al., 2016; Hachem et al., 2015). Additionally, information gathered about various environmental characteristics is communicated to the user via message alerts or trigger alerts (Liu et al., 2013). Numerous studies in the field of IoT-based agriculture have been presented during the past few decades. Therefore, it is crucial to compile, enumerate, examine, and categorize the most recent research in this field

LITERATURE REVIEW

Different IoT-based systems for the agriculture sector have been presented by researchers, enhancing production while requiring less labor from workers. To raise the standard and productivity of agriculture, researchers have worked on a variety of IoT-based agriculture projects. The literature has revealed certain IoT-based agricultural techniques, which have been compiled in this area.

In order to create a plant nursery, Carnegie Melon University used wireless sensor technologies (Junaid, 2009). A WSN-based polyhouse monitoring system that uses carbon dioxide, humidity, temperature, and light detection modules has been shown in Satyanarayana and Mazaruddin (2013). A WSN-based system has been described that monitors many agricultural parameters utilizing GPS technology and the Zigbee protocol (Sakthipriya, 2014). To boost productivity, a real-time monitoring system for rice crops has been developed (Rajesh, 2011). The crop monitoring system, which is described in Shaobo et al. (2010), gathers data on temperature and rainfall and analyses it to reduce the risk of crop loss and increase crop output.

The use of a microcontroller that doubles as a weather station in a low-cost Bluetooth-based system for tracking numerous agricultural factors, such as temperature, has been proposed in Haefke et al. (2011). The suggested technique works well for keeping track of current field data. This system's restricted communication range and requirement for Bluetooth connection with smartphones for continuous monitoring are further drawbacks. For monitoring various environmental factors such as humidity, temperature, sunshine, and pressure, Pavithra and Srinath (2014) created a smart sensing platform based on Zigbee.

In order for each node to efficiently communicate with each other, the created platform offers a high data rate, affordable hardware, and an accurate sensor that operates on a mesh network. An irrigation monitoring system based on the Global System for Mobile Communications (GSM) has been created, and it uses an Android app to measure several environmental factors like humidity, temperature, and water level control. The primary goal of this system is to create a low-cost wireless system, but one drawback is that it requires knowledge of the operational command for activating the field motor and agricultural parameters (Dinesh and Saravanan, 2011).

A system based on GSM and Field Programmable Gate Array (FPGA) has been proposed to measure greenhouse characteristics including humidity and temperature. The suggested method offers timely and reasonably priced options for crop and soil condition monitoring (Castañeda et al., 2006). A straightforward, adaptable networking low-cost system that employs a fuzzy control approach to track several greenhouse characteristics has been proposed by Ferentinos et al. (2014). A more sophisticated monitoring and controlling system for the greenhouse has been proposed in Patil et al. (2015) along with the operating and design approaches for WSN. Additionally, a number of environmental issues relating to greenhouses have been addressed, including the standardization of WSN components, wireless node packaging, and electromagnetic field interference. A system that tracks an animal's health and recognizes common ailments, whether they result from biological attacks or natural causes, is suggested by Vijayan and Suresh (2016). Watthanawisuth et al. (2009) present a low-cost system for tracking animal health that monitors body temperature, posture, and heart rate.

IOT AGRICULTURAL APPLICATION DOMAINS

This section discusses the monitoring, controlling, and tracking domains that are the main emphasis of each IoT agricultural application. Irrigation monitoring and controlling (16%), precision farming (16%), soil monitoring (13%), temperature monitoring (12%), humidity monitoring (11%), animal monitoring and tracking (11%), water monitoring and controlling (7%), disease monitoring (5%), air monitoring (5%), and fertilization monitoring (4%) are the main categories of these applications, as shown in Figure 8.1.

AIR MONITORING

This sub-domain's goal is to assess and decide on the air quality in order to guard against negative impacts. An IoT-based system for monitoring temperature, humidity, and air quality in agriculture has been suggested by Thorat et al. (2017). Based on WSNs, this system provides a real-time microclimate monitoring solution. The system comprises a temperature and humidity sensor that is powered by solar panels and supported by the Zigbee wireless technology.

SOIL MONITORING

Here, we talk about utilizing WSN to monitor soil temperature and moisture in agricultural fields. The user interacts with the systems using web applications in each of these systems, which are maintained by a variety of communication technologies including GPRS, Zigbee, and the internet (Jayaraman et al., 2015; Postolache et al., 2014).

WATER MONITORING

Here, we talk about measuring pH, temperature, and compounds that can alter the typical characteristics of water in order to monitor water quality or pollution. An

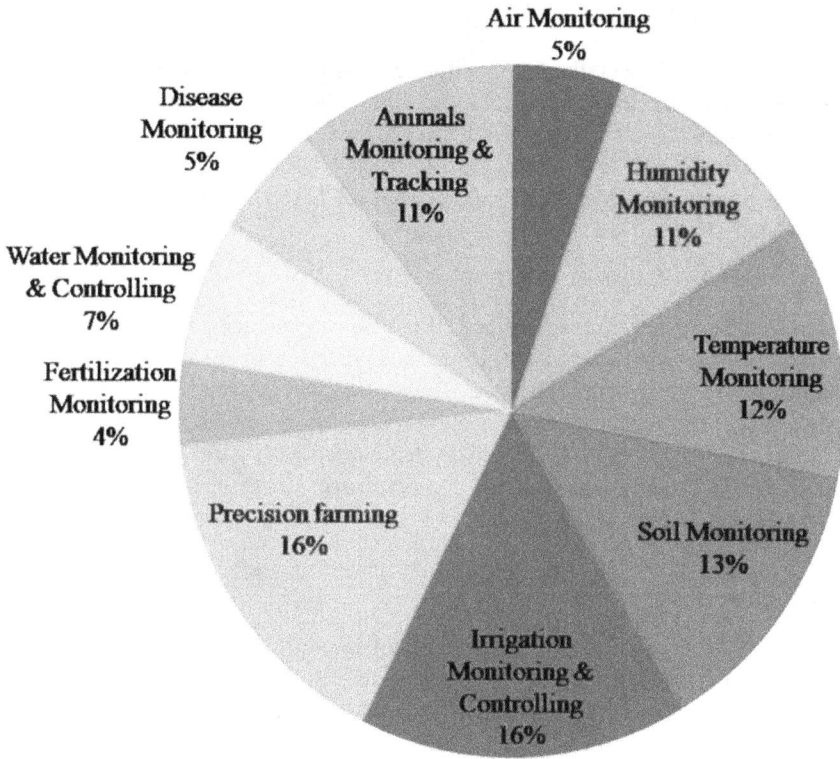

FIGURE 8.1 IoT agricultural application domains

IoT-based method to monitor water quality by gauging temperature, conductivity, and turbidity has been presented in Xijun et al. (2009). This WSN-based solution integrates sensing devices and keeps track of the various water-related metrics in metropolitan settings. Additionally, Fourati et al. (2014) created a WSN-based system to track irrigation system water levels and rainfall. In order to monitor irrigation in olive fields, a web-based decision support system that uses sensors to assess temperature, solar radiations, humidity, and rainfall has been presented by Langendoen et al. (2006).

DISEASE SURVEILLANCE

The best illustration of crop or plant monitoring is provided by the LOFAR-agro Project (Langendoen et al., 2006). By using WSN to monitor several climatic factors including temperature and humidity, this project safeguards the potato crop. The suggested system guards the crop against fungi-related diseases by analyzing the data gathered.

ENVIRONMENTAL CONDITION MONITORING

In Luan et al. (2015), a system for monitoring environmental conditions is suggested that uses WSN to measure the spatial sampling of humidity sensors. The suggested technique makes use of a historical database to ascertain the behavior of 2D correlation. Additionally, Lee et al. (2012) propose a new approach for monitoring environmental conditions that combines forecasts with IoT-based drought monitoring.

CROP AND PLANT GROWTH MONITORING

In this study, farmlands were examined utilizing the mobile sensors described in Feng et al. (2012). This system's main objective is to control plans for viticulture activities and monitor grape growth. Researchers suggested an effective and intelligent monitoring system that makes recommendations based on sensed data to keep an eye on apple orchards (Alahi et al., 2017). This system's main goals are to lower management costs, raise apple quality, and safeguard against insect infestations. It is a WSN-based system created to track the development of apples utilizing Zigbee and GPRS.

TEMPERATURE MONITORING

In terms of agricultural productivity, soil temperature is a key. A technique to track the ratio of nutrients in surface and groundwater has been suggested by Krishna et al. (2017). Electrochemical impedance was used to calculate the amount of nutrients in the soil. An inductance (L), capacitance (C), and resistance (R) (LCR) meter is used to monitor the outcomes of soil tests, and standard library readings are used to calculate the results.

HUMIDITY MONITORING

Multiple humidity sensors are used to measure the air's humidity level. Plants' ability to generate new cells is negatively impacted by excessive dampness (Malaver et al., 2015).

MONITORING GASES IN GREENHOUSE

Agriculture and greenhouse gases are related to each other. An overabundance of gases in a greenhouse raises the temperature, which has an immediate effect on agricultural yield. A WSN and solar-powered unmanned aerial vehicle (UAV) system has been presented to monitor CH_4 and greenhouse gases (Pahuja et al., 2013).

PEST CONTROL AND FERTILIZATION

An IoT solution in this area offers conservation techniques to raise crop quality and nutrient consumption rates. To keep track of pests, irrigation, fertilization, and climate in greenhouses, an online climate-monitoring system has been developed (Roy et al., 2015). To collect and process sensed data for effective analysis, the system uses WSN.

Greenhouse Illumination Control

To keep track on the development of melons and cabbages in greenhouses, an automated agricultural system is being developed (Jain al., 2008). The planned system keeps an eye on the crop-growing procedure and regulates greenhouse environmental factors including humidity, temperature, and ambient light.

Location Monitoring

We will discuss about the location tracking and movement tracing of animals as well as any unauthorized movement across the field. To protect the crop against theft and animal attacks, a variety of monitoring tools and sensors have been placed in the field. When an unauthorized movement occurs in an agricultural field, an intrusion detection system for agriculture is provided by Bryukhanov et al. (2019) that triggers an alarm and sends a text message to the farmer's mobile device. Additionally, several monitoring tools are used for livestock to keep track of the animals' whereabouts and observe their behavior. Kupriyanovsky et al. (2018) present an IoT-based animal monitoring solution that keeps tabs on swamp deer behavior. Additionally, utilizing RFID and WSN, a herd of cattle grazing in the field can be followed and observed.

THE FUNCTION OF IOT-BASED TOOLS AND SENSORS IN AGRICULTURE

For a long time, various businesses and industries have used various types of devices and sensors, but the advent of the internet of things has completely changed how far these technologies have come. An embedded system that communicates with actuators, sensors, and the necessary WSN makes up an IoT device. A CPU, memory, communication modules, and input/output components make up embedded systems. Sensors keep track of the many ambient elements, agricultural variables, and dynamic data that are gathered through facility interaction.

Sensors that are frequently utilized include those that measure temperature, humidity, soil pattern, airflow, location, CO_2, pressure, and moisture. IoT devices and sensors have important qualities that make them suitable for use in agriculture, including mobility, reliability, memory, durability, power and computational efficiency, and coverage. This comprehensive analysis demonstrates that many researchers have concentrated on soil moisture (14%), humidity (17%), and temperature monitoring (19%) (Figure 8.2).

Sensor/Devices Operation

pH Sensor

It is effective for the healthy growth of plants and crops to employ pH sensors to monitor the precise amount of nutrients in the soil (Khang et al., 2023a).

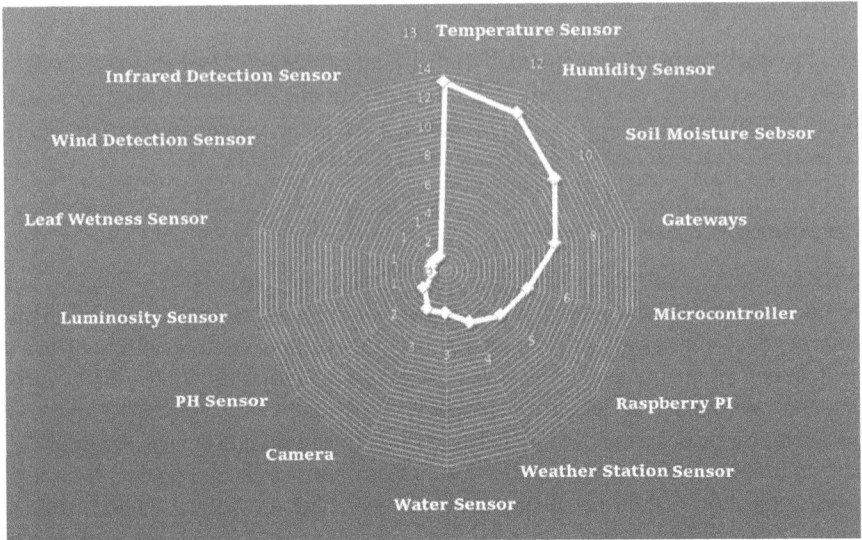

FIGURE 8.2 Sensors/devices distribution

Gas Sensor

This sensor determines the precise concentration of harmful chemicals in cattle and greenhouses by observing infrared radiations from a Waspmote Plug & Sense sensor node! The goal of Smart Agriculture Xtreme is to keep track of the water and gas levels in the soil.

Motion Detector Sensor

The sensor is used to track/trace the whereabouts of animals and fields. In addition, it may also be used to identify the motion of an unwelcome object in a field or farm and provide notifications to farmers to enable them to take prompt action and avoid crop loss.

Ultraviolet Sensor and Passive Infrared (PIR) Sensors

For efficient crop growth, an ultraviolet (UV) sensor keeps an eye on the UV rays. A motion detector that tracks a person's movement in the field is fixed within the PIR sensor. The sensor also has a light-detecting feature that converts rising temperature into voltage while tracking an item for the purpose of analyzing crop development.

Soil Moisture Sensor

The field's total amount of water and level of moisture are measured by the soil sensor. A wireless moisture sensor has been installed to track the watering system for the greenhouse. A cellular connection sensor called the CropX Starter Kit-Soil Temperature 24/7 is used to keep track of various soil conditions (Hajimahmud et al., 2022).

Temperature Sensor

The ability of the soil to absorb nutrients and moisture is impacted by changes in soil temperature. To map the precise distribution of nutrients in the soil and the sea surface, a novel sensing method has been developed. Any field area can have a 3D crop sensor array using Photosynthetically Active Radiation (PAR) technology installed to track the temperature, CO_2, and humidity.

Humidity Sensor

The development of plant leaves, photosynthesis, and pollination are all negatively impacted by humidity. As a result, this sensor directly detects the air's temperature and moisture content in order to detect the amount of humidity. Different climate-monitoring tools from Grofit can measure air temperature, air humidity, and solar radiation. The device's data transmission range is up to 200 m, and it can keep a measurement for up to 30 days.

SMART AGRICULTURE TECHNOLOGIES

Drones, farm-based mobile applications, high-tech sensors, and other new technologies are transforming the Indian agricultural environment. These technologies connect buyers and sellers with crucial farm-related information. A study predicts that the market for global smart farming would expand at a rate of about 13% CAGR.

IoT-Based Smart Farming Security Conditions

Security requirements for IoT-based smart farming are comparable to those for a typical security situation. Therefore, we must pay attention to the following security considerations in order to establish a secure farming solution:

- **Confidentiality**: Only individuals with proper authorization should have access to any relevant agricultural information or personal data.
- **Integrity**: In this context, integrity refers to the unaltered receipt and storage of data or content.
- **Authentication**: Authentication refers to the requirement that communication peers have identities.
- **Data Freshness**: This includes key freshness and data freshness because IoT agricultural networks may offer variable measurements, making it necessary to guarantee that each message is current.
- **Non-Repudiation**: A node cannot, therefore, ever refuse to send a message that has already been sent.
- **Authorization**: Authorization here refers to the notion that only authorized devices are permitted to access a network or other resources.
- **Self-Healing**: If one or more IoT-based agricultural network equipment malfunction or run out of power, other networked devices should be able to at least partially provide security.

IoT in Agriculture

The advantages of IoT applications in agriculture are as follows:

- Increasing the effectiveness with which the inputs are used such as soil, water, fertilizers, and pesticides.
- Lower production costs
- Enhanced financial success
- Sustainability
- Food security
- Continuous monitoring
- Environmental protection
- Competitive advantage
- Intelligent data collection
- Waste minimization
- Process automation
- Animal monitoring

Application of IoTs in Agriculture

The internet of things can be used in the agricultural sector in a variety of ways, which will increase productivity and financial return in addition to its other advantages.

- **IoT Sensors in Agriculture**: IoT sensors can be put in agricultural machinery to offer resource-saving production control.
- **Irrigation Demand**: Water is a crucial component of agricultural crops. However, planting performance suffers from over-application. Therefore, determining which points require more water or not has become essential. In this way, IoT devices are in charge of keeping an eye on soil quality and activating the irrigation system as needed.
- **Soil Improvement**: The same tools can be used to measure pH levels and other soil nutrients, much like water control. This allows them to pinpoint the areas that need improvement, such as fertilization.
- **Crop Evaluation**: Drone use in agriculture has produced a variety of facilities and solutions. They democratized information collection and improved decision-making because they are connected to the internet and have sensors, GPS, and other features. As a result, it is now easier to make precise diagnosis of the health of a crop. Drones have resulted in significant time and resource savings because the producer no longer needs to drive huge distances for this (Khang et al., 2022a).
- **Intelligent Greenhouses**: In greenhouses, temperature, humidity, and light levels should all be tightly regulated. Usually, this control is carried out manually using covers and other tools. You can integrate any greenhouse environment with a single system using IoT devices. Making this control becomes easier and quicker as a result.

SCOPE OF IOT

A smart city is an urban area with advanced infrastructure and communications; a smart city environment includes intelligent management of mobility, utilities, and buildings; a smart home is the integration of technologies and services through a home network to improve the quality of life; and a smart enterprise—IoT enterprise solutions are designed to support infrastructure and more versatile industry (Khang et al., 2024b).

Industry (which includes both agriculture and the agricultural processing industry), ecology, and society might be considered the three main application areas for IoT (Titovskii et al., 2019). Industry also includes pursuits associated with financial or business dealings between firms, groups, and other entities.

The usage of IoT to safeguard the environment is relevant right now. The article by Sakthipriya (2014) lists various agroecological issues, including surface and groundwater pollution in areas of intensive agricultural production, emissions from livestock complexes and agricultural machinery, generation and accumulation of agricultural waste, and pollution and degradation of the soil quality of agricultural lands.

The active application of high-tech methods in agriculture, such as those for locally differentiated, adaptive fertilizer application (e.g., the application of ameliorants using a 10-N sensor) and the use of plant protection products, precision irrigation, Earth remote sensing for in-the-moment monitoring of land and water resources, and the use of unmanned aerial vehicles to accurately analyze the soil and crops, is the solution to these issues (Jayaraman et al., 2016; Sicari et al., 2015).

The growth and blending of society, cities, and individuals are related to the social sphere. These topics are all strongly related to each other. The application areas for the internet of things are broken down into smaller application areas. Transportation, healthcare, construction, energy, agriculture, logistics, and other sectors can be distinguished. Numerous applications of IoT in agriculture have been effective. For instance, the Smart Greenhouse technology is utilized in many farms to control pests and plant diseases, offer automatic watering and an effective temperature condition, and optimize the use of fertilizers, chemicals, and power. Sensor data is processed on a computer running specialized software that is intended to analyze input data and make decisions about actions that are carried out using various electro-mechanical devices (Hahanov et al., 2022).

Utilizing smart sensors, gadgets, and monitoring software (also known as "Smart Farm technology") can increase animal productivity and product quality. Experts claim that automated systems for feeding, milking, and monitoring the health of livestock and poultry can boost egg and milk production by 20% and 30%, respectively (Yang et al., 2017). Experts estimate that GPS monitoring of cars could result in fuel consumption reductions of up to 20%. It also offers chances for personnel scheduling and route optimization (Kumar and Patel, 2014). In Russia, the issue of raw material safety during collection and transportation is important because the appropriate sensors can track the weight and location of the transported raw materials, almost eliminating the possibility of fraud.

SECURITY OF IOT-BASED SMART FARMING

Three fundamental requirements—authentication, access control, and stakeholder confidentiality—make up the security of IoT-based smart farming. While the network must be protected from outside attacks at the perception layer and the data aggregation must be protected at the network layer, it is ensured that only authorized users can access data from the application layer by authorizing specified entities. The most common security issue in the perception layer is physical security, which comprises hardware and information acquisition security (Khang et al., 2022b). Because all the technology is placed up in a public area, physical security is essential.

Since IoT devices can be utilized in a number of contexts, a single security standard is insufficient. Another important security concern is information leakage, which involves revealing of private information and locations. A single security standard is insufficient since IoT devices may be used in a variety of settings. Information leakage, which includes the disclosure of sensitive information and locations, is another significant security risk. Data encryption, jamming, the use of blocker tags, changing the frequency of tags, and tag destruction techniques are all examples of security countermeasures. Since sensor nodes and RFID tags differ from each other, it is important to keep this in mind while developing encryption algorithms, intrusion detection policies, key distribution and routing regulations, and hardware limits.

In the internet of things, information travels from a gateway to an end device while also being uploaded to other platforms, such as cloud infrastructure. For sensor nodes, there are numerous security regulations in place, such as identity verification, data filtering, cryptographic algorithms, and data flow control mechanisms. Security threats also include tampering, wiretapping, replay attacks, and cheating. Therefore, throughout the data-collecting phase, secrecy, authentication, and integrity must be used.

CONCLUSION

Single farmers might be able to deliver their product directly to customers, thanks to the internet of things, and not just in a limited area like in direct marketing or shops. This could result in a more direct, shorter route between producers and customers, changing the entire supply chain, which is currently mostly under the control of major corporations. The corporate sector would be able to offer all necessary services to farmers in remote locations at reasonable prices, thanks to cloud computing.

Researchers are looking into IoT-based technological solutions to improve farm productivity in a way that complements current services all over the world. In order to achieve this, we talk about the topology, platform, and design of agricultural networks that enable access to the internet of things (IoT) backbone and enable farmers to increase crop productivity. A thorough review of recent and ongoing developments in IoT agricultural applications, gadgets and sensors, communication protocols, and numerous cutting-edge technology is discussed here.

REFERENCES

Alahi, M.E.E.; Xie, L.; Mukhopadhyay, S.; Burkitt, L. A temperature compensated smart nitrate-sensor for agricultural industry. *IEEE Trans. Ind. Electron.* 2017, *64*(9), 7333–7341. https://ieeexplore.ieee.org/abstract/document/7906489/

Anh, P.T.N.; Hahanov, Vladimir; Triwiyanto, Ragimova Nazila Ali; İsmibeyli, Rashad; Abdullayev, Vugar; Alyar, Abuzarova Vusala AI models for disease diagnosis and prediction of heart disease with artificial neural networks. *Computer Vision and AI-Integrated IoT Technologies in Medical Ecosystem* (1st ed.) 2024. CRC Press. https://doi.org/10.1201/9781003429609-9

Bryukhanov, A.Y. et al. Technologies and technical means of mechanized production of crop production and animal husbandry. 2019, *1*(98), 257–268. https://iopscience.iop.org/article/10.1088/1742-6596/1515/2/022009/meta

Castañeda-Miranda, R.; Ventura-Ramos, E., Jr.; del RocíoPeniche-Vera, R.; Herrera-Ruiz, G. Fuzzy greenhouse climate control system based on a field programmable gate array. *Biosyst. Eng.* 2006, *94*(2), 165–177. https://www.sciencedirect.com/science/article/pii/S1537511006000729

Dinesh, M.; Saravanan, P. FPGA based real time monitoring system for agricultural field. *Int. J. Electron. Comput. Sci. Eng.* 2011, *1*, 1514–1519. https://www.mdpi.com/2073-4395/11/9/1881

Ferentinos, K.P.; Katsoulas, N.; Tzounis, A.; Kittas, C.; Bartzanas, T. A climate control methodology based on wireless sensor networks in greenhouses. In *Proceedings of the XXIX International Horticultural Congress on Horticulture: Sustaining Lives, Livelihoods and Landscapes (IHC2014)*, Brisbane, 17–22 August 2014, pp. 75–82. https://www.actahort.org/books/1107/1107_9.htm

Feng, C.; Wu, H.R.; Zhu, H.J.; Sun, X. The design and realization of apple orchard intelligent monitoring system based on internet of things technology. In *Advanced Materials Research; Trans Tech Publications*. Stafa-Zurich, 2012, Volume 546, pp. 898–902. https://www.scientific.net/AMR.546-547.898

Fourati, M.A.; Chebbi, W.; Kamoun, A. Development of a web-based weather station for irrigation scheduling. *In Proceedings of the, 2014 Third IEEE International Colloquium in Information Science and Technology (CIST)*. Tetouan, 20–22 October 2014, pp. 37–42. https://ieeexplore.ieee.org/abstract/document/7016591/

Giorgetti, A.; Lucchi, M.; Tavelli, E.; Barla, M.; Gigli, G.; Casagli, N.; Dardari, D. A robust wireless sensor network for landslide risk analysis: System design, deployment, and field testing. *IEEE Sens. J.* 2016, *16*, 6374–6386. https://ieeexplore.ieee.org/abstract/document/7488208/

Hachem, S.; Mallet, V.; Ventura, R.; Pathak, A.; Issarny, V.; Raverdy, P.G.; Bhatia, R. Monitoring noise pollution using the urban civics middleware. In *Proceedings of the 2015 IEEE First International Conference on Big Data Computing Service and Applications*, Redwood City, CA, 30 March–2 April 2015, pp. 52–61. https://ieeexplore.ieee.org/abstract/document/7184864/

Haefke, M.; Mukhopadhyay, S.C.; Ewald, H. A Zigbee based smart sensing platform for monitoring environmental parameters. In *Proceedings of the 2011 IEEE International Instrumentation and Measurement Technology Conference*, Binjiang, 10–12 May 2011, pp. 1–8. https://ieeexplore.ieee.org/abstract/document/5944154/

Hajimahmud, V.A.; Khang, A.; Hahanov, V.; Litvinova, E.; Chumachenko, S.; Alyar, V.A. Autonomous Robots for Smart City: Closer to Augmented Humanity. In *AI-Centric Smart City Ecosystems: Technologies, Design and Implementation* (1st ed.), 7(12), 2022. CRC Press. https://doi.org/10.1201/9781003252542-7

Hahanov, V.; Khang, A.; Litvinova, E.; Chumachenko, S.; Hajimahmud, V.A.; Alyar, V.A. The Key Assistant of Smart City - Sensors and Tools. *AI-Centric Smart City Ecosystems: Technologies, Design and Implementation* (1st ed.), 17(10), 2022. CRC Press. https://doi .org/10.1201/9781003252542-17

Jain, V.R.; Bagree, R.; Kumar, A.; Ranjan, P. wildCENSE: GPS based animal tracking system. In *Proceedings of the 2008 International Conference on Intelligent Sensors, Sensor Networks and Information Processing*, Sydney, 15–18 December 2008, pp. 617–622. https://ieeexplore.ieee.org/abstract/document/4762058/

Jayaraman, P.P.; Palmer, D.; Zaslavsky, A.; Salehi, A.; Georgakopoulos, D. Addressing information processing needs of digital agriculture with OpenIoT platform. In *Interoperability and Open*, 2015. https://link.springer.com/chapter/10.1007/978-3-319 -16546-2_11

Jayaraman, P.P.; Yavari, A.; Georgakopoulos, D.; Morshed, A.; Zaslavsky, A. Internet of Things platform for smart farming: Experiences and lessons learnt. *Sensors (Basel)* 2016. *16*(11), 1884. https://www.mdpi.com/1424-8220/16/11/1884

Junaid, A. *Application of Modern High Performance Networks*. Oak Park, IL, Bentham Science Publishers Ltd., 2009, pp. 120–129. https://www.google.com/books?hl=en&lr =Google& id=GK0TDgAAQBAJ&oi=fnd&pg=PP1&dq=Application+of+Modern+ High+Performance+Networks%3B+Bentham+Science+Publishers+Ltd.:+Oak+Park ,+IL&ots=bW4oYSOmYN&sig=4pY6zbSnXNt4XX6IwZNIV2CJ1Ug

Khandani, S.K.; Kalantari, M. Using field data to design a sensor network. In *Proceedings of the 2009 43rd Annual Conference on Information Sciences and Systems*, Baltimore, MD, 18–20 March 2009, pp. 219–223. https://link.springer.com/chapter/10.1007/978-3 -319-16546-2_11

Khang, A.; Ragimova, N.A.; Hajimahmud, V.A.; Alyar, V.A. Advanced technologies and data management in the smart healthcare system. In *AI-Centric Smart City Ecosystems: Technologies, Design and Implementation* (1st ed.), 16 (10), 2022a. CRC Press. https:// doi.org/10.1201/9781003252542-16

Khang, A.; Hahanov, V.; Abbas, G.L.; Hajimahmud, V.A. Cyber-Physical-social system and incident management. In *AI-Centric Smart City Ecosystems: Technologies, Design and Implementation* (1st ed.), 2 (15), 2022b. CRC Press. https://doi.org/10.1201 /9781003252542-2

Khang, A. *Advanced Technologies and AI-Equipped IoT Applications in High-Tech Agriculture* (1st ed.), 2023a. IGI Global Press. https://doi.org/10.4018/978-1-6684 -9231-4

Khang, A.; Rana, G.; Tailor, R. K., Hajimahmud, V. A. *Data-Centric AI Solutions and Emerging Technologies in the Healthcare Ecosystem* (Eds.) (2023b). CRC Press. https://doi.org/10.1201/9781003356189

Khang, A.; Abdullayev, V.; Hahanov, V.; Shah, V. *Advanced IoT Technologies and Applications in the Industry 4.0 Digital Economy* (1st ed.), 2024a. CRC Press. https:// doi.org/10.1201/9781003434269

Krishna, K.L.; Silver, O.; Malende, W.F.; Anuradha, K. Internet of Things application for implementation of smart agriculture system. In *Proceedings of the 2017 International Conference on I-SMAC (IoT in Social, Mobile, Analytics and Cloud) (I-SMAC)*, Palladam, 10–11 February 2017, pp. 54–59. https://ieeexplore.ieee.org/abstract/docu ment/8058236/

Kumar, J.S.; Patel, D.R. A survey on Internet of Things: Security and privacy issues. *Int. J. Comput. Appl.* 2014. *90*(11), 1–7. https://course.ccs.neu.edu/cs7680su18/resources/ pxc3894454.pdf

Kupriyanovsky, V.P. et al. *Int. J. Open Inf. Technol.* 2018, *6*(10), 46–67. https://doi.org/10 .1201/9781003434269

Langendoen, K.; Baggio, A.; Visser, O. Murphy loves potatoes: Experiences from a pilot sensor network deployment in precision agriculture. In *Proceedings of the 20th IEEE International Parallel Distributed Processing Symposium*, Rhodes Island, 25–29 April 2006. https://ieeexplore.ieee.org/abstract/document/1639412/

Luan, Q.; Fang, X.; Ye, C.; Liu, Y. An integrated service system for agricultural drought monitoring and forecasting and irrigation amount forecasting. In *Proceedings of the 2015 23rd International Conference on Geo Informatics*, Wuhan, 19–21 June 2015, pp. 1–7. https://ieeexplore.ieee.org/abstract/document/7378617/

Lee, J.; Kang, H.; Bang, H.; Kang, S. Dynamic crop field analysis using mobile sensor node. In *Proceedings of the 2012 International Conference on ICT Convergence (ICTC)*, Jeju Island, 15–17 October 2012, pp. 7–11. https://ieeexplore.ieee.org/abstract/document/6386766/

Liu, Z.; Huang, J.; Wang, Q.; Wang, Y.; Fu, J. Real-time barrier lakes monitoring and warning system based on wireless sensor network. In *Proceedings of the 2013 Fourth International Conference on Intelligent Control and Information Processing (ICICIP)*, Beijing, 9–11 June 2013, pp. 551–554. https://ieeexplore.ieee.org/abstract/document/6568136/

Malaver Rojas, J.A.; Gonzalez, L.F.; Motta, N.; Villa, T.F.; Etse, V.K.; Puig, E. Design and flight testing of an integrated solar powered UAV and WSN for greenhouse gas monitoring emissions in agricultural farms. In *Proceedings of the 2015 IEEE/RSJ International Conference on Intelligent Robots and Systems, Hamburg, Germany, 28 September–2 October 2015*, Volume 1, No. 1, pp. 1–6. https://core.ac.uk/download/pdf/33503789.pdf

Medela, A.; Cendón, B.; González, L.; Crespo, R.; Nevares, I. IoT multiplatform networking to monitor and control wineries and vineyards. In *Proceedings of the 2013 Future Network Mobile Summit*, Lisboa, 3–5 July 2013, pp. 1–10. https://ieeexplore.ieee.org/abstract/document/6633525/

Pahuja, R.; Verma, H.K.; Uddin, M. A wireless sensor network for greenhouse climate control. *IEEE Pervasive Comput.* 2013, *12*(2), 49–58. https://ieeexplore.ieee.org/abstract/document/6504857/

Patil, A.; Pawar, C.; Patil, N.; Tambe, R. Smart health monitoring system for animals. In *Proceedings of the 2015 International Conference on Green Computing and Internet of Things (ICGCIoT)*, Noida, 8–10 October 2015, pp. 1560–1564. https://ieeexplore.ieee.org/abstract/ document/7380715/

Pavithra, D.S.; Srinath, M.S. GSM based automatic irrigation control system for efficient use of resources and crop planning by using an android mobile. *IOSR J. Mech. Civ. Eng.* 2014, *11*(4), 49–55. http://smartfasal.in/wp/wp-content/uploads/2019/09/GSM-based -Automatic-Irrigation-Control-System-for-Efficient-Use-of-Resources-and-Crop -Planning-by-Using-an-Android-Mobile.pdf

Postolache, O.; Pereira, J.D.; Girão, P.S. Wireless sensor network-based solution for environmental monitoring: Water quality assessment case study. *IET Sci. Meas. Technol.* 2014, *8*(6), 610–616. https://ietresearch.onlinelibrary.wiley.com/doi/abs/10.1049/iet-smt.2013 .0136

Rajesh, D. Application of spatial data mining for agriculture. *Int. J. Comput. Appl.* 2011, *15*(2), 7–9. https://www.researchgate.net/profile/Rajesh-D-2/publication/50946211 _Application_of_Spatial_Data_mining_for_Agriculture/links/60a5fc1192851c4 3da031ec0/Application-of-Spatial-Data-mining-for-Agriculture.pdf

Rani, S.; Chauhan, M.; Kataria, A.; Khang, A. IoT equipped intelligent distributed framework for smart healthcare systems. *Networking and Internet Architecture.* 2021, *2*, 30. https://doi.org/10.48550/arXiv.2110.04997

Roy, S.K.; Roy, A.; Misra, S.; Raghuwanshi, N.S.; Obaidat, M.S. AID: A prototype for agri-cultural intrusion detection using wireless sensor network. In *Proceedings of the 2015 IEEE International Conference on Communications (ICC)*, London, 8–12 June 2015, pp. 7059–7064. https://ieeexplore.ieee.org/abstract/document/7249452/

Satyanarayana, G.V.; Mazaruddin, S.D. Wireless sensor based remote monitoring system for agriculture using ZigBee and GPS. In *Proceedings of the Conference on Advances in Communication and Control Systems-2013*, Makka Wala, 6–8 April 2013. https://www.atlantis-press.com/proceedings/cac2s-13/6287

Sakthipriya, N. An effective method for crop monitoring using wireless sensor network. *Middle East J. Sci. Res.* 2014, *20*, 1127–1132. https://www.academia.edu/download/53733474/ Crop_Monitoring.pdf

Shaobo, Y.; Zhenjianng, C.; Xuesong, S.; Qingjia, M.; Jiejing, L.; Tingjiao, L.; Kezheng, W. The appliacation of Bluetooth module on the agriculture expert System. In *Proceedings of the 2010 2nd International Conference on Industrial and Information Systems*, Dalian, 10–11 July 2010, Volume 1, pp. 109–112. https://ieeexplore.ieee.org/abstract/document/5565902/

Sicari, S.; Rizzardi, A.; Grieco, L.A.; Coen-Porisini, A. Security privacy and trust in Internet of Things: The road ahead. *Comput. Netw. 76*, 146–164, January 2015. https://pdfs.semanticscholar.org/5085/517ab5803c3ab2bc6fa984e10e541a5f2f37.pdf

Thorat, A.; Kumari, S.; Valakunde, N.D. An IoT based smart solution for leaf disease detec-tion. In *Proceedings of the 2017 International Conference on Big Data, IoT and Data Science (BID)*, Pune, 20–22 December 2017, pp. 193–198. https://ieeexplore.ieee.org/abstract/ document/8336597/

Tikhonov, A.A. et al., 2018 Mining information and analytical Bulletin, *12*, pp. 192–198. https://www.giab-online.ru/en/catalog/issledovanie-mezhsloevyh-deformaciy-vozni-kayushchih-pri-posadke-

Torres-Ruiz, M.; Juárez-Hipólito, J.H.; Lytras, M.D.; Moreno-Ibarra, M. Environmental noise sensing approach based on volunteered geographic information and spatio-tem-poral analysis with machine learning. In *Proceedings of the International Conference on Computational Science and Its Applications*, Beijing, 4–7 July 2016, pp. 95–110. https://link.springer.com/chapter/10.1007/978-3-319-42089-9_7

Titovskii, S.N. Science and education: Experience, problems, prospects for development. In *Proceedings of the of the International Scientific and Practical Conference (Krasnoyarsk: KrasGAU)*, 2019, pp. 304–307. https://doi.org/10.1201/9781003434269

Vijayan, A.; Suresh, M. Wearable sensors for animal health monitoring using Zigbee. *Int. Adv. Res. J. Sci. Eng. Technol.* 2016, *3*, 369–373. https://ieeexplore.ieee.org/abstract/ document/8553844/

Watthanawisuth, N.; Tuantranont, A.; Kerdcharoen, T. Microclimate real-time monitor-ing based on ZigBee sensor network. In *Proceedings of the Sensors*, 2009 IEEE, Christchurch, 25–28 October 2009, pp. 1814–1818. https://ieeexplore.ieee.org/abstract/document/ 5398587/

Xijun, Y.; Limei, L.; Lizhong, X. The application of wireless sensor network in the irriga-tion area automatic system. In *Proceedings of the 2009 International Conference on Networks Security, Wireless Communications and Trusted Computing*, Wuhan, 25–26 April 2009, Volume 1, pp. 21–24. https://ieeexplore.ieee.org/abstract/document/4908205/

Yang, Y.; Wu, L.; Yin, G.; Li, L.; Zhao, H. A survey on security and privacy issues in Internet-of-Things. *IEEE Internet Things J., 4*(5), 1250–1258, October 2017. https://ieeexplore.ieee.org/abstract/document/7902207/

Zheng, R.; Zhang, T.; Liu, Z.; Wang, H.; An EIoT system designed for ecological and environ-mental management of the Xianghe Segment of China's Grand Canal. *Int. J. Sustain. Dev. World Ecol.* 2016, *23*(4), 372–380. https://www.tandfonline.com/doi/abs/10.1080/13504509.2015.1124470

9 Application of Internet of Things (IoT) Technologies for Agriculture

Ankit Jain, Anita Shukla, Imran Ullah Khan, and Raghvendra Singh

INTRODUCTION

Agriculture in India has a long history that extends back to the Indus Valley Civilization, and in some regions of South India, much earlier (Indian Geography, 2019). India is currently in second place globally in farm output. With India's overall economic growth, agriculture's economic contribution to GDP is continuously shrinking. Nevertheless, agriculture is India's largest economic sector by population and contributes significantly to the country's overall socioeconomic structure (Khang, 2023).

India was the sixth largest net exporter and the seventh largest agricultural exporter in the world in 2013 with $38 billion worth of agricultural exports (Wikipedia, 2019). The emerging and least developed countries receive the majority of their agricultural exports (Wikipedia, 2019). More than 120 countries receive Indian agricultural, horticultural, and processed food exports, mostly from the Middle East, Southeast Asia, SAARC nations, the European Union, and the United States (Agriculture in India, 2019) (Yalla et al., 2013).

Although we live in a world where everything can be managed and operated automatically, there are still some significant industries in our nation where automation has not yet been fully adopted or utilized. Agriculture is one of those fields. Automated farming entails keeping track of and making adjustments to the climatic factors that either directly or indirectly affect plant growth and the end product.

If remedial measures are not put in place before the issue becomes unsolvable or not worth solving, groundwater overexploitation issues in India will inevitably worsen and expand. Other issues with farming, such as keeping the right temperature, humidity level, and light available for advanced plants, as well as keeping an eye on unauthorized entry into farmed fields, call for technology solutions. On the other hand, a timely and adequate supply of electricity is the most crucial factor for agriculture (Upinder & Manjit et al., 2023). In addition to decreasing efficiency, a highly unpredictable power supply with frequent power outages has also posed issues for farmers, who spend their time ensuring that power is available since without it,

DOI: 10.1201/9781003434269-9

their jobs cannot begin. The node MCU Wi-Fi module used to connect this system to the internet allows it to show real-time data that can be seen and accessed via the internet utilizing IoT technology from anywhere in the globe (Subhashini & Khang et al., 2023).

The identified problem in the Indian farming system is addressed by this hardware in an effective and efficient manner. The offered remedy is cost-effective, environmentally benign, and electronically operated, making the Indian agricultural system more farmer-friendly.

RELATED WORK

New approaches and technologies are frequently put out and adopted in order to address the demands of humanity as they exist today on a global scale. The internet of things (IoT) has emerged as a result (Quy et al., 2022a; Sinche et al., 2020) and is described as a network of all objects that are integrated within devices, sensors, machines, software, and people through the internet environment to communicate, share information, and interact in order to offer a holistic solution between the real world and the virtual world (Rani et al., 2021). IoT has been used in a variety of fields recently, including smart homes (Li et al., 2019; Shin et al., 2019), smart cities (An et al., 2019; Cirillo et al., 2020), smart energy (Ammad et al., 2020; Metallidou et al., 2020), autonomous cars (Quy et al., 2022b; Kiani et al., 2022), smart agriculture (Patle et al., 2022; Vangala et al., 2020), campus management (Chang and Lai, 2020; Sutjarittham et al., 2019), healthcare (Rani et al., 2019; Zhou et al., 2020), and logistics (Humayun et al., 2020; Song et al., 2021; Shafique et al., 2020) and described a number of different IoT applications.

According to Yalla et al. (2013), information regarding the automatic supply of water to fields is explained. The automation of the system is given with modules and a soil moisture sensor, and the source to generate power from renewable resources prefers sunlight as the primary source. The PV farmers pump was created and atomized by Pradeep et al. (2011) while taking into account the power supply, direct current (DC), alternating current (AC), inverter frequency, GSM technology, a well, the water level in the well, and submersible pump. Here, the researchers present a cutting-edge method using a GSM module. An irrigation system based on a variable-rate automated microcontroller has been proposed by Uddin et al. (2012). Only solar power is employed as a source of power to operate the entire system. On the paddy field, sensors are positioned. These sensors continuously sense the water level and communicate this information to the farmer.

Seal et al. (2014) explored the design of a solar tracking system to capture the most solar energy, which is then turned into electrical energy and used to power the irrigation system. Kumar et al. (2014) discuss a moisture sensor-based automatic water flow control system with a solar-powered vehicle tracker. It is the suggested remedy for the current energy crisis affecting Indian farmers. Patel and Mathurkar (2013) highlight several monitoring and controlling systems that have been built to raise yield. One of the main causes of the yield's decline is diseases. In order to predict when a disease would begin to spread, a monitoring system has been designed.

A sensor module is used to detect various environmental conditions across the farm, and a microprocessor displays the sensed data on an LCD.

Irrigation, as defined by Pasha and Yogesha (2014), is the artificial application of water to land or soil for the purpose of assisting in the growth of agricultural crops, maintaining landscapes, and re-greening damaged soils in arid climates and during dry spells. Dhanne et al. (2014) explored details pertaining to the automatic provision of water to fields, automation of the system is given with modules and soil moisture sensors, and the source to generate power from renewable resources favors sunshine as the major source.

SMART FARMING SYSTEM

The term "smart agriculture" refers to a broad range of internet of things (IoT)-related uses in farming, agro-technology, and other areas of agriculture. The use of sensors, data collecting, wireless networks, cloud platforms, and data analysis is already revolutionizing the farming and agricultural industries. It can be interpreted as anything that improves animal and crop production through more precise and controlled farming practices (Kamran & Noman et al., 2023). The utilization of IoT and different tools like sensors, control systems, robots, autonomous vehicles, automated hardware, variable-rate technologies, and other things is a crucial element in this method of farm management (Jebaraj & Khang et al., 2023). The primary benefits include:

- Increased production.
- Lower production costs.
- Operational efficiencies.
- Real-time and intelligent cost management.

These products have the ability to lessen the impact on the environment while also optimizing your earnings by either increasing revenues or decreasing costs. Consequently, the IoT agricultural applications enable ranchers and farmers to gather useful data. Large landowners and small farmers must recognize the potential of the IoT market for agriculture and implement smart technology to boost their operations' competitiveness and sustainability. Demand can be successfully met given the population's rapid growth if ranchers and small farmers successfully use agricultural IoT technologies (Khang & Gupta et al., 2023).

FEATURES OF THE PROPOSED SYSTEM

Numerous features have been added to this system to increase its functionality and make it more and more farmer-friendly. In general, greenhouses are utilized for high-tech crops or for specific crops that are used for irrigation research. This system is connected to the internet using a node MCU Wi-Fi module, which may show real-time data that is viewable and retrieved via internet utilizing IoT technology from anywhere in the globe via a Blynk server. The attributes are listed as follows:

- This model includes a feature where the water pump automatically turns on when the soil needs water and shows real-time data that can be viewed via IoT from anywhere in the world.
- For advanced crops, an automatic artificial lighting system is turned "ON" if light levels drop below a set threshold. With the help of IoT, real-time data on light intensity might be retrieved from any location in the world.
- This method uses a single push button to manually or automatically open or close the green house's roof using a servo motor depending on the amount of light present via the Blynk server.
- A rain sensor is used to track down rainfall, providing us with data on the amount of precipitation. Using the idea that ultrasound behaves like radar, an ultrasonic sensor is used to measure the water level in the tank.
- When the humidity reaches a certain level, the water-distributing jet motor automatically activates. The % humidity level is available online.
- For complex crops, the technology automatically switches "ON" the fan when the temperature rises and updates the temperature values online. Pressure sensors gauge the environment's absolute pressure.
- When there is an alarming circumstance, such as an unauthorized occupant, the buzzer begins to beep, and we receive the latest information online.
- It has solar panels that, in the event of a power outage, give the system a backup source of electricity.
- It has a real-time display system through the internet that shows the data gathered by all sensors together with the choice or action made by the microcontroller.
- It includes the option for both automated and manual relay operation using switches on the control panel over the internet, which is accessible from anywhere in the world.

ALGORITHM AND IMPLEMENTATION

In this work, we suggest creating a model through the use of control engineering in order to solve the stated issue and also to include additional beneficial characteristics, the methodical approach taken to create the system's hardware in stages. Building a broad system to gather input from outside sources (sensors) and alter it to produce a certain output is the main goal. The sensors and circuitry listed below make up the hardware created for the current system.

- ESP 8266 node MCU Wi-Fi module
- LCD 16×2
- DHT11 temperature and humidity sensor
- Light-dependent register (LDR) sensor
- Pressure sensor (BMP18)
- Rain sensor
- Soil moisture sensor
- Ultrasonic sensor (HC-SR04)

- Servo motor
- Buzzer
- ULN 2003A relay driver IC & relay
- Solar panel with power backup battery

Node MCU ESP8266 Pin Out

The node MCU ESP8266 Wi-Fi module, as shown in Figure 9.1, is a development board with an open-source Lua-based firmware that has been created specifically for internet of things (IoT) applications. Although node MCU ESP8266-based boards originally shipped with the LUA scripting language for programming (Lua is a lightweight, high-level, multi-paradigm programming language designed primarily for embedded use in applications), ESP-based boards now require the ESP8266 package to be installed into Arduino IDE.

Main Components

The main components related to the proposed system are explained development of circuit, algorithm and display panel development for Internet of things as below.

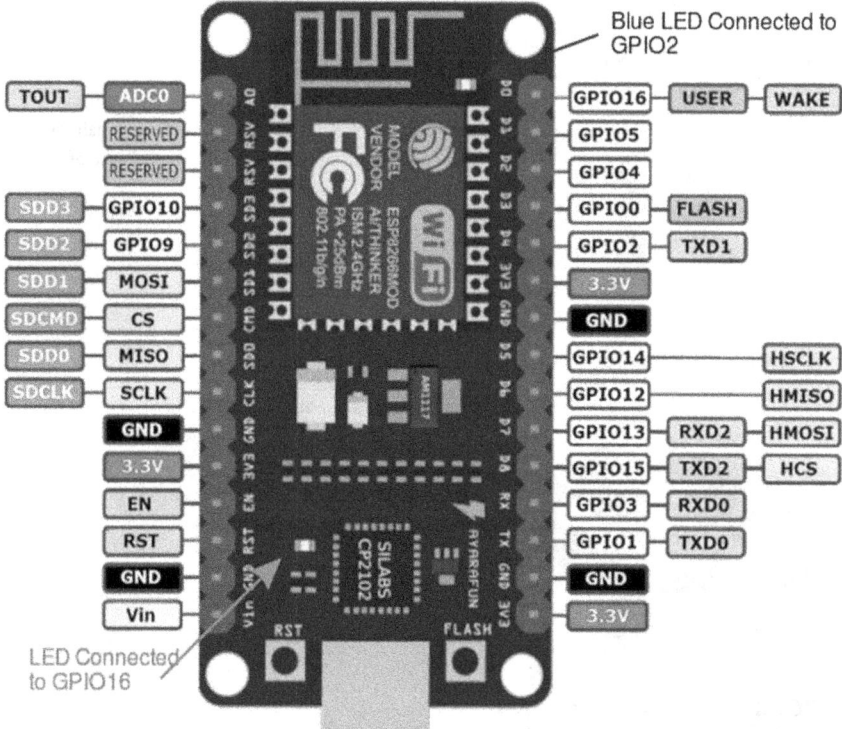

FIGURE 9.1 Pin mapping of node MCU

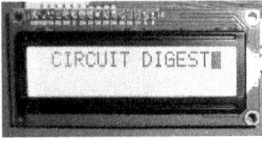

FIGURE 9.2   16 × 2 LCD

A relatively basic module that is frequently used in many different devices and circuits is the 16 × 2 LCD. LCDs are inexpensive, simple to program, and have no restrictions on the display of special and even customized characters and animations (unlike in seven segments) (Figure 9.2).

FIGURE 9.3   DHT11 sensor.

The DHT11 digital temperature and humidity sensor is a composite sensor that produces a calibrated digital signal for both temperature and humidity (Figure 9.3).

FIGURE 9.4   LDR module

The LDR uses a light-sensitive resistor, and as light hits it, the resistance changes. As light levels increase, resistance values decrease (Figure 9.4), ranging in magnitude by several orders of magnitude.

FIGURE 9.5   Pressure sensor

The BMP180 instrument measures the environment's absolute pressure. The BMP180 is delivered, calibrated, and prepared for usage. Due to the device's use of I2C (Figure 9.5).

FIGURE 9.6   Rain sensor

A rain sensor (water drops) is used to track when it rains. It functions like a switch. It is made up of two components: a sensor module and a sensing pad. While water drops touch the sensing pad surface, the switch closes (Figure 9.6), closing it while it is normally open.

FIGURE 9.7 Soil moisture sensor

This sensor employs two probes to carry current through the soil and measure the resistance in order to calculate the quantity of moisture present. The soil transmits electricity more readily when it is moist (less resistance) and less effectively when it is dry (more resistance) (Figure 9.7).

FIGURE 9.8 Ultrasonic sensor

Sonar waves are used by the HC-SR04 ultrasonic sensor to calculate distance. From 2 cm to 400 cm, or 1 inch to 13 feet, it offers remarkable non-contact range detection with high precision and consistent results (Figure 9.8).

FIGURE 9.9 Buzzer

An audio signaling device, such as a buzzer or beeper, can be mechanical, electromechanical, or piezoelectric. Buzzers and beepers are frequently used as timers, alarm clocks, and confirmation of user input (Figure 9.9).

FIGURE 9.10 ULN 2003A relay driver IC

The primary use of ULN is the interface of low-level circuits with several peripheral power loads. The driver uses the ULN2003Adriver IC, which has seven power Darlington arrays with a combined 500 mA driving capacity. Typical loads over 230 V can be regulated simultaneously at a rough duty cycle, depending on the input sensor data number of drivers turned on (Figure 9.10).

FIGURE 9.11 Relay

Through the use of a relay, one circuit can switch a completely independent second circuit. The mechanical and magnetic connection between the two circuits within the relay is the only connection available. The magnetic field produced as a current passes through the coil draws an armature that is mechanically connected to a moving contact. A connection with a fixed contact is either made or broken by the movement (Figure 9.11).

FIGURE 9.12   Solar panel PV cell

Using the photovoltaic effect, a solar cell or photovoltaic (PV) cell transforms light into an electric current. Direct current (DC) power generated by solar cells varies with the brightness of the sun. Inside modules, several solar cells are linked. An inverter, which generates power at the desired voltage and, for AC, the desired frequency/phase, is connected to an array of modules that have been linked together. A solar cell can generate 1.5 W of power. A solar panel, which may generate 3–110 W of power, is made up of individual cells that are connected to one another. Typically, panels are made to provide energy at 12 V (Figure 9.12).

DEVELOPMENT OF CIRCUIT

The circuit diagram is shown in Figure 9.13; the system is built with node MCU microcontroller board. Green house is generally used for sophisticated crops or for certain crops that are used for research purposes in the field of irrigation. By the use of I2C protocol, we have interfaced 16 × 2 LCD and BMP180 pressure sensors using Serial data (SDA) and Serial clock (SCL) pin of node MCU D2 and D1, respectively. The DHT11 sensor is employed to measure the temperature and humidity. If the temperature range goes below the threshold level, it will automatically switch on the fan using a relay for cooling purposes.

A digital light sensor is employed to measure the light intensity; it will automatically switch on the artificial light if the light intensity goes below a certain threshold level. This circuitry has one push button to open or close the roof of the greenhouse using a servo motor, manually as well as automatically according to the intensity of light through the Blynk server. A rain sensor, which also serves as a switch, is used to gauge rainfall. It is made up of two parts: a sensor pad and a sensor module. The switch closes from its normally open position when a drop of water touches the sensing pad's surface, providing us with information about the amount of rain that has fallen.

To measure the amount of moisture in the soil, a soil moisture sensor is used. Using the idea that ultrasonic is similar to radar, an ultrasonic sensor is used to determine the amount of water in the tank. When the tank is empty, it senses the water level and turns on the water pump. An ultrasonic sensor is also used to detect any unwanted occupancy through the entry area. A buzzer is also provided for alarm. Node MCU is not able to provide enough current, which is required to operate relay and that is why ULN2003A is used as a relay driver module. BMP180 measures the absolute pressure of the environment which is continuously displayed on LCD.

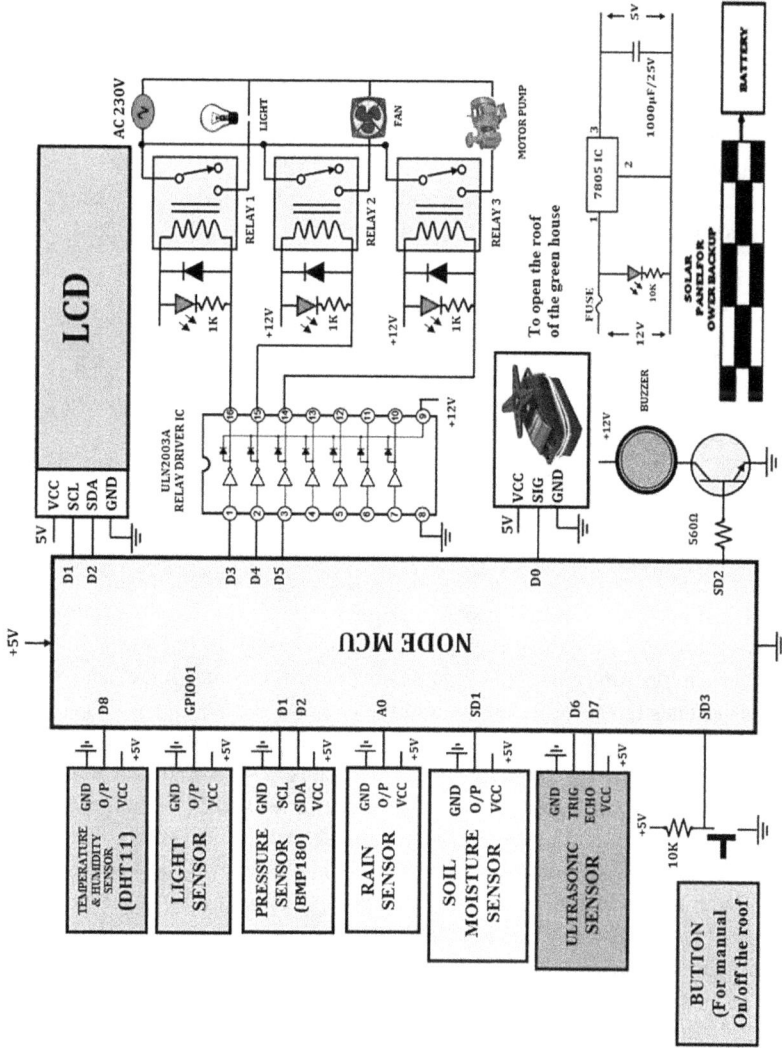

FIGURE 9.13 Circuit diagram of the proposed model

ALGORITHM AND DISPLAY PANEL DEVELOPMENT FOR INTERNET OF THINGS

The hardware system, which is based on IoT and node MCU Module, is run by a program created in the Arduino IDE (Figure 9.14 (a, b, c, d, e)).

SETTING UP BLYNK APPLICATION

Just go to https://blynk.io and click on start free as shown in Figure 9.15. After that, register with Gmail ID. Enter your email address and password to join up for the Blynk IoT cloud server (Figure 9.16), and then hit the icon of new project to assign project name.

Next, choose the node MCU board, and then pick Wi-Fi as the connection type. Lastly, press the Create button as displayed in Figure 9.17.

IMPLEMENTATION OF PRINTED CIRCUIT BOARD (PCB)

Dip trace was used to create this PCB (Figure 9.18).

RESULTS AND DISCUSSION

In this work, we have designed a green house in order to understand former's problem related to field area in real-time situation. Here, we examined an IoT-based weather monitoring system for irrigation. As we know, green house is generally used for sophisticated crops or for certain crops that are used for research purposes in the field of irrigation. The proposed system was put to the test in real time following its development, and the outcomes were good and appropriate. In order to maintain the green house's temperature in both manual and automatic modes, it is crucial to often check the system's temperature and humidity.

We know that sunlight is necessary for the research plants used in green house, so according to the intensity of light during day time our greenhouse roof will be opened with the help of manual switch or by using the server as well by operating the servo motor. During dark an artificial light will be automatically switched on. The pressure and moisture level of the soil give proper reading and whenever the soil needs water we can switch on the water pump. The water pump can be turned on when the tank is empty and we can use an ultrasonic sensor to monitor the water level (Figure 9.19).

The system was found to work successfully and all sensors also worked satisfactorily. The user would be able to see all sensor data on LCD and transfer the data through IoT to the Blynk server. The developed technology offers an electronic and environmentally-friendly answer to the issues with the Indian agricultural sector. This will lead to fewer issues for farmers, particularly with irrigation systems. This system leads to various pleasant results (Rani & Chauhan et al., 2021), which are as follows:

- **Solution to Water Problem**: An automatic water pump is included in the proposed model and activates when the soil requires moisture.

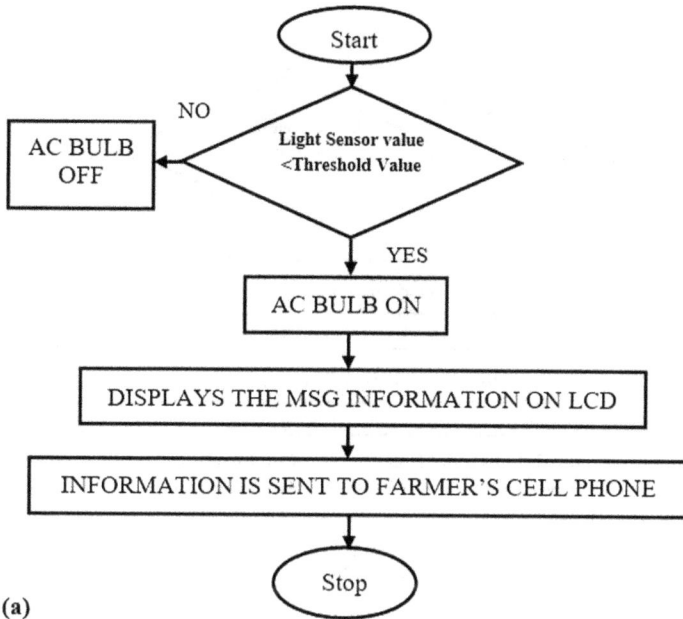

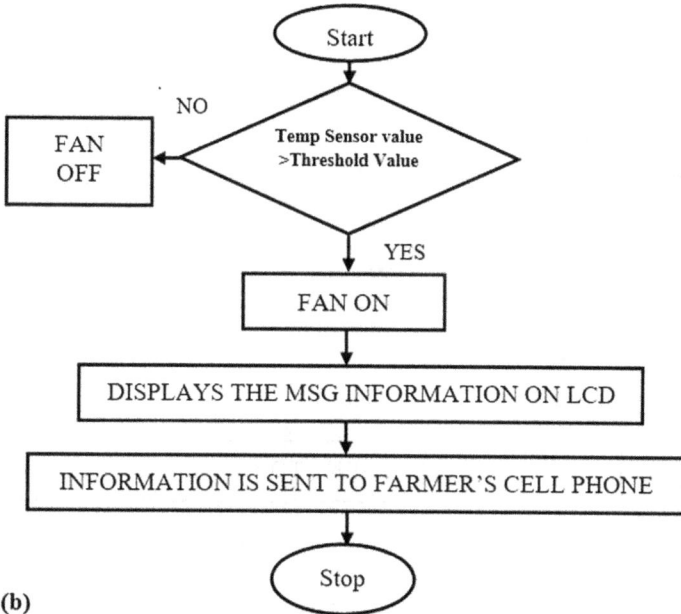

FIGURE 9.14   (a)–(e) The flowchart of the system and the primary sensor systems that are connected to it, upon which the program has been built

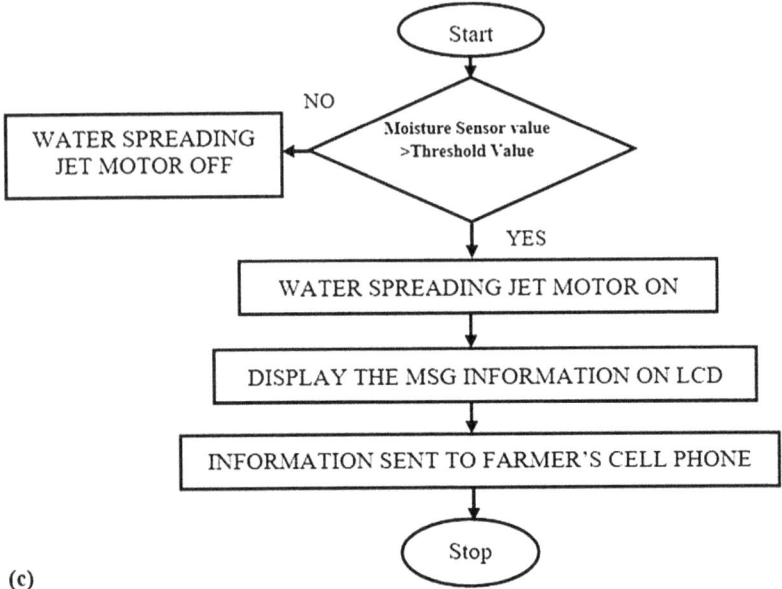

(c)

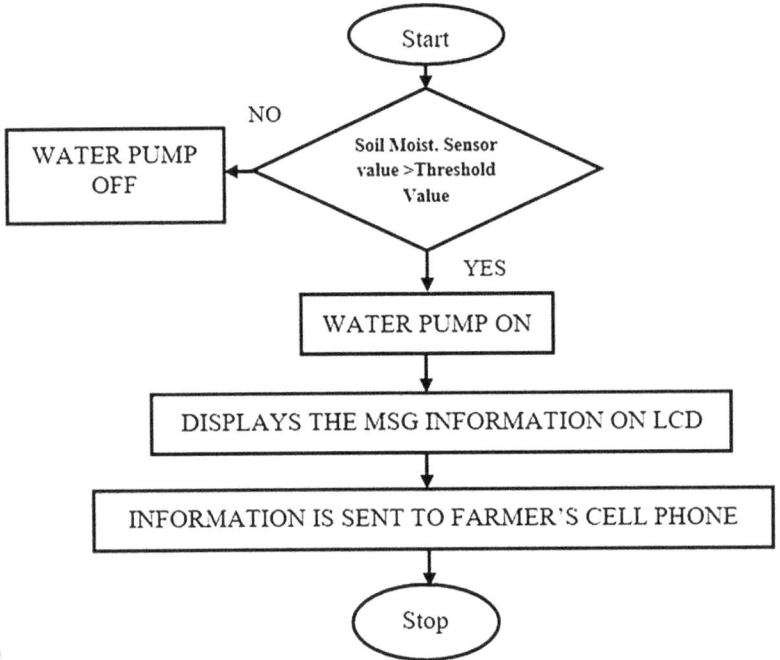

(d)

FIGURE 9.14 Continued

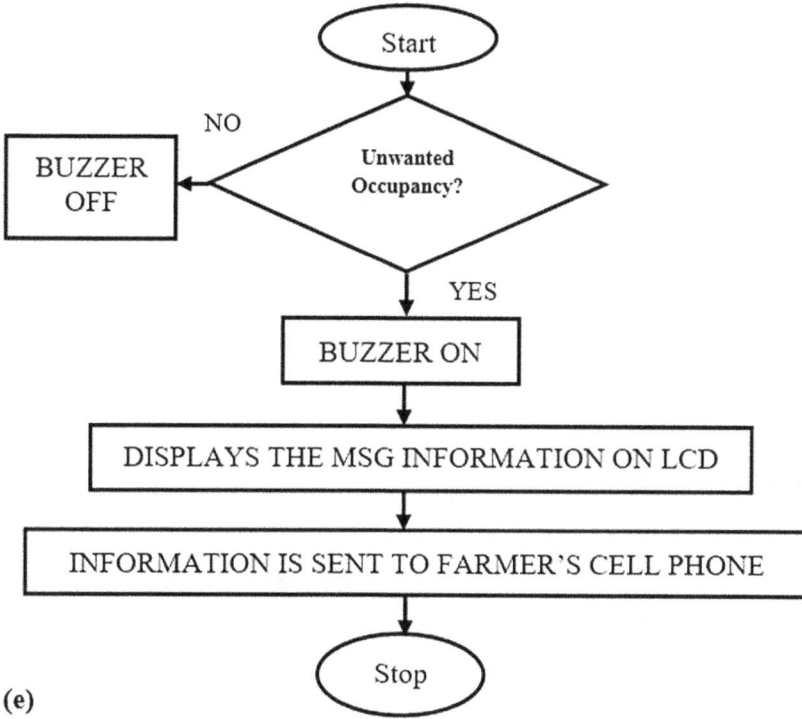

(e)

FIGURE 9.14 Continued

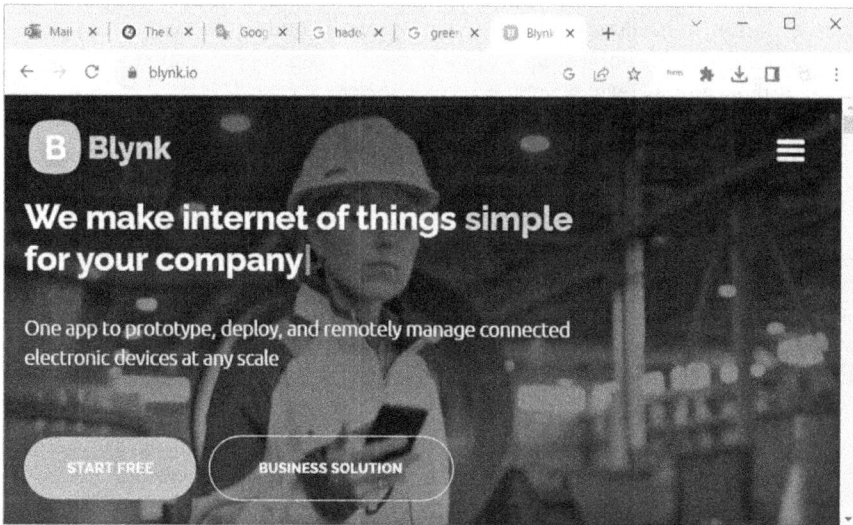

FIGURE 9.15 Blynk cloud user interface

FIGURE 9.16   Log in with registered mail ID

- **Solution to Unwanted Occupancy**: Using an ultrasonic sensor at the field's entrance helps combat the issues of unauthorized habitation and crop vandalism. If an unauthorized entry or occupant is discovered at the farmer's land entrance, an automatic buzzer system will sound an alarm.
- **Solution to Humidity Level Maintenance**: Using water spreading jet motor that starts automatically when humidity rises above a set threshold solves the problem of maintaining humidity levels.
- **Solution to Temperature Maintenance**: When the temperature rises, the system automatically turns "ON" the fan for complex crops to solve the temperature maintenance problem.
- **Solution to Electricity Supply**: The system is outfitted with a solar panel and battery for uninterrupted power backup and supply, which keeps the system running even when there is no power source.
- **Solution to Light Arrangement for Sophisticated Plants**: The automatic artificial lighting system, which is turned "ON" if light levels drop below a preset limit for crops, is used to provide the right amount of light for advanced crops.
- **Solution to Rain Problem and Weather Condition**: Proposed model is provided with a rain sensor, pressure sensor, and light sensor. Sunlight is necessary for the research plants used in greenhouse, so according to the rain and weather condition or intensity of light during day time, our greenhouse roof will be open with the help of a manual switch or by using a server as well by operating aservo motor.
- **Solution to Communication Problem**: An IoT-based module is utilized in the system to automatically alert the farmer about the temperature, humidity, wetness, light, fire, and unwelcome occupancy through real-time monitoring over the internet, which can be accessed from anywhere in the globe, to keep them informed about various field conditions.

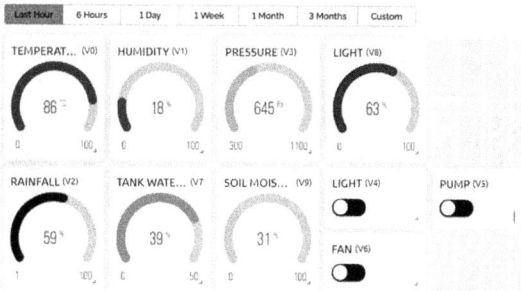

(a) Laptop dashboard

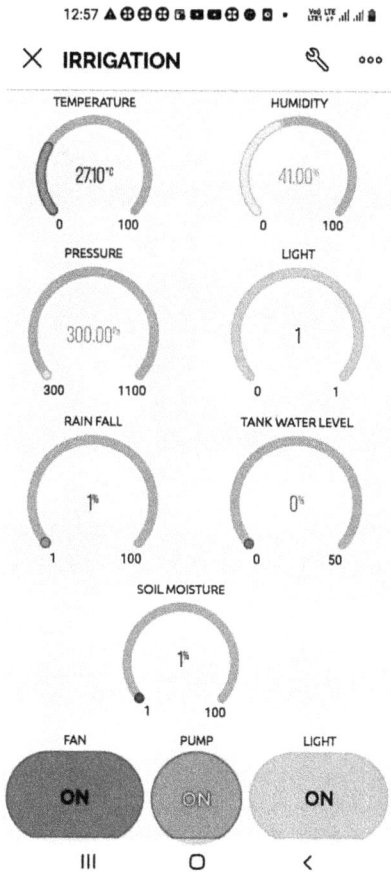

(b) Mobile dashboard

FIGURE 9.17 Blynk server dashboard. (a) Laptop dashboard. (b) Mobile dashboard

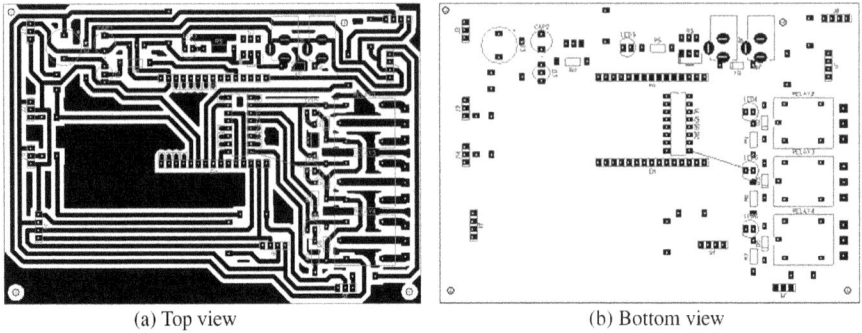

(a) Top view   (b) Bottom view

FIGURE 9.18   PCB prepared using dip trace. (a) Top view. (b) Bottom view

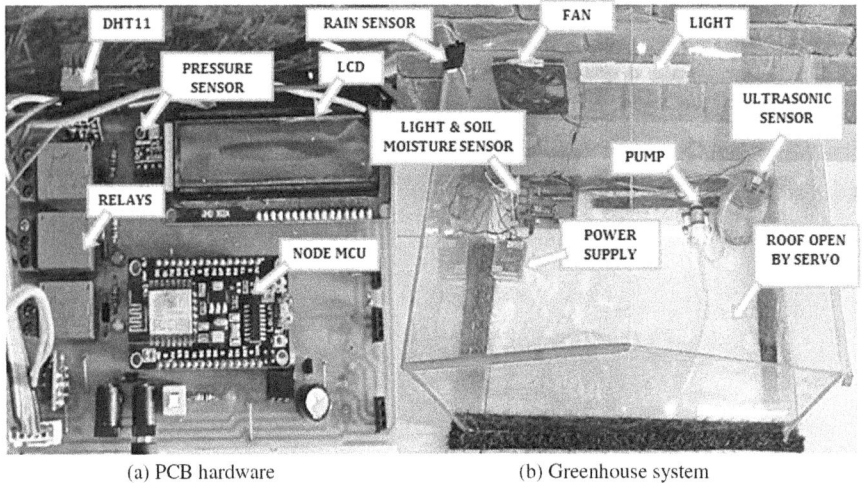

(a) PCB hardware   (b) Greenhouse system

FIGURE 9.19   Original picture of proposed system. (a) PCB hardware. (b) Greenhouse system

CONCLUSION

By careful observation and real-time testing, we conclude that our system is secure and we are able to measure and display all readings of a sensor on LCD as well on the Blynk server successfully. If any security flaws are detected, then our system will be able to detect and recover it automatically. All the sensors work on low power, so energy saving is also taken care. This approach produces a number of pleasing outcomes.

The system offers technical solutions for issues including water issues, unauthorized habitation, humidity level maintenance, temperature maintenance, electrical supply, light setup for advanced plants, and to keep the farmer updated on various field situations. In this system, a Wi-Fi module built on the internet of things (IoT) automatically alerts the farmer to changes in temperature, humidity, moisture, light, and unwelcome occupants by showing real-time data from around the globe. The relay can be run automatically via internet or manually.

FUTURE SCOPE

The suggested hardware can serve the following purposes with a few modifications:

- **Implementation of GSM Module**: The system has incorporated an IoT system, but if Wi-Fi signals are not available, then we can implement a GSM module with our system. This can give us the real-time sensor data all over the world using a cellular technique of GSM (Khang & Vugar et al., 2024).
- **Implementation of Fire Sensor**: Despite the system having a number of sensors, there is still room for more sensors. The utility of the system can be improved by adding one significant sensor, such as a gas sensor, which will be beneficial for protecting crops from fire damage (Rath & Khang et al., 2024).

REFERENCES

Agriculture in India. Wikipedia. Accessed 31 January 2019. https://en.wikipedia.org/wiki/Agriculture_in_India

Ammad, M.; Shah, M.A.; Islam, S.U.; Maple, C.; Alaulamie, A.A.; Rodrigues, J.J.P.C.; Mussadiq, S.; Tariq, U. A Novel Fog-Based Multi-level Energy-Efficient Framework for IoT-Enabled Smart Environments. *IEEE Access* 2020, 8, 150010–150026.

An, J.G.; Le Gall, F.; Kim, J.; Yun, J.; Hwang, J.; Bauer, M.; Zhao, M.; Song, J.S. Toward Global IoT-Enabled Smart Cities Interworking Using Adaptive Semantic Adapter. *IEEE Internet of Things Journal* 2019, 6(3), 5753–5765. https://ieeexplore.ieee.org/abstract/document/8667627/

Chang, Y.; Lai, Y. Campus Edge Computing Network Based on IoT Street Lighting Nodes. *IEEE Systems Journal* 2020, 14(1), 164–171. https://ieeexplore.ieee.org/abstract/document/8490873/

Cirillo, F.; Gomez, D.; Diez, L.; Maestro, I.E.; Gilbert, T.B.J.; Akhavan, R.; Smart City IoT Services Creation through Large-Scale Collaboration. *IEEE Internet of Things Journal* 2020, 7(6), 5267–5275. https://ieeexplore.ieee.org/abstract/document/9026979/

Dhanne, B.S.; Kedare, S.; Dhanne, S.S. Modern Solar Power Irrigation System by Using ARM. *International Journal of Research in Engineering and Technology* 2014, 3, 20–25. https://www.academia.edu/download/42020835/MODERN_SOLAR_POWERED_IRRIGATION_SYSYTEM_BY_USING_ARM.pdf

Humayun, M.; Jhanjhi, N.; Hamid, B.; Ahmed, G. Emerging Smart Logistics and Transportation Using IoT and Blockchain. *IEEE Internet Things Mag.* 2020, 3(2), 58–62. https://ieeexplore.ieee.org/abstract/document/9125435/

Ikram, K.; Buttar, N. A.; Waqas, M. M.; Muthmainnah, M. M. O.; Niaz, Y.; Khang, A. *Robotic Innovations in Agriculture: Maximizing Production and Sustainability* Advanced Technologies and AI-Equipped IoT Applications in High-Tech Agriculture (1st ed.) (2023). IGI Global Press. https://doi.org/10.4018/978-1-6684-9231-4.ch007

Indian Geography. Accessed 31 January 2019. https://www.learnapt.com/lesson-player/415-indian geography/ sections /727/items/47248

Jebaraj, L.; Khang, A.; Chandrasekar, V.; Pravin, A. R.; Sriram, K. Smart City Concepts, Models, Technologies and Applications. *Smart Cities: IoT Technologies, Big Data Solutions, Cloud Platforms, and Cybersecurity Techniques* (Eds.) (2023). CRC Press. https://doi.org/10.1201/9781003376064-1

Kaur, U.; Manjit Kaur, K. A. Futuristic Technologies in Agriculture Challenges and Future Prospects *Advanced Technologies and AI-Equipped IoT Applications in High-Tech Agriculture* (1st ed.) (2023). IGI Global Press. https://doi.org/10.4018/978-1-6684-9231 -4.ch020

Kiani, F.; Seyyedabbasi, A.; Nematzadeh, S.; Candan, F.; Çevik, T.; Anka, F.A.; Randazzo, G.; Lanza, S.; Muzirafuti, A. Adaptive Metaheuristic-Based Methods for Autonomous Robot Path Planning: Sustainable Ag-Ricultural Applications. *Applied Sciences* 2022, 12(3), 943. https://www.mdpi.com/2076-3417/12/3/943

Khang, A. *Advanced Technologies and AI-Equipped IoT Applications in High-Tech Agriculture* (1st ed.) (2023). IGI Global Press. https://doi.org/10.4018/978-1-6684 -9231-4

Khang, A.; Abdullayev, V.; Hahanov, V.; Shah, V. *Advanced IoT Technologies and Applications in the Industry 4.0 Digital Economy* (1st ed.) (2024). CRC Press. https://doi.org/10.1201/9781003434269

Khang, A.; Gupta, S.K.; Rani, S.; Karras, D.A.; *Smart Cities: IoT Technologies, Big Data Solutions, Cloud Platforms, and Cybersecurity Techniques* (Eds.) (2023). CRC Press. https://doi.org/10.1201/9781003376064

Kumar, I.; Vij, J.; Patel, L.V.; Sharma, M.; Haldar, D. Solar Powered Auto Irrigation System. *International Journal of Engineering Research & Management Technology* 2014, 1, 61–66. https://www.agrisolarclearinghouse.org/wp-content/uploads/info-library/ agrisolar-info/solar-irrigation/solar-powered-auto-irrigation-system.pdf

Li, W.; Logenthiran, T.; Phan, V.; Woo, W.L. A Novel Smart Energy Theft System (SETS) for IoT-Based Smart Home. *IEEE Internet of Things Journal* 2019, 6(3), 5531–5539. https:// ieeexplore.ieee.org/abstract/document/8661504/

Metallidou, C.K.; Psannis, K.E.; Egyptiadou, E.A. Energy Efficiency in Smart Buildings: IoT Approaches. *IEEE Access* 2020, 8, 63679–63699. https://ieeexplore.ieee.org/abstract/ document/9050775/

Patle, K.S.; Saini, R.; Kumar, A.; Palaparthy, V.S. Field Evaluation of Smart Sensor System for Plant Disease Prediction Using LSTM Network. *IEEE Sensors Journal* 2022, 22(4), 3715–3725. https://ieeexplore.ieee.org/abstract/document/9667350/

Patel, N.R.; Mathurkar, S.S. Microcontroller Based Drip Irrigation System Using Smart Sensor. Annual IEEE India Conference, 978–982. 2013. https://ieeexplore.ieee.org/ abstract/document/6726064/

Pasha, B.R.; Yogesha, B. Microcontroller Based Automatic Irrigation System. *International Journal of Engineering and Science* 2014, 3, 6–9. https://theijes.com/papers/v3-i7/ Version-1/B03710609.pdf

Pradeep, E.; Ganeshmurthy, R.; Sekar, K.; Arun, E. Automation of PV Farmer Pump. 2nd International Conference on Sustainable Energy and Intelligent System 1, 163–166 (2011). https://digital-library.theiet.org/content/conferences/10.1049/cp.2011.0354

Quy, V.K.; Nam, V.H.; Linh, D.M.; Ban, N.T.; Han, N.D. Communication Solutions for Vehicle Ad-hoc Network in Smart Cities Environment: A Comprehensive Survey. *Wirel. Pers. Commun.* 2022a, 122(3), 2791–2815. https://link.springer.com/article/10 .1007/s11277-021-09030-w

Quy, V.K.; Nam, V.H.; Linh, D.M.; Ngoc, L.A.; Gwanggil, J. Wireless Communication Technologies for IoT in 5G: Vision, Applications, and Challenges. *Wireless Communications and Mobile Computing* 2022b, 3229294. https://www.hindawi.com /journals/wcmc/2022/3229294/

Rani, S.; Ahmed, S.H.; Shah, S.C.; Smart Health: A Novel Paradigm to Control the Chickungunya Virus. *IEEE Internet of Things Journal* 2019, 6(2), 1306–1311. https://ieeexplore.ieee.org/abstract/document/8283593/

Rani, S.; Chauhan, M.; Kataria, A.; Khang, A. IoT Equipped Intelligent Distributed Framework for Smart Healthcare Systems. *Networking and Internet Architecture* (Eds.) 2021, 2, 30. https://doi.org/10.48550/arXiv.2110.04997

Rath, K. C.; Khang, A.; Roy, D. The Role of Internet of Things (IoT) Technology in Industry 4.0. *Advanced IoT Technologies and Applications in the Industry 4.0 Digital Economy* (1st ed.) (2024). CRC Press. https://doi.org/10.1201/9781003434269-1

Sinche, S.; Raposo, D.; Armando, N.; Rodrigues, A.; Boavida, F.; Pereira, V.; Silva, J.S. A Survey of IoT Management Protocols and Frameworks. *IEEE Communications Surveys and Tutorials* 2020, 22(2), 1168–1190. https://ieeexplore.ieee.org/abstract/document/8848791/

Shin, D.; Yun, K.; Kim, J.; Astillo, P.V.; Kim, J.-N.; You, I. A Security Protocol for Route Optimization in DMM-Based Smart Home IoT Networks. *IEEE Access* 2019, 7, 142531–142550. https://ieeexplore.ieee.org/document/8850082/

Song, Y.; Yu, F.R.; Zhou, L.; Yang, X.; He, Z. Applications of the Internet of Things (IoT) in Smart Logistics: A Comprehensive Survey. *IEEE Internet of Things Journal* 2021, 8(6), 4250–4274. https://ieeexplore.ieee.org/abstract/document/9241736/

Shafique, K.; Khawaja, B.A.; Sabir, F.; Qazi, S.; Mustaqim, M. Internet of Things (IoT) for Next-Generation Smart Systems: A Review of Current Challenges, Future Trends & Prospects for Emerging 5G-IoT Scenarios. *IEEE Access* 2020, 8, 23022–23040. https://ieeexplore.ieee.org/abstract/document/8972389/

Seal, S. O.; Shewale, S.; Sirsikar, A.; Hankare, P. Solar Based Automatic Irrigation System. *International Journal of Research in Advent Technology* 2014, 2, 186–189. https://www.academia.edu/download/36607014/paper_id-24201498.pdf

Subhashini, R.; Khang, A. The Role of Internet of Things (IoT) in Smart City Framework. *Smart Cities: IoT Technologies, Big Data Solutions, Cloud Platforms, and Cybersecurity Techniques* (Eds.) (2023). CRC Press. https://doi.org/10.1201/9781003376064-3

Sutjarittham, T.; Habibi Gharakheili, H.; Kanhere, S.S.; Sivaraman, V. Experiences with IoT and AI in a Smart Campus for Optimizing Classroom Usage. *IEEE Internet of Things Journal* 2019, 6(5), 7595–7607. https://ieeexplore.ieee.org/abstract/document/8656525/

Uddin, J.; Reja, S.M.T.; Newaz, Q.; Islam, T.; Kim, J.M. Automated Irrigation System Using Solar Power. 7th International Conference on Electrical and Computer Engineering, 228–23. 2012. https://ieeexplore.ieee.org/abstract/document/6471527/

Vangala, A.; Das, A.K.; Kumar, N.; Alazab, M. Smart Secure Sensing for IoT-Based Agriculture: Blockchain Perspective. *IEEE Sensors Journal* 2020, 21(16), 17591–17607. https://ieeexplore.ieee.org/abstract/document/9149915/

Wikipedia. Accessed 31 January 2019. https://en.wikipedia.org/wiki/Agriculture_in_India

Yalla, S.P.; Ramesh, B.; Ramesh, A. Autonomous Solar Powered Irrigation System. *International Journal of Engineering Research and Applications* 2013, 3(1), 060–065. https://citeseerx.ist.psu.edu/document?repid=rep1&type=pdf&doi=2e07d0008a3d9c0b929a6b3eca7d9539cb4d0821

Zhou, Z.; Yu, H.; Shi, H. Human Activity Recognition Based on Improved Bayesian Convolution Network to Analyze Health Care Data Using Wearable IoT Device. *IEEE Access* 2020, 8, 86411–86418. https://ieeexplore.ieee.org/abstract/document/9086799/

10 Internet of Things (IoT)-based Water Conservation and Monitoring System

Hanuman Prasad and Rupendra Kumar Pachuri

INTRODUCTION

Water management is recognized as a process to plan, design, distribute, and handle the optimal utilization of water resources. The conservation of water is a key feature of water management and it is very difficult to keep a record of water consumption in urban areas. In India, 50% water is wasted between source and consumer, i.e., leakage of water. Water leakage causes wastage of water and money, and it shows the poor plan and management. Water deficit is one of the biggest issues in the world. The main issues of water conservation systems are monitoring of water level, leakage detection, water quality, and water flow through various channels. IoT-based appliances are becoming more useful and prominent devices in various fields such as domestic, agriculture, and smart cities.

The IoT-based water conservation system is designed to remove the above-mentioned problems and it is a well-proof solution for water pipeline surveillance. The internet of things (IoT) links humans, other humans, and other things to share information. An IoT-based water management system has many advantages such as real-time monitoring, identification of water leakages, recaptures revenue, manages the water wastage, and also reduces the human effort (Boobalan et al., 2018; Devare and Hajare, 2019; Jisha et al., 2019; Johar et al., 2018; Kumar et al., 2017). This technology provides real-time data through the internet and shares the information to the consumer. These technologies are radio frequency identification tags, embedded systems, actuators, sensors, and nanotechnology that collaborate things to communicate together via distributed networks (Hahanov & Khang et al., 2022).

Wireless sensor networks (WSNs) are used to compile real-time data and verify the current situation in the pipe. Control unit is connected to the power source and sends real information to the nodes. WSN is a cutting-edge information-gathering technique used to create the communication system and its applications in different areas such as military, medical, environmental control, and monitoring of the

DOI: 10.1201/9781003434269-10

system. In order to provide proper transportation of water without water loss, we need complex pipeline infrastructures (Khang & Hajimahmud et al., 2024).

In the past ten years, IoT has emerged as a novel and promising approach for water monitoring in the smart cities. By leveraging technologies like sensors, wireless communications and networking, cloud computing, and other related ones, the internet of things technology aims to connect conventional objects to the internet and make them intelligent (Bhambri & Rani et al., 2022). There are many approaches used to develop water conservation systems. The paper by Raj kumar et al. (2017) illustrates an IoT-based intelligent irrigation system. The agricultural field is associated with various issues such as limited natural resources of water and labor supply. The IoT framework-based approach technique can be used in water and food areas through ZigBee protocol for data acquisition (Mekonnen et al., 2018). IoT cloud services are recommended for water treatment facility smart water management (Sayed et al., 2019). Cloud computing platforms like ThingSpeak have the potential to examine and envisage big data analytics to send data to wireless sensor networks (WSNs). The IoT-based method for monitoring and leak detection in water pipelines utilizing soil moisture sensors is described in this study (Elleuchi et al., 2019).

The induction of smart water monitoring and controlling through big data analytics and IoT-based operation is explained for green smart society (Ayaz et al., 2019; Gloria et al., 2019; Nie et al., 2020; Nizetica et al., 2020; Dubey et al., 2020). This article discusses the development and testing of an IoT-based solution for water conservation and also detecting leakage and location in water pipelines through moisture sensor and data acquisition in speak cloud (Figure 10.1). These sensors are accomplished by measuring moisture variations due to leakage and the moisture sketch can be used to find the leaks as well.

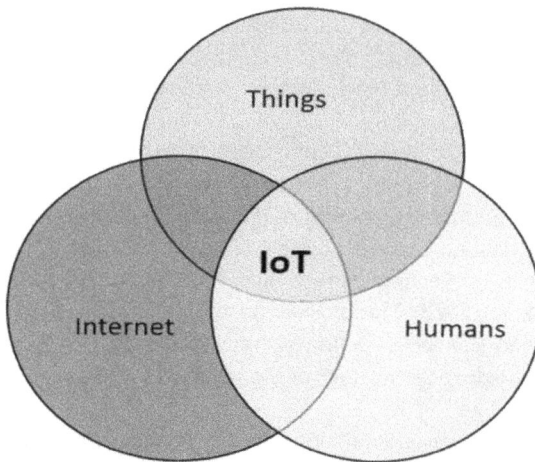

FIGURE 10.1   Relationship between internet, humans, and things (Bouali et al., 2022)

LITERATURE STUDY

There are many research works that have been dictated to implement an IoT-based smart water conservation and monitoring system for freshwater biodiversity (Wang et al., 2022). It elaborated the real-time application of IoT-based water monitoring system and effective water quality regulation of the system using artificial intelligence (AI)-based approach to estimate parameters using the different database. The article by Bouali et al. (2022) leverages the three main axes using IoT technologies such as smart water metering, renewable energy-integrated energy-efficient agriculture by water-level pumping, and smart irrigation-based ecosystem in a real-world smart farm test bed.

The paper by Ali et al. (2022) discusses the clarification for the water management and distribution issues. The literature is divided into different sections like water distribution network development, sensor networks installations, and sensor applications for reading in IoT test bed setup. The proposed system is used in smart homes enabling water quality monitoring, consumption metering, and leak detection. In addition to that, it provides users with consumer awareness highlights and a monitoring platform for users and administrators to detect leaks. The paper by Olatinwo and Joubert (2019) elaborated wireless sensor water monitoring systems with energy-efficient solutions. The authors are mainly focused on the area of dedicated wireless monitoring systems and water quality parameter calculations. The authors reviewed many research papers and concluded the strength and weakness of the papers with different energy-efficient solution techniques like EH techniques and directed the future research work.

The literature by Pan et al. (2018) has proposed a low-cost water surveillance system for water monitoring systems. The main issues of the traditional system are time consuming, complex, and costly. These problems can be compensated by using the proposed techniques. The proposed methods are designed from replacement of the manual system and provide the higher accuracy and stability of the system. Abolghasemi and Anisi (2021) have proposed a solution for flood monitoring systems with efficient energy controls. The authors approached the techniques in three sections like random block sample; gradient-based sensing approach, and other parameters.

The article by Singh et al. (2018) contributes to the applications of smart sensors for monitoring intelligent buildings. The article also mentioned the different features of the sensors and comparative studies of the traditional sensors and actuators. The smart monitoring system focuses on air quality, ventilation, electricity and water management, lighting, heating/cooling, and building health monitoring systems (Khang & Rana et al., 2023). The smart multistoried buildings use IoT for intelligent communication and leakage detection in water pipelines. It can manage the water supply system and reduce the wastage of the water system (Khang & Gadirova et al., 2023).

In the modern era, sensor-actuators and IT-based technology have played a significant role in the water monitoring management system. The smart water monitoring and conservation system is shown in Figure 10.2.

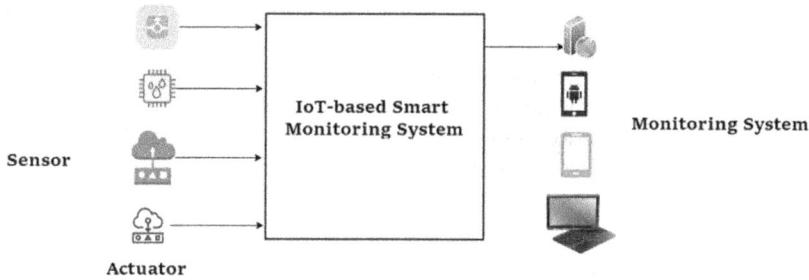

FIGURE 10.2   Smart water monitoring system

Arduino is an extensive electronics platform, which can interface between hardware and software. Arduino Uno is a microcontroller based on the ATmega328P. It has 14 digital input/output pins, 6 analog inputs, 16 MHz frequency, USB connection, a power jack, and a reset button. The Arduino board is capable of examining inputs like light on a sensor, a finger on a button and turning it into an output activating a motor, turning on an LED, and publishing something online. The instructions can be sent using the Arduino programming language and based on processing. The Arduino software (IDE) runs on Windows, Macintosh OSX, and Linux operating systems. Beginners may utilize IDE with ease, and advanced users can benefit from its flexibility as well.

Water leak detection is used for larger, integrated systems installed in modern buildings for the notification of water leakage. Water leak detection systems may be incorporated with building management structures the use of a couple of protocols inclusive of Modbus. Simple Network Management Protocol (SNMP) protocols leak detection systems can inform the IT team of workers in charge of monitoring the statistics center and server rooms. Water leak detection structures were particularly designed to identify and on some products turn off the water supply while a leak is detected. There are two main causes of water damage: leaking pipes and burst water pipes. The water detection system alerts you to even the slightest leak so you can fix it quickly.

ESP8266 is a Wi-Fi embedded system, which is used for the development of IoT applications. The ESP8266 model is a low price standalone Wi-Fi transceiver that may be used for end factor IoT developments. It has the following capabilities and features are 2.4 GHz Wi-Fi, 16 general-purpose input/output, 10 bit Analog-to-digital-converter (ADC), Pulse-width modulation (PWM) 32 KB instruction RAM, 80 KB user data RAM, 32 KB instruction cache RAM, and 16 KB Erlang Term Storage (ETS) system data RAM and Universal Asynchronous Receiver/Transmitter (UART).

The Arduino Integrated Development Environment (IDE) is a cross-platform program like Windows, macOS, and Linux that uses C and C++ functions. It is used to create and upload applications to boards that are compatible with Arduino as well as other vendor development boards with the aid of third-party cores. The Arduino IDE has specific code organization guidelines to support the languages C and C++.

A software library from the Wiring project, which offers numerous standard input and output operations, is provided by the Arduino IDE.

An IoT-based water management system has many advantages and efficiently responds to and avoids emergencies, collects revenue, reduces energy costs, reduces emergency repair scenarios, moisture loss detection and management, predicts potential pipe failures, manages water pressure, and distribution of maintenance and repair costs.

PROPOSED SYSTEM

The proposed method is based on a structural model and it consists of a power supply system, Arduino microcontroller, water leakage detection sensor, Wi-Fi module, and LCD display. The proposed model is shown in Figure 10.3. Water leak detection system by Elleuchi et al. (2019) is extensively used in large, integrated systems, installed in multistory buildings, aircraft, and others. It is connected to Ethernet. The moisture sensor senses the water leakage and sends the text message to the Arduino microcontroller using Wi-Fi trap. The ESP8266 is an economical-based Wi-Fi microchip and external hardware interface controller to the Ethernet device. Arduino unified development context platform is used for compiling, editing, and uploading the code in the Arduino device. It is integrated with the Arduino and hardware to upload programs and communicate with them (Figure 10.3).

ThingSpeak is an IoT integrated service that allows you to imagine, aggregate, and explore live data streams in the cloud (Rani & Bhambri et al., 2023). It can be used to store and retrieve data from ThingSpeak using the Hypertext Transfer

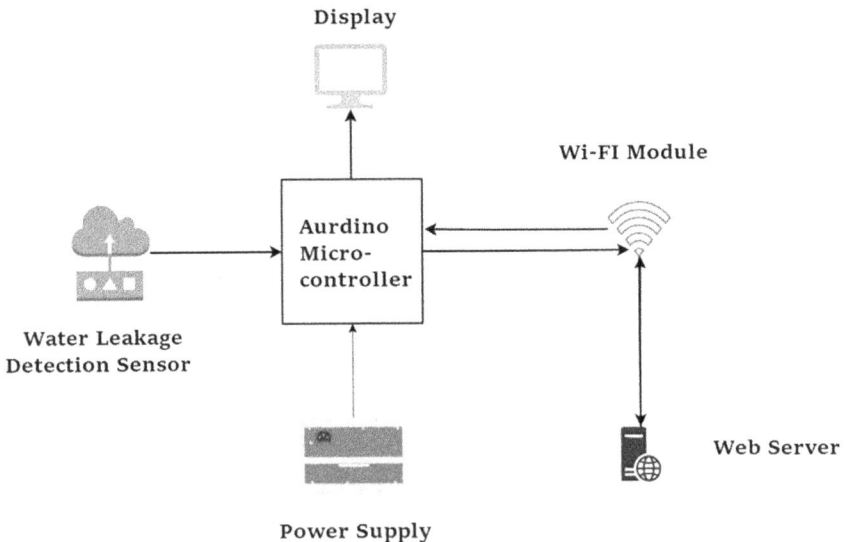

FIGURE 10.3 The proposed system model

Protocol (HTTP) and Message Queuing Telemetry Transport (MQTT) protocol over the internet (Figure 10.4). Its applications are sensor logging, social network, and also soft computing.

RESULTS AND DISCUSSION

The proposed setup consists of Arduino controller, water leakage detection sensor, LCD display, Esp8266 Wi-Fi module, and power supply. The software system is a setup of Arduino IDE, Embedded C, and ThingSpeak cloud. The hardware setup model is shown in Figure 10.5(a). Table 10.1 shows the hardware descriptions of the proposed system (Ushaa & Khang, 2024).

As above proposed solution, the wireless sensor network is integrated with a set of hardware nodes, accountable for the collection of sensor data that communicate to others.

The software component is used to store and analyze the collected data. Figure 10.5(b) shows the water leakage detection system. The WSN is integrated with two

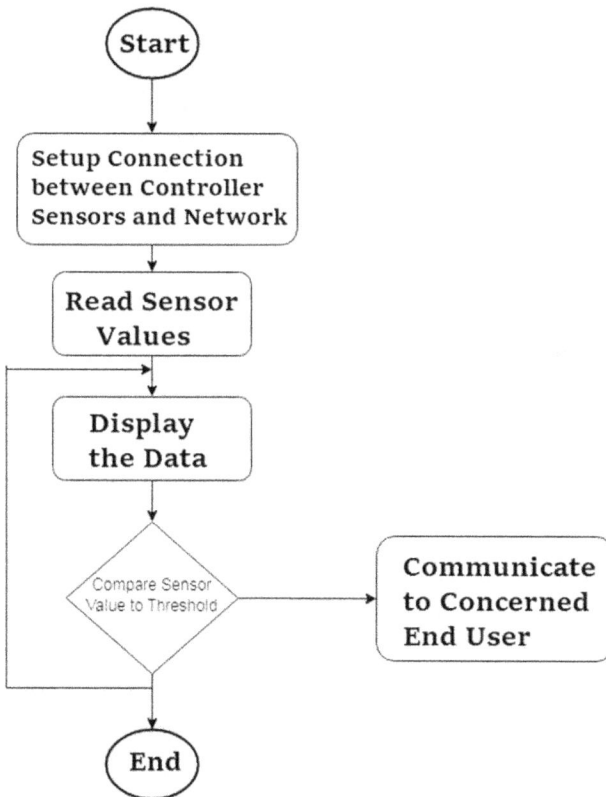

FIGURE 10.4   Operating flow chart of the proposed system

(a): Hardware setup of the proposed system

(b): Water leakage detection system

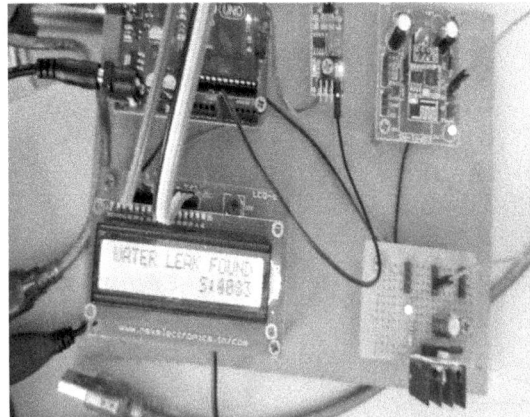

(c): Testing and implementation of the proposed system

FIGURE 10.5 WSN is used in this project to integrate the various nodes and synchronize the data: (a) hardware setup of the proposed system, (b) water leakage detection system, and (c) testing and implementation of the proposed system

TABLE 10.1

Hardware Component of the Proposed System

S. No	Name of the Components	Descriptions
1	Arduino controller	ATmega328, 5 V, input voltage 6–20 V, Current 40 mA/input–output pin, Flash memory 32 KB, clock speed frequency 16 MHz
2	Moisture sensor	Soil hygrometer humidity detection module, Vcc 3.3–5 V, LM 393 comparator chip
3	Esp8266 Wi-Fi module	Remote control GPIO/PWM, voltage 3–4 V, operating current and voltage are 80 MA and 3–4 V, respectively, operating temperature −40°C to 125°C
4	Power supply	5 V

different nodes and its features are aggregation of nodes for the network connection and communication with the server and sensor node for collection of information transmitted to the server.

When the water leakage scenarios are executed during the field testing, the moisture sensor is sending the command to the controller. Testing and implementation of the proposed system are shown in Figure 10.5(c).

CONCLUSION

In this era, any business is dependent on data processing and information technology using modern tools. This chapter recommended the design and implementation of an IoT-based smart water system. The proposed scheme is used for better water management and using soil moisture sensors included inthe Arduino controller, pipeline monitoring, and leak detection are performed. It is an economical and efficient platform for automation in the agriculture sector, multistory buildings, and industrial applications.

The traditional process system has a number of shortcomings, including a lack of real-time reactivity that makes it difficult to control water processes or identify leaks effectively. The IoT-based solution is a flexible approach for the water management system and it will ensure the optimum use of the existing resources. The proposed Arduino-based microcontroller is used to monitor and control manpower and efforts. This idea of a smart city that does not require human interaction and less operating expenses is supported by the system.

REFERENCES

Abolghasemi V. and M. H. Anisi, "Compressive sensing for remote flood monitoring," *IEEE Sensors Letters*, vol. 5(no. 4) (2021). DOI: 10.1109/LSENS.2021.3066342

Ali A. S., M. N. Abdelmoez et al., "A Solution for water management and leakage detection problem using IOT based Approach," *Internet of Things Journal*, vol. 18 (2022). DOI: 10.1016/j.iot.2022.100504

Ayaz M., M. Ammad-uddin, Z. Sharif, A. Mansour and M. Aggoune, "Internet of things (IOT) based smart agriculture: towards making the fields talk," *IEEE Access New Technologies for Smart Framing 4.O Research challenges and opportunities*, vol. 7, pp. 1–34 (2019). DOI: 10.1109/ACCESS.2019.2932609

Bhambri P., S. Rani, G. Gupta and A. Khang, *Cloud and Fog Computing Platforms for Internet of Things* (2022). CRC Press. DOI: 10.1201/9781003213888

Boobalan J., V. Jacintha, J. Nagarajan, K. Thangayogesh and S. Tamilarasu, "An IOT based Agriculture Monitoring System," *International Conference on Communication and Signal Processing*, pp. 594–598 (2018). DOI: 10.1109/ICCSP.2018.8524490

Bouali E. T., M. R. Abid, E. M. Boufounas, T. A. Hamed and D. Benhaddou, "Renewable Energy integration into cloud and IOT based smart agriculture," *IEEE Access Journal*, vol. 10 (2022). DOI: 10.1109/ACCESS.2021.3138160

Devare J. and Neha Hajare, "A survey on IOT based agricultural crop growth monitoring and quality control," *IEEE International Conference on Communication and Electronics Systems (ICCES 2019)*, pp. 1624–1630. DOI: 10.1109/ICCES45898.2019.9002533

Dubey S., P. Singh, P. Yadav and K. Singh, " Household waste management system using IoT and machine," *International Conference on Computational Intelligence and Data Science (ICCDS 2019), Procedia Computer Science*, vol. 167, pp. 1950–1959 (2020). DOI: 10.1016/j.procs.2020.03.222

Elleuchi M., R. Khelif, M. Kharrat, M. Aseeri and M. Abid, "Water pipeline monitoring and leak detection using soil moisture sensors: IoT based solutions," *IEEE International Multi-Conference on Systems, Signals and Devices*, pp. 772–775 (2019). DOI: 10.1109/ SSD.2019.8893200

Gloria A., C. Dionision, G. Simoes, J. Cardoso and P. Sebastiao, "Water management for sustainable irrigation systems using internet of things," *Sensing and Instrumentation in IoT Era*, pp. 1402: 1–14 (2019). DOI: 10.3390/s20051402

Hahanov V., A. Khang, E. Litvinova, S. Chumachenko, V. A. Hajimahmud and V. A. Alyar, "The key assistant of smart city - sensors and tools," *AI-Centric Smart City Ecosystems: Technologies, Design and Implementation* (1st ed.), 17(10) (2022). CRC Press. DOI: 10.1201/9781003252542-17

Jisha R. C., G. Vignesh and D. Deekshti, "IOT Based water level monitoring and implementation on both agriculture and domestic areas," *IEEE International Conference on Intelligent Computing, Instrumentation and Control Technologies (ICICICT)*, pp. 1119–1123 (2019). DOI: 10.1109/ICICICT46008.2019.8993272

Johar R., A. Bensenouci and M. A. Bensenouci, "IOT based smart sprinkling system," *IEEE 15th Learning and Technical Conference*, pp. 147–152 (2018). DOI: 10.1109/ LT.2018.8368499

Khang A., G. Rana, R. K. Tailor and V. A. Hajimahmud, *Data-Centric AI Solutions and Emerging Technologies in the Healthcare Ecosystem* (Eds.) (2023). CRC Press. DOI: 10.1201/9781003356189

Khang A., G. E. Musrat and V. A. Hajimahmud, "Role of photochemical reactions in the treatment of water used in the high-tech agriculture," *Advanced Technologies and AI-Equipped IoT Applications in High-Tech Agriculture* (1st ed.) (2023). IGI Global Press. DOI: 10.4018/978-1-6684-9231-4.ch018

Khang A., V. A. Abdullayev, Olena Hrybiuk and Arvind Kumar Shukla, *Computer Vision and AI-Integrated IoT Technologies in Medical Ecosystem* (1st ed.) (2024). CRC Press. DOI: 10.1201/9781003429609

Kumar A., A. Surendra, H. Mohan, M. Valliappan and N. Kirthika, "Internet of things based smart irrigation using regression algorithm," *IEEE International Conference on Intelligent Computing, Instrumentation and Control Technologies (ICICICT)*, pp. 1652–1657 (2017). DOI: 10.1109/ICICICT1.2017.8342819

Mekonnen Y., L. Burton, A. Sarwat and S. Bhansali, "IoT sensor network approach for smart framing: An application in food, energy and water system," *IEEE Global Humanitarian Technology Conference*, pp. 1–5 (2018). DOI: 10.1109/GHTC.2018.8601701

Nie X., T. Fan, B. Wang, Z. Li and, A. Manickam, "Big data analytics and IOT in operation safety management in underwater management," *Computer Communications*, pp. 188–196, vol. 154 (2020). DOI: 10.1016/j.comcom.2020.02.052

Nizetica S., P. Solic, D. Lopez-de-Ipina, González-de-Artaza and L. Patrono, "Internet of things (IoT) opportunities, issues and challenges towards a smart and sustainable future," *Journal of Cleaner Production*, 274 (2020). DOI: 10.1016/j.jclepro.2020.122877

Olatinwo S. O. and T. H. Joubert, "Energy efficient solutions in wireless sensor systems for water quality," *IEEE Sensors Journal*, vol. 19(no. 5), pp. 1596–1625 (2019). DOI: 10.1109/JSEN.2018.2882424

Pan J., Y. Yin, J. Xiong et al., "Deep learning-based unmanned surveillance systems for observing water levels," *IEEE Access*, vol. 6, pp. 73561–73571 (2018). DOI: 10.1109/ACCESS.2018.2883702

Rani S., P. Bhambri, A. Kataria, A. Khang and A. K. Sivaraman, *Big Data, Cloud Computing and IoT: Tools and Applications* (1st ed.) (2023). Chapman and Hall/CRC. DOI: 10.1201/9781003298335

Sayed H., M. AI-Kady and Y. Siddik, "Management of smart water treatment plant using IoT cloud services," *IEEE International Conference on Smart Applications, Communications and Networking*, pp. 1–5 (2019). DOI: 10.1109/SmartNets48225.2019.9069763

Singh A., Anshul Gaur, A. Kumar et al., "Sensing technologies for monitoring intelligent buildings: A review," *IEEE Sensors Journal*, vol. 18(no. 12), pp. 4847–4860 (2018). DOI: 10.1109/JSEN.2018.2829268

Ushaa Eswaran, Khang A., "Application of Internet of things (IoT) and sensors technologies for agriculture," *Advanced IoT Technologies and Applications in the Industry 4.0 Digital Economy* (1st ed.) (2024). CRC Press. DOI: 10.1201/9781003434269-8

Wang Y., I. W.-H. Ho, Y. Chen, Y. Wang and Y. Lin, "Real-time water quality monitoring and estimation in AIoT for freshwater biodiversity conservation," *IEEE Internet of Things Journal*, vol. 9(no. 16), pp. 14366–14374 (2022). DOI: 10.1109/JIOT.2021.3078166

11 Application of Internet of Things (IoT) in Logistics and Supply Chain Management

Lokpriya M. Gaikwad, Vivek Sunnapwar,
Sandip Kanase, and Alex Khang

INTRODUCTION

Managers of organizations realize that seriousness isn't just accomplished by streamlining the production lines; in addition, it is essential to further develop the supply chain network, upgrading the efficiency development. Hearty ongoing data is an early stage for the activity of any association. Specifically, tracking and a hint of merchandise, knowing where everything is whenever, can be useful to decrease capacity costs, the chance of misfortune or burglary, and so forth. Consequently, it has expanded consideration put on the exhibition, plan, and examination of the supply network to optimize for productivity and keep viability.

Strategies assume a crucial part in financial development and are a driver of nations' and firms' seriousness . Be that as it may, by virtue of the intricate inventory chains and high work costs, the expenses of strategies are currently at a generally significant level. High operation costs will influence the productivity of the assembling worldwide worth chains and the seriousness of a nation's economy. Certainly, creating more brilliant ways to deal with further develops coordinated factors' productivity and diminishes strategies' costs, and in both, scholarly community and industry is an ideal and significant theme these days. As of late, the idea of smart operations has been proposed.

Smart logistics depend on cutting-edge progressed data and innovation of Information and Communications Technology (ICT). It can understand the cutting-edge coordinated operations framework in a keen manner by continuously handling and completely examining the data of all parts of strategies. Smart coordinated factors can work on the method of planned operations transportation, warehousing, circulation handling, dispersion, data administrations, etc., and can reduce time and cost reserve funds of the journey of passage of goods from departure and through the multiple transit points to the last destination.

DOI: 10.1201/9781003434269-11

In addition, it can possibly decrease the ecological contamination brought about by coordinated operations. Nonetheless, many testing issues actually should be tended to during the time spent acknowledging cleverly coordinated factors. The major questions incorporate how to make it conceivable to understand the full interoperability of interconnected gadgets, and how to empower the transformation and independence of smart coordinated factors frameworks to give it a consistently more significant level of smartness.

As one of the significant innovations of ICT, the internet of things (IoT) is perceived as one of the main areas of future advances. Particularly, with the fast advancement of remote correspondence advancements, IoT is quickly making progress in the situations of present-day remote media communications. The meaning of IoT is continually developing from a unique spotlight on machine-to-machine (M2M) association and applications to the "pervasive collection" of information. That is, IoT has made expanses of information, and the complicated connections between the exchanges addressed by this information can be investigated consistently with the assistance of different numerical analysis technologies.

Undeniably, IoT will assume a vital part in the execution of smart logistics which will change the strategies activity mode, and the design of the framework of the operation significantly. In any case, there actually are many issues that should be considered during the most common way of making IoT-based brilliant strategies a reality, like material situations, existing difficulties, and future directions. It leads this work to help the people who are keen on the turn of events and improvement of this space. Albeit some reviews connected with smart logistics and IoT have previously been introduced in the literature, they for the most part center on only one of these two regions or one application part of IoT in coordinated operations.

A roadmap considers the connected issues in the combination of these two regions, all the more exhaustively. The primary commitments of this chapter are as per the following.

- We examine the cutting-edge investigations on smart logistics and IoT innovations, sum up and dissect the concentrations and insufficiencies of various studies on IoT-based smart logistics, and talk about their cutoff points.
- It presents the foundation information on smart logistics operations, including its idea, advancement, situations and essential capabilities, and key innovations of smart logistics. We further talk about empowering advances for IoT in smart logistics, including radio frequency identification (RFID), wireless sensor networks (WSNs), remote communication innovations, and middleware innovation.
- It features the jobs of the elements engaged with smart logistics of transportation, warehousing, loading/unloading, conveying, bundling, distribution processing, appropriation, and data handling when incorporated with IoT advances.
- It talks about the difficulties of IoT-based smart logistics in the part of information security, information protection, and asset the board.

A roadmap of our methodology is given in Figure 11.1, where it centers around the connected work, background information on brilliantly planned operations, empowering advancements for IoT in smart logistics, the utilization of IoT in smart logistics, difficulties, and future course. We accept that our conversation and investigation will permit readers to comprehend this field all the more thoroughly and advance the connected ensuing studies on this issue.

Especially, following and tracing products as a course of deciding the current and past areas of a remarkable thing or property, detailing the appearance or takeoff of the item, and recording the distinguishing proof of the item, the area, the time, and the status are a run-of-the-mill organization's issue. In this line, internet of things (IoT) innovation can assist with working on each part of the store network with the executives, further developing interest in the board, customization, and programmed renewal of products. In addition, it permits making new protected and proficient plans that add QoS and the board detectability of moved products. IoT licenses oversee omnipresent data about the shipped merchandise through various kinds of correspondences and devices (Lee et al., 2017) coordinated in smart urban communities.

This chapter proposes the use of IoT to contrast the moved merchandise and the conveyance note, decide their starting point and objective, and track and follow them from the distance. Such data will help the specialists as well as the exporters and merchants, who have some control over their product, guarantee its unwavering quality, improve its transportation through versatile travel course tasks, and offer an added benefit to the clients. Here, we depict an imaginative answer for the administration of the total production network process, which utilizes various current IoT advancements, for example, RFID, EPC, Wi-Fi, GPS, QR codes, and so on in a protected and effective manner.

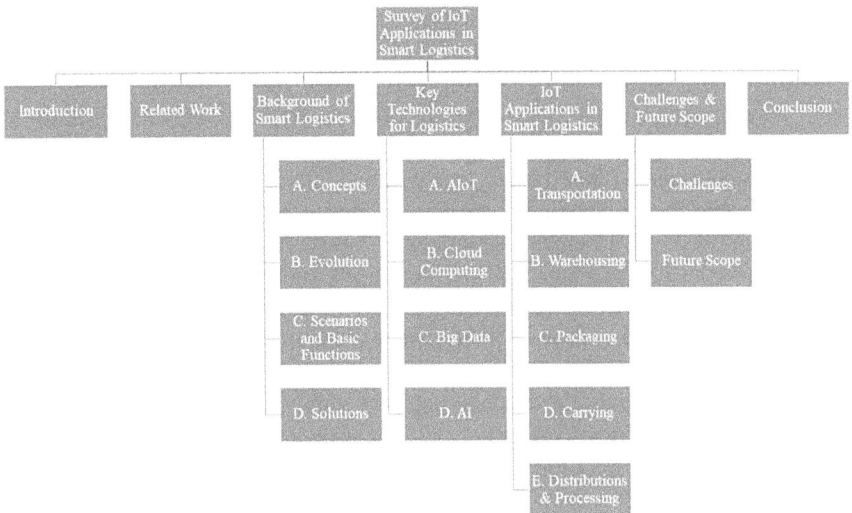

FIGURE 11.1   Roadmap of IoT in smart logistics

LITERATURE SURVEY

There are some reviews about IoT in smart logistics in recent years. The greater part of them principally centers on IoT design and a few application parts of IoT in smart logistics. Issues of remote correspondence innovations, smart logistics improvement, and difficulties are just introduced as a little piece of these studies.

Liu et al. (2010) assessed the primary advancements of IoT related to automate backing of business processes in logistics. They present brilliant things, including RFID and WSN. Then, they center on IoT-based help of plan and runtime changes considering dynamic changes in business processes. In this review, IoT is itemized exclusively from brilliantly coordinated factors in data handling (e.g., business cycles), and remote correspondence advances, other important applications, difficulties, and future headings of IoT in smart logistics are disregarded.

Khang et al. (2024) talked about the new prerequisites and valuable open doors for cyber physical logistics system (CPLSs) that depend on IoT advancements. In this article, they show the idea of CPLS, list the essential mechanical layers, and utilize a dissemination place as a genuine situation to delineate the improvement of CPLS. Albeit the study outlines the nitty-gritty arrangements of IoT in smart logistics joining with viable application situations, it gives no specific consideration to the difficulties of IoT in smart logistics and the effect of new key advancements on brilliantly coordinated operations.

Liu et al. (2012) studied the literatures connected with the advancements of IoT that incorporate RFID, WSN, and cloud computing. The development and the utilization of a few critical advances of IoT in smart logistics are dissected by the patent guide perception approach. The creators don't talk about challenge issues and their conversations of future examination bearings are restricted to shallow inquiries.

Rath et al. (2024) presents the use of IoT in the connected fields based on the hypotheses of IoT, examines the impact of IoT on coordinated operations data in strategies administration production network (LSSC), assembles the design of LSSC in light of IoT, and gauges the application prospect. This study was depicted the impact of IoT on the planned operations data issue, including strategies/administration stream, data stream, and asset stream. However, no specific consideration is paid to uses of IoT in different parts of smart logistics and difficulties need to face in the real world.

Liu et al. (2010) investigated the attributes of cloud computing and IoT, giving arrangements in light of cloud computing and IoT to acknowledge coordinated factors data trade and information trade. Tragically, the authors just spotlight on the examination and utilization of the critical innovations of IoT, ignoring numerous other significant key advances of IoT and their effect on smart logistics, like RFID, WSN, and simulated intelligence (AI).

Lee et al. (2017) introduced five IoT advancements (RFID, WSNN, cloud computing, and so on) that are fundamental in the organization of effective IoT-based items and administrations. Simultaneously, the specialized and administrative difficulties are broke down. This chapter presents IoT-related advances all the more completely and talk about the difficulties of IoT; however, they don't concede sufficient consideration regarding wireless communication advancements and the utilizations of IoT in logistics that are just been referenced momentarily.

Prasse (2014) presents the utilization of IoT in smart logistics operations as a part of strategies circulation and data processing, breaking down the difficulties and the fate of IoT-based smart logistics operations. Albeit the eventual fate of IoT-based smart logistics is examined, the creators just spotlight on the effect of strategy viewpoints. This implies that they don't focus harder on the effect of cutting-edge innovations. This study also gave new difficulties in the strategies space of the development of the business IoT (IIoT) and address a few reflections with respect to the main aspects expected for full execution of the logistics 4.0 worldview. This chapter proposes the idea of smart logistics and presents the mechanical applications in the part of asset arranging, distribution center administration, transportation, and data security. Then, at that point, they examine the future course. Albeit this article makes reference to the difficulties of IoT in smart logistics, it doesn't lead to a top-to-bottom examination.

There are other reviews that predominantly intend to introduce the IoT worldview, for example, various utilizations of IoT in different fields are presented, like spaces of transportation, coordinated operations, and smart city. A portion of those literatures examines the difficulties of IoT, including security and protection. Some examine the future bearings of IoT consolidating other trend-setting innovations. For instance, the rudiments and advancement of smart cities, zeroed in on the establishments and standards required for propelling the science, designing, and innovation of brilliant urban areas, including IoT, and covered uses of smart cities as they connect with smart transportation/connected vehicle (CV) and smart transportation frameworks (STSs). Be that as it may, the primary impediments of the work are the absence of far-reaching information about smart strategies and remote correspondence advances, particularly the utilization of IoT advances in smart logistics factors is just recorded basically.

BACKGROUND KNOWLEDGE OF SMART LOGISTICS

Smart logistics is an unavoidable pattern in the improvement of current modern logistics, the exploration subject of which has drawn in a ton of consideration from the scholarly world and industry. In this segment, we present information about smart logistics operations momentarily from the definition, advancement, fundamental capabilities, and arrangements.

CONCEPT

Since "smart logistics" is proposed, there is no agreement on the idea of it that has been arrived at in the scholarly world. This chapter proposed the idea of smart items and smart administrations from the beginning, and that implies that people can assign a portion of their control exercises to smart items and shrewd administrations. Smart logistics are characterized in light of the idea of brilliant product and smart administrations. This chapter also explains on qualities of brilliant strategies, which are utilized to characterize smart logistics as the given measures and characterized smart logistics as a strategies framework, which can upgrade adaptability, the acclimation to market changes, and will cause the organization to be nearer to the client's needs. This chapter introduced the comprehension of coordinated operations as the incorporated preparation, control, acknowledgment, and observation of all inward and extensive material, part, and item streams along the total worth added chain. With the cutting-edge innovations improvement, the strategic presentation is turning out to be increasingly more reliant upon mechanical advancement, and the meaning of smart logistics has likewise grown as needs are. Smart logistics is frequently used to allude to various coordinated factors tasks that are arranged, made due, or controlled in a more smart way contrasted with regular arrangements.

In any case, regardless of how the analysts characterize the idea of smart logistics, we can find that they share, practically speaking, that smart logistics join the high-level data advances and communication advances. It can incorporate and enhance the framework of the operation by complete examination, opportune handling, and self-change in accordance with making the logistics system framework more brilliant.

EVOLUTION

The improvement of logistics is firmly connected with the headway of innovation. We sum up the advancement of operations according to the viewpoint of innovative turn of events. The advancement of present-day smart logistics has gone through four phases that are logistics motorization, strategies computerization, coordinated operations mix, and planned operations knowledge. Strategies of motorization can be followed back to the 1920s when the principal truck work vehicle that was utilized to convey the freight was worked by the CLARK equipment Organization in 1917.

At the main stage, because of the advancement of the IC engine, mechanical manufacturing, and so forth, the labor in logistics operations exercises is supplanted by mechanical hardware. With the development of sensors, scanner tags, RFID innovation, and so on, present-day operations have advanced from motorization to robotization progressively since the 1960s. Delegate kinds of gear are the computerized directed vehicle (AGV) and the robotized stockpiling and recovery frameworks (AS/RS, etc.). This chapter presented an AGV framework that is disintegrated into non-overlapping, single-vehicle circles working couple. This chapter fostered a heuristic method to take care of the issue that the limit of the AS/RS was lacking to store all things. During the 1990s–2000s, logistics advanced from computerization to collaboration with the improvement of RFID, organization, and correspondence innovation.

This chapter examined the impacts of data accessibility on logistics integration and talked about how the internet is being utilized in dealing with significant parts of supply chains, including transportation, buying, client care, etc. This chapter also concentrated on the connection between data, authoritative design, and the fruitful execution of the incorporated conveyance idea. This chapter has introduced a system for an internet business local area organization, which broadens the customary business-to-business online business to internet business at the business level and understands the web-based incorporation of deals. In the 21st century, the quick advancement of technology, for instance, IoT, artificial intelligence, and big data, has advanced the improvement of smart logistics.

Gregor et al. (2017) examine a unique production logistics synchronization framework, which incorporates cloud fabricating and IoT foundations deliberately to empower a smart logistics services control instrument with multilevel dynamic flexibility. This chapter proposed a robot control framework that can sort divides with a setting mindful component for producing data to control the development of a robot. This chapter fostered the goal of aligning the different global positioning framework that looks into object area from continuous perspective, ensures the precision, unwavering operations of area, stability of the operating area, and protect the quality of assets by combining the capability of the various global positioning frameworks.

SCENARIOS AND BASIC FUNCTIONS

Logistics scenarios and the essential elements of smart logistics are displayed in Figure 11.2.

- **Transportation**: Logistics transportation is delivering things starting with one place and then onto the next place utilizing offices and tools. It is the main financial action among the parts of operations frameworks.
- **Warehousing**: Logistics warehousing is exercises that control, group, and deal with the stock, which is a huge, powerful component in the smart supply chain network.

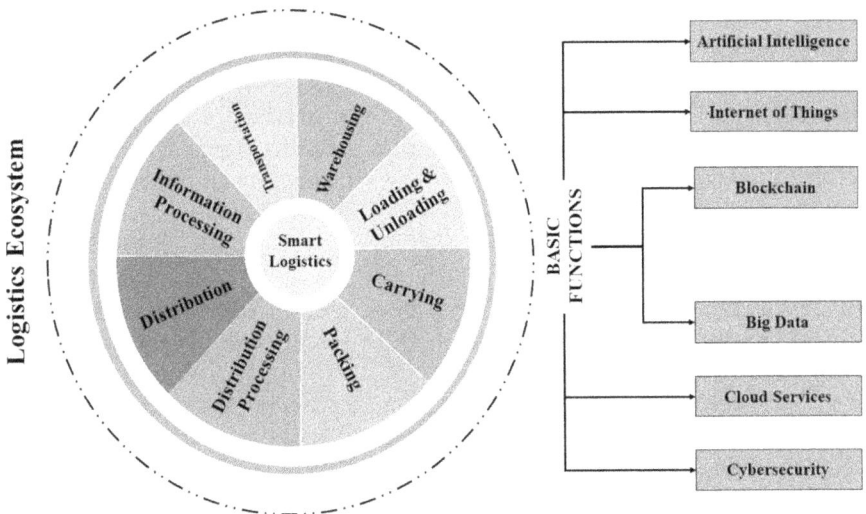

FIGURE 11.2 Smart logistics scenarios, basic functions, and key technologies (Khang, 2021)

- **Loading/Unloading**: It implies stacking/dumping things at the assigned area by human or mechanical means.
- **Conveying**: It is the principal logistics activity for moving things on a level plane in a similar spot.
- **Packaging**: It is to safeguard items during conveyance, to work with capacity and transportation. It is one of the main areas of logistics.
- **Conveyance Processing**: It implies basic activities, like bundling, division, metering, arranging, naming, and so on, as indicated by the requirements of the production spot to the utilization place.
- **Distribution**: It is a planned operations strategy that takes care of business for the representative as indicated by the custom's organization prerequisites. An exhaustive and wonderful planned operations dissemination arrangement has turned into a significant element influencing coordinated factors costs.
- **Data Handling**: The powerful data on production, markets, costs, and so forth is gathered and handled to make logistics-related forecast and plans, which empower logistics exercises to be done all the more productively and easily.

The essential elements of smart logistics incorporate sensing, combination, insightful investigation, advancement choice, system support, and opportune input for every logistics situation.

- **Sensing**: Acknowledging smart perception is utilized. It utilizes different trend-setting innovations to gather a lot of exact data on transportation, warehousing, stacking, dumping, data administrations, and different viewpoints.
- **Integration**: It is to accomplish information availability, transparency, and elements by the normalization of information and cycles. The gathered data is sent to the server farm through the organization for information documenting to lay out a strong data set.
- **Smart Investigation**: It is to dissect coordinated operations issues utilizing keen test system models and calculations. During the activity, the framework of the operation can call the first experience information to investigate. Then, at that point, consolidate the recently gathered information to track down provisos or failure points in the logistics exercises.
- **Enhancement Choice**: It is a brilliant choice capability. The smart logistics framework can propose the most sensible and viable arrangements by the prescient investigation. Then, at that point, it pursues more exact and logical choices as per various circumstances.
- **Framework Backing**: Smart logistics aren't free of every situation. Every situation can be associated with one another to share information and enhance asset portion, which gives the most impressive framework support for all situations of logistics.

- **Programmed Amendment**: In view of the past capabilities, the smart logistics framework can work following the best arrangement. At the point when a few issues are found, they will be fixed naturally.
- **Opportune Input**: The smart logistics framework is a real-time update framework. Criticism is a fundamental piece of executing framework remedy and framework improvement, which gives areas of strength for taking care of the framework issues in time.

SOLUTIONS

Numerous analysts examine smart logistics arrangements according to the viewpoint of innovation improvement. For instance, this chapter dissected the chances of Industry 4.0 with regard to the different logistics scenarios. This chapter planned operations situated Industry 4.0 application model as well as the center parts of Industry 4.0 and delineated the expected ramifications of industry 4.0 on various strategic situations. Some "smart" arrangements in logistics and production networks on the board as per new innovation advancement of IoT, large information, and Industry 4.0. This chapter delineated the use of trend-setting innovations in the Business 4.0 period on smart logistics from the parts of warehousing, smart transportation, and data security. (Khang et al. (2022a) portrayed a few utilizations of big data frameworks and choice of emotionally supportive networks that can be utilized to upgrade the plan and assessment of city logistics plans.

A few specialists examine the effect of consumer behavior on smart logistics solution and propose arrangements consolidating with innovation improvement. For instance, on the job of the client in strategies tasks and proposed a calculated model for client direction in intelligent logistics. This chapter inspected the impact of buyers on the effectiveness of working of the framework of the planned operation and proposed models that permit deciding the association impacts of the logistics system and purchaser.

KEY TECHNOLOGIES IN SMART LOGISTICS

As displayed in Figure 11.3, it needs four vital advancements as help to acknowledge smart logistics, that is, IoT, cloud computing, big data, and artificial intelligence. To give an image of the job they will probably play in savvy planned operations, and we present these advances in this segment. Besides, we feature IoT and a few critical empowering innovations of it (Khang et al., 2024).

INTERNET OF THINGS

IoT is the premise of the improvement of smart logistics. The fundamental thought of IoT is that different items can collaborate with one another and help out their neighbors to arrive at the principal objective of comprehending data without the guide of human intercession, through one of a kind tending to plans. Utilizing RFID labels, sensors, actuators, cell phones, and so on, the smart logistics framework can get

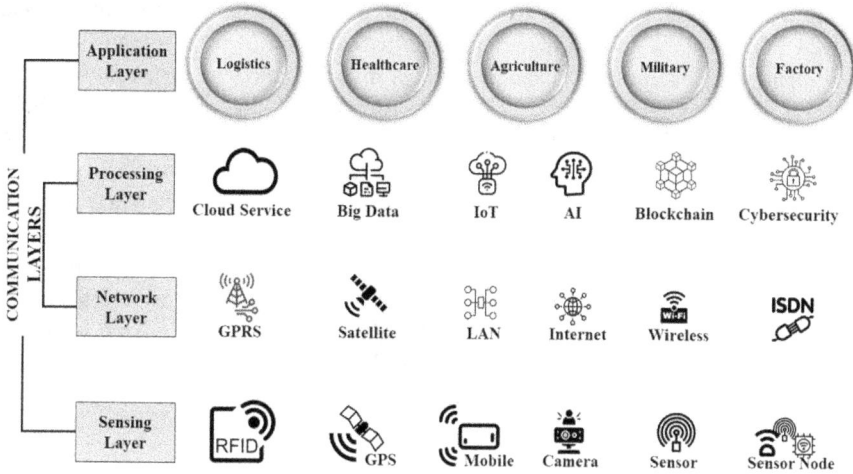

FIGURE 11.3 IoT design predominantly composed of four layers (Khang, 2021)

data on cargoes, strategies vehicles, transportation courses, warehousing, and so on, whenever and anyplace, and can accomplish start to finish data interconnection by network communication advancements. In view of the IoT stage, the smart logistics framework can break down and cycle enormous measures of strategies information and data, and pursue a choice to acknowledge shrewd control of items joined with distributed computing, big data, and computer-based intelligence, and other trend-setting innovations as displayed in Figure 11.3.

The detecting layer is utilized to gather and detect different actual boundaries, logos, sound, video, and different information in the actual world by RFID, camera, 2D code, and other high-level sensors. The organization layer comprises information correspondence and systems administration frameworks for conveying information accumulated from gadgets at the detecting layer to higher layers. The handling layer gives an office to information access, stockpiling, and handling joined with equip-ment stages and insightful calculations, like the cloud computing, big data innova-tion, and artificial intelligence. The application layer gives access administrations to IoT clients.

With regard to empowering innovation of IoT, it needs a lot of innovations, includ-ing sensor innovation, wireless network, communication innovation, etc. In this part, we center on a few empowering innovations that are generally utilized for the send-ing of effective IoT-based items and administrations in smart logistics (Hahanov et al., 2022).

Radio Frequency Identification (RFID)

RFID assumes a vital part in smart logistics to distinguish and catch the information, and it is broadly utilized in different logistics situations.

It utilizes electromagnetic fields to naturally recognize and follow labels con-nected to objects. Dissimilar to a standardized tag, RFID needn't bother with being inside the view of the per-user, so it empowers recognizable proof in good ways. RFID labels support a bigger arrangement of remarkable IDs than standardized identifications and can consolidate extra information. There are four kinds of RFID labels as per different recurrence ranges, including low frequency (LF) labels, high-frequency (HF) labels, ultrahigh-frequency (UHF) labels, and ultra-wide band (UWB) labels for various application objects in brilliantly coordinated operations.

For instance, the run-of-the-mill applications for LF RFID labels are product ID and information assortment that can be broadly utilized in shrewd warehousing, savvy dispersion, and brilliant bundling. HF labels can be handily framed into a card shape and are utilized in electronic tickets, electronic ID cards, and so on, so they are normal in smart warehousing. UHF labels are essentially utilized for smart transportation, and UWB labels can accomplish exact situation inside a portion of a meter, which works with the administration of significant instruments and staff the board in brilliantly smart logistics.

Wireless Sensor Networks (WSN)

Due to a large number of terminal sensor nodes in application scenarios of smart logistics, it is very important to organize and combine these terminal nodes freely. Therefore, WSN has gained considerable popularity due to its flexibility in solving problems in smart logistics, such as monitoring of transport vehicle status in smart transportation and item status monitoring in smart warehousing. WSN (Khang et al., 2022b) refers to a self-organizing network, which is built of tens to thousands of spatially dispersed and dedicated "sensor nodes" for monitoring, recording, and organizing the acquisition data at a central location by wireless connectivity and spontaneous formation of networks (Figure 11.4).

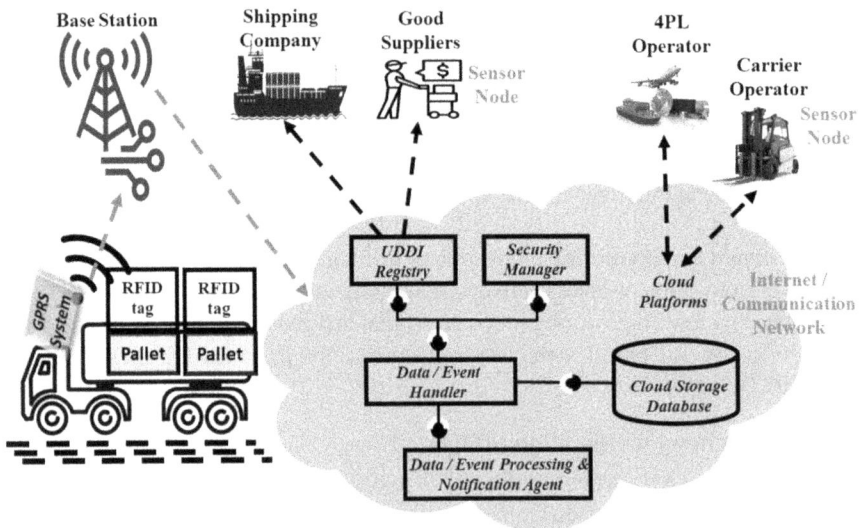

FIGURE 11.4   The architecture of WSN (Khang, 2021)

Wireless Communication Technologies

Wireless communication advances are urgent advances to the field of the organization used to gather all sensor node information and ship off the base station and the assigned client. These advances empower gadgets to speak with others without being actually associated. With the development of radio frequency (RF) innovation, cell organization, and other trend-setting innovations, remote communication advances save in step with a tremendous interest for IoT applications in smart logistics.

CLOUD COMPUTING

Logistics system framework joining is to take care of the issue of a "data disengaged island" during the time spent on the inventory network. In any case, the ongoing coordinated operations framework is a dynamic, heterogeneous, disseminated, and huge framework. It has the hindrances of unfortunate dynamics, slow reaction, low contagiousness, and high upkeep and development costs. With the nonstop advancement of brilliant coordinated factors innovation, it is the agreement of the business to incorporate and share the operations data assets of enormous frameworks and give a wide range of on-request strategies administrations for different clients in the method of "cloud computing."

Cloud computing is a model for empowering omnipresent, helpful, on-request network admittance to a common pool of configurable registering assets (e.g., networks, servers, capacity, applications, and administrations) that can be quickly provisioned and delivered with negligible administration exertion or specialist co-op connection (Liawatimena et al., 2011). The principal objective of cloud computing is to utilize enormous figuring and stockpiling assets under concentrated administration. Its fundamental qualities are as per the following: on-request self-administration, expansive organization access, asset pooling, quick flexibility, and estimated administration. Because of its qualities, distributed computing straightforwardly divides between client's versatile flexible assets over a boundless organization, which makes it more like the human mind and the operational hub of a brilliant logistics system as shown in Figure 11.5.

BIG DATA

As referenced above, IoT innovation can understand the interconnection of things in the logistics system and acquire data from the associated hubs. Cloud computing gives an excellent innovation stage for asset joining. While big data innovation can mine new coordinated operations business esteem through huge scope strategies, information gathered by IoT joined with distributed computing innovation, which will advance the improvement of smart logistics.

Big data is an innovation that gets ways to investigate and deliberately extricate data from enormous informational collections. These informational collections are excessively enormous or complex to be managed by conventional information handling application programming (Lewandowski et al., 2013). Current utilization of the term large information will in general allude to the utilization of prescient

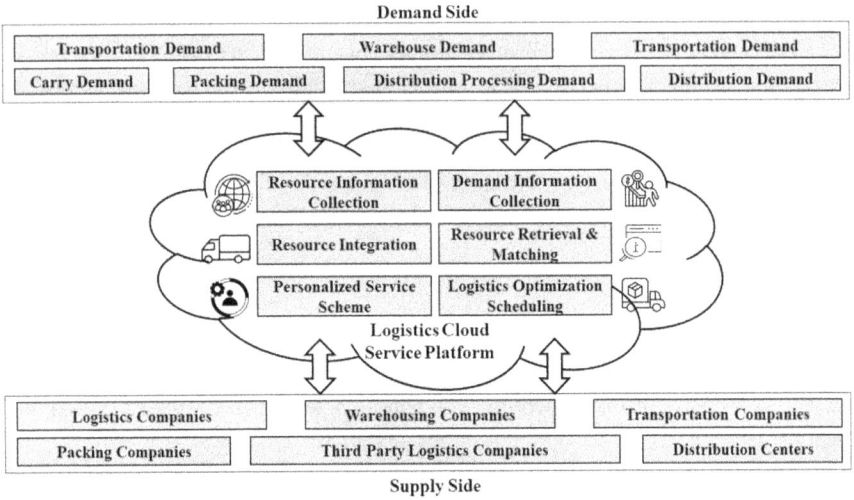

FIGURE 11.5 Logistics cloud service mode (Khang, 2021)

examination, client conduct investigation, or certain other high-level information investigation strategies that concentrate esteem from information, and sometimes to a specific size of the informational collection. Large information can be portrayed by the accompanying attributes: the enormous amount of created and put away information, the range of type and nature of the information, continuous information handling speed, and the high information quality and worth (Mohr and Khan, 2015). The meaning of large information innovation isn't to hold immense information, but to remove viable data contained in the information by particular information handling. It is the reason for the attention to smart logistics system (Rana et al., 2021).

ARTIFICIAL INTELLIGENCE (AI)

Artificial intelligence is a significant innovation that has a framework's capacity to accurately decipher outside information, gain from such information, and utilize those figuring out how to accomplish explicit objectives and errands through adaptable variation (Kaplan and Haenlein, 2019). There are three significant delicate figuring standards in simulated intelligence innovation (Zhang et al., 2011), or at least, artificial neural network (ANNs) (Silva et al., 2017), fuzzy logic (Lee and Cheng, 2012), and developmental computation (Fogel, 2000), (Xue et al., 2016). The use of computer-based intelligence innovation will speed up the improvement of smart logistics, so it has drawn in the consideration of specialists as a critical innovation for smart logistics (Figure 11.6).

- **Operation Administration**: With artificial intelligence innovation, the activity board community will have the capacity of self-learning and self-versatile through AI, and can settle on an autonomous choice in the wake of seeing business conditions. For instance, utilizing AI, the planned operations

FIGURE 11.6 Fusion scenarios of AI technology in smart logistics (Khang, 2021)

booking framework can become familiar with the order and booking experience and slowly acknowledge assistant and programmed direction.

- **Warehouse Administration**: Later on, the automated distribution center will be acknowledged by consolidating artificial intelligence innovation with IoT, enormous information, distributed computing, and so on. It can group and oversee things by smart algorithm.
- **Unmanned Dispersion**: Simulated intelligence is valuable for transportation. Through the man-made intelligence calculation, the automated dispersion robot can understand way arranging, clever hindrance aversion, etc., which will change the dissemination mode and assist with decreasing coordinated factors costs (Hajimahmud et al., 2022).
- **Sorting**: Through the man-made intelligence framework, various cameras and sensors can catch constant information about millions of products, and afterward recognize things through brand logos, names, and three-dimensional shapes. It implies the arranging framework never again needs transport machines, examining gear, manual handling hardware, and staff to sort individually, which will lessen costs incredibly.

- **Packing**: By computing the volume information and pressing box size of items utilizing the simulated intelligence calculation, the pressing framework can insightfully work out and suggest pressing material and pressing arranging, to organize the crate type and product position plot sensibly.
- **Site Determination**: The area of the warehousing and dispersion focus straightforwardly influences the effectiveness of the planned operation. Simulated intelligence calculation can give an area model that is near the ideal arrangement by completely learning and enhancing as indicated by different limitations of the genuine climate, for example, the geological area of clients, providers and makers, transportation economy, work accessibility, development cost, tax collection framework, and so on.
- **Client Assistance**: In brilliantly coordinated operations, computer-based, intelligence-based discourse acknowledgment has become one of the significant applications. The client care specialist in light of discourse acknowledgment innovation can understand the perception and clever examination of the client's voice, to help the manual specialist rapidly look and match the key information focuses. It will further develop work proficiency and administration quality. What's more, purchaser conduct expectation is additionally a significant piece of man-made intelligence client assistance. In short, artificial intelligence will be applied to planned operations transportation, warehousing, appropriation, the board, etc. to accomplish an effective strategies framework. In spite of the fact that man-made intelligence innovation is as yet not completely mature as of now, the advancement of smart strategies can't be without it. It will lead the future improvement course of brilliantly coordinated operations.

APPLICATIONS OF IOT IN SMART LOGISTICS

In this part, we center on the uses of IoT advances in smart logistics according to the viewpoint of wireless communication.

SMART LOGISTICS TRANSPORTATION

In the smart logistics transportation, IoT advances are in many cases used to finish the constant checking of vehicles, freight, and driver. During the strategies transportation process, data of the vehicle, freight, driver circumstance, and so forth are productively joined to further develop transportation proficiency, lessen transportation costs, decrease freight misfortune, and obviously comprehend everything in the planned operations transportation process (Kali et al., 2024).

Vehicle Status Observing

For vehicle checking, it fundamentally incorporates area following and constant circumstances of vehicles, for example, the speed of the vehicle, the tire pressure, fuel utilization, and the number of brakes.

Freight Status Observing

Through freight checking, the area and the situation with freight can be acquired continuously. A smart freight global positioning framework in view of IoT is proposed by Nassereddine and Khang (2024). This freight global positioning framework utilizes RFID, GIS, 3G communication, middleware innovation, and artificial intelligence innovation. It understands freight following through ongoing sign obtaining, information correspondence, and data handling. In Rani et al. (2023), dynamic street transport of the perilous merchandise observing framework is proposed, which depends on IoT and RFID innovation.

This framework helps out the interstate foundation and data sharing framework data sets by cell correspondence and gets more data about perilous merchandise by data handling. A wise freight arrangement EURIDICE is introduced in Bhambri et al. (2022). The arrangement embraces an IoT approach in appropriating knowledge on portable and conveyed gadgets. It permits them to speak with one another as well similarly as with a focal stage. As such, EURIDICE gives various administrations, including freight restriction, rerouting, and observing of freight conditions, to be performed without human mediation.

Driver Monitoring

In terms of driver monitoring, it can be realized by detecting the driver's healthy and driving behaviors through IoT technology (Khang et al., 2023).

Choudhari and Giripunje (2016) described the wearable sensor network design for a low-power healthcare real-time processing and other application of IoT, which uses the wireless technology to transmit the physiological parameter of the people at remote locations. This design is useful for the driver's community to take some preventive measures and thus road accidents can be reduced to some extends.

In Aazam and Fernando (2017), an architecture for driver behavior monitoring based on IoT and fog computing is presented. In this architecture, all the environmental, vehicle, and driver influential factors are considered by using multi-sensors. Data communication can be realized through various communication technologies depending on the different types of sensors and data frequency, such as RFID, bluetooth, Wi-Fi, and so on.

Smart Warehousing

Warehousing the operations is turning out to be more complicated and basic as business and innovation keep on evolving. Presently with IoT innovations, we can streamline the usage of stockroom space, screen the distribution center climate, and further develop the item-the-board interaction in logistics ecosystem.

Warehouse Space Advancement

As far as distribution center space streamlines, it chiefly incorporates item area arranging and distribution center space structure enhancement. Trab et al. (2018) proposed multi-agent engineering for item distribution arranging with similarity requirements, which involves a choice system for the item's situation, in light of exchanges between specialists related to similarity tests. This exchange component depends on an IoT foundation and multi-agent frameworks are characterized to tackle the security issue of item designation tasks.

Warehouse Climate Monitoring

Checking warehouse climate for the most part incorporates climate temperature and dampness, power appropriation of the board, and so on. In Jiang et al. (2015), an IoT design that is reasonable for cotton capacity is built. It utilizes IoT advances to further develop cotton warehousing the executives. this study planned a computerized distribution center observing framework in light of IoT to take care of the issues of checking slacking and non-intelligent. The plan can be utilized to screen the temperature, moistness, and the instance of fire in the distribution center.

Warehouse Management

As one of the warehousing of the board frameworks, further developing the warehousing of the executive's interaction is vital. Lee et al. (2017) proposed an IoT-based stockroom-the-board framework with a high-level information insightful methodology utilizing computational knowledge strategies to empower savvy planned operations for Industry 4.0 (Figure 11.7).

SMART LOADING/UNLOADING

The essential activities of stacking/dumping exercises incorporate stacking (transport), dumping (transport), stacking, capacity, outbound transportation, and so forth, which principally allude to dealing with the upward course. They are the essential

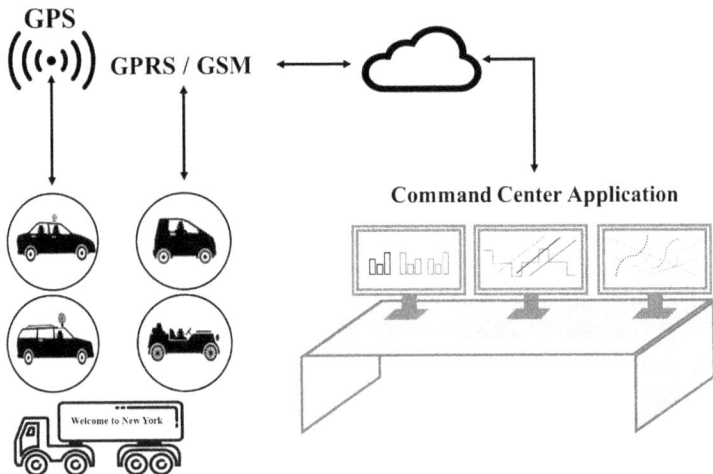

FIGURE 11.7 Framework of the proposed IoT-based warehouse management system (Khang, 2021)

exercises emerging from transportation and capacity exercises. In the coordinated factors process, stacking/dumping exercises are continually showing up and rehashing, so they frequently become the keys to deciding the strategy's speed and expenses. Besides, the utilization of stacking/dumping supplies (like a forklift) in view of IoT can significantly further develop planned operations effectiveness.

SMART CARRYING

Carrying is the exercise expected for the transportation and capacity of merchandise, which predominantly alludes to the treatment of the level course. Lately, the keen improvement of conveying is primarily reflected in the turn of events and utilization of AGV that is utilized for the inside and outside transport of materials (Liawatimena et al., 2011). With the quick improvement of IoT advancements, organizations in view of wireless communication innovations are progressively applied in AGVs, which generally give various benefits to AGVs, for example, low expenses and high proficiency (Zhang and Yu, 2018).

SMART PACKAGING

The packaging is the innovation of encasing or safeguarding items for dispersion, capacity, deal, and use. With the improvement of IoT advances, the packaging market is changing from traditional bundling to intuitive, mindful, and savvy bundling. All in all, brilliant bundling can use IoT and large information to lay out a unique connection with sensors on the bundling, like RFID, NFC, Bluetooth, and smart labels (Maksimovic et al., 2015).

SMART DISTRIBUTION HANDLING

Appropriation handling alludes to the handling that is all performed at the stockroom or planned operations community before the items are transported. The objective is to expand the additional worth of the item. There are numerous sorts of exercises relating to conveyance handling, like bundling, arranging, marking (applying names), division, metering, etc. For IoT advancements, more are applied in arranging and marking.

Smart Distribution

As of late, current coordinated factors appropriation keeps on creating toward informatization, digitization, organizing, joining, insight, and adaptability. The uses of IoT advances in keen coordinated operations conveyance are mostly reflected in the wise administration of circulation focuses and the improvement of smart conveyance draws near.

Smart Information Handling

Strategies data alludes to all data connected with operations exercises (transportation, distribution center, etc.). In smart logistics, operations data handling is fundamental.

Strategies for data handling exist in each coordinated operations situation, as a matter of fact. Notwithstanding the application situations portrayed above, operations data handling in light of IoT advances is likewise applied to the design of logistics models and logistics information frameworks.

CHALLENGES AND FUTURE DIRECTIONS

The IoT innovations can assist with understanding the vision of future smart logistics, yet additionally face challenges and different difficulties that it brings. For instance, because of the blast of information produced by IoT gadgets during the network operations, information security, protection, and asset that network operators will be faced difficulty coordinating combine technological elements together. In this segment, we examine these difficulties according to a specialized point of view and present some future exploration headings.

CHALLENGES

This part talks about three specialized difficulties: (1) data security; (2) information protection; and (3) resource administration.

Data Security: During the time for applying IoT innovations to accomplish keen strategies, the data should be crossed various authoritative limits and can be utilized for different purposes whenever, in any event, for obscure purposes. As a developing number and assortment of associated gadgets are brought into IoT organizations, the potential security danger heightens.

Information Protection: In the operations cycle, it contains a ton of individual data, item data, and undertaking data, which is imparted and traded over the web. Hence, protecting information security is a delicate subject and turns into a significant examination heading for scientists.

Resource Administration: Because of the multiplication of IoT advances in brilliant strategies, it requires the administration of an enormous assortment of conventions, information designs, and actual detecting assets in the activity cycle. This brings up the issue of how the assets given by the gadgets can be effectively overseen and provisioned (Kliem and Kao, 2015), that is to say, asset the board is one more test in shrewd coordinated factors and it has likewise turned into a hotly debated issue for specialists.

FUTURE SCOPE

Since shrewd coordinated operations in light of IoT advancements have drawn broad consideration and have been concentrated generally, its improvement can be impacted by a ton of different advances. Notwithstanding the previously mentioned innovations, including simulated intelligence, distributed computing, and large information, there are blockchains (Khanh and Khang, 2021), digital actual framework (CPS), three-dimensional printing, and so on. For instance, Tijan et al. (2019)

has researched the chance of blockchain innovation in reasonable operations and production networks the executives.

In Lewandowski et al. (2013), CSP-based engineering is represented, which is for dynamic and secluded control of single material taking care of gear inside a strategies framework. Mohr and Khan (2015) talked about three-dimensional printing and its troublesome effects on supply chains representing things to come.

Overall, with the ceaseless improvement of IoT, remote correspondence innovation, man-made intelligence, and other trend-setting innovations, an ever-increasing number of scientists and undertakings, have expanded their exploration endeavors on shrewd coordinated factors in view of the multi-innovation mix, which will likewise significantly advance the advancement speed of brilliantly planned operations (Khang et al., 2022b).

CONCLUSION

This chapter gives a review of the ongoing IoT innovations applied to smart logistics. We started our conversation with certainly connected papers and find information on smart logistics. Then, at that point, we zeroed in on empowering advances for IoT in smart logistics. Moreover, how IoT advances are applied in the domain of brilliantly coordinated operations was examined exhaustively, according to the viewpoints of transportation, warehousing, stacking/dumping, conveying, circulation handling, appropriation, and data handling.

We additionally talked about some critical exploration difficulties and future research directions in IoT-based smart logistics. In summary, research on applying IoT innovations in smart logistics is very expansive and various exploration issues and difficulties lay ahead. By and by, it is agreeable to the local area to address these difficulties in smart logistics quickly. This chapter endeavors to momentarily investigate how IoT advances work and when they ought to be utilized to take care of issues in smart logistics. We trust that our conversation can assist with advancing the improvement of smart innovation logistics.

REFERENCES

Aazam, M., and Fernando, X., "Fog assisted driver behavior monitoring for intelligent transportation system," in Proc. IEEE 86th Veh. Technol. Conf. (VTC-Fall), Toronto, ON, Sep. 2017, pp. 1–5. https://ieeexplore.ieee.org/abstract/document/8288317/

Bhambri, P., Rani, S., Gupta, G., and Khang, A., *Cloud and Fog Computing Platforms for Internet of Things*. CRC Press, 2022. https://doi.org/ 10.1201/9781003213888

Choudhari, S., and Giripunje, V., "Remote healthcare monitoring system for drivers community based on IoT," *Int. J. Emerg. Technol. Eng. Res.*, vol. 4, no. 7, pp. 118–121, Jul. 2016. https://www.ijeter.everscience.org/Manuscripts/Volume-4/Issue-7/Vol-4-issue-7-M-26.pdf

Fogel, D., "What is evolutionary computation?," *IEEE Spectr.*, vol. 37, no. 2, pp. 26–32, Feb. 2000. https://ieeexplore.ieee.org/abstract/document/819926/

Hahanov, V., Khang, A., Litvinova, E., Chumachenko, S., Hajimahmud, V. A., and Alyar, V. A., "The key assistant of smart city - sensors and tools," *AI-Centric Smart City Ecosystems: Technologies, Design and Implementation* (1st ed.), 17(10). CRC Press, 2022. https://doi.org/10.1201/9781003252542-17

Hajimahmud, V. A., Khang, A., Hahanov, V., Litvinova, E., Chumachenko, S., and Alyar, V. A., "Autonomous robots for smart city: Closer to Augmented Humanity," *AI-Centric Smart City Ecosystems: Technologies, Design and Implementation* (1st ed.), 7(12). CRC Press, 2022. https://doi.org/10.1201/9781003252542-7

Jiang, J., Yang, D., and Gao, Z., "Study on application of IoT in the cotton warehousing environment," *Int. J. Grid Distrib. Comput.*, vol. 8, no. 4, pp. 91–104, Sep. 2015. https://www.earticle.net/Article/A253944

Kaplan, A., and Haenlein, M., "Siri, Siri, in my hand: Who's the fairest in the land? on the interpretations, illustrations, and implications of artificial intelligence," *Bus. Horiz.*, vol. 62, no. 1, pp. 15–25, Jan./Feb. 2019. https://ieeexplore.ieee.org/abstract/document/8528677/

Khang, A., "Material4Studies," *Material of Computer Science, Artificial Intelligence, Data Science, IoT, Blockchain, Cloud, Metaverse, Cybersecurity for Studies*, 2021. https://www.researchgate.net/publication/370156102_Material4Studies

Khang, A., Abdullayev, Vugar, Hahanov, Vladimir, and Shah, Vrushank, *Advanced IoT Technologies and Applications in the Industry 4.0 Digital Economy* (1st ed.). CRC Press, 2024. https://doi.org/10.1201/9781003434269

Khang, A., Chowdhury, S., and Sharma, S., *The Data-Driven Blockchain Ecosystem: Fundamentals, Applications, and Emerging Technologies* (1st ed.). CRC Press, 2022b. https://doi.org/10.1201/9781003269281

Khang, A., Hahanov, V., Abbas, G. L., and Hajimahmud, V. A., "Cyber-physical-social system and incident management," *AI-Centric Smart City Ecosystems: Technologies, Design and Implementation* (1st ed.), 2(15). CRC Press, 2022a. https://doi.org/10.1201/9781003252542-2

Khang, A., Rana, G., Tailor, R. K., and Hajimahmud, V. A. (Eds.), *Data-Centric AI Solutions and Emerging Technologies in the Healthcare Ecosystem*. CRC Press, 2023. https://doi.org/10.1201/9781003356189

Khanh, H. H., and Khang, A., "The role of artificial intelligence in blockchain applications," *Reinventing Manufacturing and Business Processes through Artificial Intelligence*, 2 (pp. 20–40). CRC Press, 2021. https://doi.org/10.1201/9781003145011-2

Kliem, A., and Kao, O., "The internet of things resource management challenge," in Proc. IEEE Int. Conf. Data Sci. Data Intensive Syst., Sydney, NSW, Dec. 2015, pp. 483–490. https://ieeexplore.ieee.org/abstract/document/7396547/

Lee, C., Lv, Y., Ng, K., Ho, W., and Choy, K., "Design and application of internet of things based warehouse management system for smart logistics," *Int. J. Prod. Res.*, vol. 56, no. 8, pp. 2753–2768, Oct. 2017. https://www.tandfonline.com/doi/abs/10.1080/00207543.2017.1394592

Lee, J., and Cheng, W., "Fuzzy-logic-based clustering approach for wireless sensor networks using energy predication," *IEEE Sens. J.*, vol. 12, no. 9, pp. 2891–2897, Sep. 2012. https://ieeexplore.ieee.org/abstract/document/6217269/

Liu, T., Liu, J., and Liu, B., "Design of intelligent warehouse measure and control system based on Zigbee WSN," in Proc. IEEE Int. Conf. Mechatron. Autom., Xi'an., China, Aug. 2010, pp. 888–893. https://ieeexplore.ieee.org/abstract/document/5588999/

Liu, W., Zheng, A., Li, H., Qian, M., and Wang, R., "Dangerous goods dynamic monitoring and controlling system based on IoT and RFID," in Proc. 24th Chin. Control Decis. Conf. (CCDC), Taiyuan, China, May 2012, pp. 4171–4175. https://ieeexplore.ieee.org/abstract/document/6243113/

Liawatimena, S., Felix, B., Nugraha, A., and Evans, R., "A mini forklift robot," in Proc. 2nd Int. Conf. Next Gener. Inf. Technol., Gyeongju, Jun. 2011, pp. 127–131. https://ieeexplore.ieee.org/abstract/document/9241736/

Lewandowski, M., Gath, M., Werthmann, D., and Lawo, M., "Agent-based control for material handling systems in in-house logistics—Towards cyber-physical systems in in-house logistics utilizing real size," in Proc. Eur. Conf. Smart Obj. Syst. Technol. (Smart SysTech), Erlangen, Jun. 2013, pp. 1–5. https://ieeexplore.ieee.org/abstract/document /6525255/

Maksimovic, M., Vujovi, V., Omanovi, E., and Mikli, C., "Application of Internet of Things in food packaging and transportation," *Int., J. Sustain. Agr. Manag. Informat.*, vol. 1, no. 4, pp. 333–350, 2015. https://www.inderscienceonline.com/doi/abs/10.1504/ IJSAMI.2015.075053

Mohr, S., and Khan, O., "3D printing and its disruptive impacts on supply chains of the future," *Technol., Innovat. Manag. Rev.*, vol. 5, no. 11, pp. 20–25, Nov. 2015. https:// timreview.ca/sites/default/files/Issue_PDF/TIMReview_November2015.pdf#page=20

Nassereddine, Mohamed, and Khang, A., "Applications of internet of things (IoT) in smart cities," *Advanced IoT Technologies and Applications in the Industry 4.0 Digital Economy* (1st ed.). CRC Press, 2024. https://doi.org/10.1201/9781003434269-6

Rana, G., Khang, A., Sharma, R., Goel, A. K., and Dubey, A. K. (Eds.), *Reinventing Manufacturing and Business Processes through Artificial Intelligence*. CRC Press, 2021. https://doi.org/10.1201/9781003145011

Rani, S., Bhambri, P., Kataria, A., Khang, A., and Sivaraman, A. K., *Big Data, Cloud Computing and IoT: Tools and Applications* (1st ed.). Chapman and Hall/CRC, 2023. https://doi.org/10.1201/9781003298335

Rath, Kali Charan, Khang, A., and Roy, Debanik, "The role of Internet of things (IoT) technology in Industry 4.0," *Advanced IoT Technologies and Applications in the Industry 4.0 Digital Economy* (1st ed.). CRC Press, 2024. https://doi.org/10.1201/9781003434269-1

Silva, I., Spatti, D., Flauzino, R., Liboni, L., and Alves, S., *Artificial Neural Networks*. Springer, 2017. https://link.springer.com/chapter/10.1007/978-3-319-43162-8_2

Trab, S., Bajic, E., Zouinkhi, A., Abdelkrim, M., and Cheki, H., "RFID IoT-enabled warehouse for safety management using product class based storage and potential fields methods," *Int. J. Embedded Syst.*, vol. 10, no. 1, pp. 71–88, 2018. https://www.inderscienceonline.com/doi/abs/10.1504/IJES.2018.089436

Tijan, E., Aksentijevic, S., Ivanic, K., and Jardas, M., "Blockchain technology implementation in logistics," *Sustainability*, vol. 11, no. 4, pp. 1–13, Feb. 2019. https://www.mdpi .com/2071-1050/11/4/1185

Xue, B., Zhang, M., Browne, W., and Yao, X., "A survey on evolutionary computation approaches to feature selection," *IEEE Trans. Evol. Comput.*, vol. 20, no. 4, pp. 606–626, Aug. 2016. https://ieeexplore.ieee.org/abstract/document/7339682/

Zhang, B., Wu, Y., Lu, J., and Du, K., "Evolutionary computation and its applications in neural and fuzzy systems," *Appl. Comput. Intell. Soft Comput.*, vol. 2011, pp. 1–20, Aug. 2011. https://www.hindawi.com/journals/acisc/2011/938240/abs/

Zhang, M., and Yu, K., "Wireless communication technologies in automated guided vehicles: Survey and analysis," in Proc. IEEE 44th Annu. Conf. Ind. Electron, Soc. (IECON), Washington, DC, Oct. 2018, pp. 4155–4161. https://ieeexplore.ieee.org/abstract/document/8592782/

12 Application of IoT in Drug Supply Chain Management (SCM) during COVID-19 Outbreak

Mina Bahadori, Masoumeh Soleimani,
Morteza Soltani, and Mehdi Davari

INTRODUCTION

These days, the internet of things (IoT) has a significant impact on various aspects of the medical field. Researchers are actively exploring ways to leverage IoT in their work to enhance the primary medical objective, which ultimately benefits patients. To take a closer look at some of these researches, see Nanda et al. (2023), Kali et al. (2024), and Kumar and Pundir (2020). The COVID-19 pandemic has revealed the critical importance of efficient drug supply chain management. This chapter explores how the internet of things (IoT) can enhance drug supply chain management in the pharmaceutical industry during a pandemic. By leveraging IoT technologies, pharmaceutical companies can overcome challenges, seize opportunities, and improve their supply chain operations.

The pandemic has severely impacted the global healthcare supply chain, including the drug supply chain. Healthcare organizations and pharmaceutical companies have faced challenges in ensuring timely delivery of medical equipment and pharmaceutical products. Disruptions in global supply chains due to travel restrictions and border closures have caused shortages of critical supplies. Increased demand has strained supply chains, making it difficult to maintain adequate inventory levels. Additionally, the lack of real-time data on location and status hampers supply chain management, leading to inefficiencies and delays. Counterfeit and substandard products entering the supply chain pose risks to public health (Yan and Huang, 2009).

To address these challenges, healthcare organizations and pharmaceutical companies are adopting new strategies and technologies. This includes working with new suppliers, implementing risk management strategies, and investing in IoT and blockchain to enhance supply chain visibility and transparency (Jia et al., 2012).

DOI: 10.1201/9781003434269-12

As the world continues to grapple with the pandemic, robust and resilient healthcare supply chains are crucial. This requires identifying and addressing the vulnerabilities and challenges in the current drug supply chain management system. By exploring the application of IoT in drug supply chain management, we can uncover potential solutions that improve efficiency and ensure the availability of essential medical supplies and pharmaceutical products. Embracing IoT technologies in drug supply chain management is crucial for building a resilient and robust system capable of meeting the challenges posed by pandemics, ultimately safeguarding public health on a global scale (Aich et al., 2019). This chapter contains nine sections that collectively explore the application and impact of IoT in drug supply chain management during a pandemic. It begins with a compelling case study that highlights real-world examples and outcomes, providing a practical understanding of IoT-enabled drug supply chain management. The subsequent sections delve into key aspects of the topic, including the difference between IoT technologies and platforms, data-driven demand forecasting, the economic implications of IoT-enabled drug supply chain management, and the ethical and privacy considerations associated with its implementation. Furthermore, practical recommendations are provided for companies looking to adopt IoT in their drug supply chain management practices. Finally, the conclusion section summarizes the key findings and insights presented throughout the paper, emphasizing the significance of IoT in improving efficiency, reducing costs, and enhancing patient care (Brown and Russell, 2007).

LITERATURE REVIEW

There have been many examples of IoT healthcare applications of IoT, and this section presents some of them. For instance, Philips developed devices that are able to alert patients when it is time to take their pills and can provide pre-filled cups with medication[1]; Eversense provides a glucose monitoring system that uses a sensor, implanted below the patient's skin, to measure the blood glucose level and send it to patient's doctor using a mobile phone app[2] (Khanh and Khang, 2021). In 2019, Farooq et al. analyzed supply chain risks in IoT systems and their unique aspects, discussed research challenges in supply chain security, and identified future research directions. In 2020, Tu (2018) published a paper that highlights the potential applications of IoT in pharmaceutical manufacturing (Khang et al., 2024b) and supply chain management to enhance product quality, increase productivity, and reduce errors during different stages of a pharmaceutical product (Nassereddine and Khang, 2024).

In 2023, Javaid and Khan, 2021 introduced the telemedicine, including the advantages of telemedicine and the telemedicine in China. Telemedicine uses computers, communication networks, medical technology, and equipment so that the patient and experts, medical experts, and grassroots medical men can consult the patient's condition face to face despite they are in different places (Mohammadi and Harouni, 2018). The studies on IoT are continued in recent years; in 2022, an improved method "Low power and Lossy Network" is proposed to control and monitor the Alzheimer's patient in the cloud robot on the internet of things in smart homes (Karimi et al., 2021). Increasing the accuracy of diagnosis, reducing costs, and reducing human resources in the medical sector have been proven by researchers (Harouni et al., 2021).

CASE STUDY

Here we conduct a case study on a pharmaceutical industry supply chain to illustrate how IoT can be applied to improve drug supply chain management during a pandemic, identify the key challenges and opportunities for implementing IoT in drug supply chain management, and propose recommendations for companies to adopt IoT-enabled solutions.

CASE STUDY OVERVIEW

- **Company**: AbbVie Pharmaceuticals.
- **Scenario**: AbbVie Pharmaceuticals operates a global supply chain network responsible for distributing essential medications during the COVID-19 pandemic.

The company faces several critical challenges that need to be addressed. Actually, traditional supply chain management systems fail to provide real-time visibility into the location, condition, and status of pharmaceutical products. This deficiency leads to delays in delivery and inaccuracies in inventory management. Also, effectively predicting and responding to fluctuating demand patterns during a pandemic is crucial. However, existing supply chain systems often struggle to adapt, leading to stockouts or excess inventory, both of which can have significant financial implications.

Another challenge is regarding the storage and transportation of temperature-sensitive pharmaceuticals, the drugs that proper storage and transportation conditions for them are essential to maintain product efficacy and safety. Unfortunately, current supply chain systems face challenges in adequately monitoring and maintaining the required temperature throughout the supply chain. The last challenge that we mention here is that the proliferation of counterfeit drugs during a pandemic poses a grave risk to public health. Conventional supply chain systems encounter difficulties in effectively tracking and authenticating products, leaving room for counterfeit drugs to infiltrate the market.

Addressing the above challenges is crucial for the company's success and to ensure the safety, efficiency, and effectiveness of the pharmaceutical supply chain. Here IoT has some opportunities for implementation. First of all, by utilizing IoT devices such as radio frequency identification (RFID) tags and sensors, pharmaceutical companies can track and monitor pharmaceutical products in real time. This improves visibility, reduces delivery time, and enables proactive decision-making. Second, the integration of IoT-enabled temperature sensors into packaging, storage facilities, and transportation vehicles allows for continuous monitoring and regulation of temperature conditions. Real-time alerts can be generated if temperature thresholds are breached, ensuring product quality and safety. Also, combining IoT solutions with technologies like blockchain enables the creation of tamper-proof digital records for each product's journey. This ensures authenticity and prevents the entry of counterfeit drugs into the supply chain (Figure 12.1).

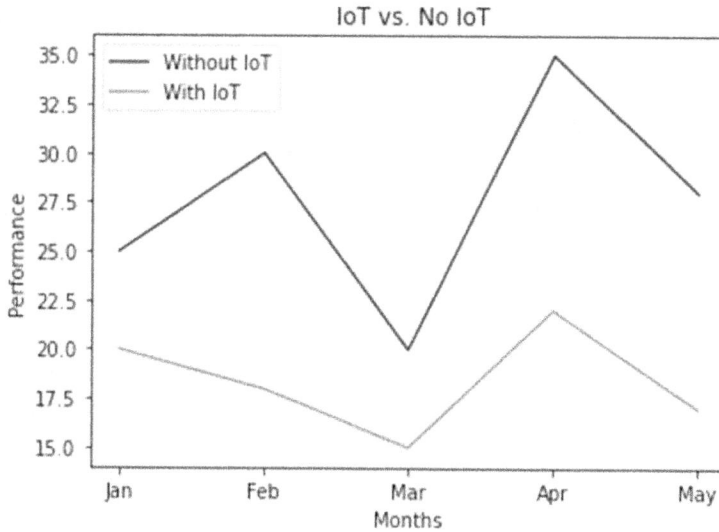

FIGURE 12.1   The graph demonstrates the difference between using IoT and not using it on a trend related to the case study on drug supply chain management

In contrast, the performance with IoT (orange line) starts at a slightly lower value but shows a more consistent and steady improvement over time. The comparison suggests that the utilization of IoT in drug supply chain management has a positive impact on performance. The trend with IoT showcases a more controlled and stable performance trajectory, indicating that IoT implementation enhances efficiency, real-time monitoring, and decision-making capabilities. The difference between the two trends highlights the potential benefits of adopting IoT in the pharmaceutical industry's supply chain, emphasizing the need for leveraging IoT technologies to optimize drug supply chain management during a pandemic.

COMPARISON OF IOT TECHNOLOGIES AND OTHER DRUG SUPPLY CHAIN MANAGEMENT PLATFORMS

There are different IoT technologies and platforms that offer distinct advantages and limitations for drug supply chain management. In this section, we explore the various commonly used IoT technologies and platforms in this context. In the following, an introduction to these technologies and platforms is provided, accompanied by a comprehensive examination of their respective benefits and drawbacks.

RFID TAGS

- **Advantages**: Radio frequency identification (RFID) tags are cost-effective and provide real-time tracking and identification of pharmaceutical products throughout the supply chain. They can store product information, such as batch numbers and expiration dates, and enable quick and accurate inventory management.

- **Limitations**: RFID tags require proximity to an RFID reader for data capture, which limits their effectiveness in large-scale tracking scenarios. Additionally, their range may be affected by physical obstructions and environmental factors of sensors:
- **Advantages**: Sensors, such as temperature sensors, humidity sensors, and vibration sensors, play a crucial role in ensuring the quality and integrity of pharmaceutical products. They can monitor environmental conditions, detect deviations, and provide real-time alerts to prevent spoilage or damage.
- **Limitations**: Sensor accuracy and reliability can vary, requiring regular calibration and maintenance. They may also have limited battery life, necessitating periodic replacement or recharging. Additionally, different types of sensors may be required to monitor specific parameters, increasing complexity and cost (Karimi et al., 2022).

CLOUD-BASED PLATFORMS

- **Advantages**: Cloud-based platforms provide a scalable and centralized infrastructure for data storage, analysis, and management. They enable real-time data access from various devices and locations, facilitating collaboration and decision-making. Cloud platforms also offer advanced analytics capabilities for data-driven insights and predictive modeling.
- **Limitations**: Cloud-based platforms depend on internet connectivity, and any disruptions may hinder data accessibility and real-time monitoring. Concerns related to data security and privacy arise when storing sensitive information on external servers, requiring robust cybersecurity measures and compliance with regulations (Khang et al., 2024a).

BLOCKCHAIN

- **Advantages**: Blockchain technology provides an immutable and decentralized ledger that can enhance transparency, traceability, and security in the drug supply chain. It enables tamper-proof records of product transactions, authentication, and provenance verification, thereby reducing the risk of counterfeit drugs and ensuring supply chain integrity (Georgios et al., 2019).
- **Limitations**: Blockchain implementation requires coordination and collaboration among multiple stakeholders, as all participants must agree on the rules and protocols governing the blockchain network (Lu and Liu, 2011).

By understanding the advantages and limitations of these IoT technologies and platforms, stakeholders in the drug supply chain can make informed decisions to optimize their operations and improve efficiency while addressing the specific requirements and challenges they face (Khang et al., 2022).

To determine the most suitable IoT technologies and platforms for drug supply chain management, organizations should consider their specific needs, budget, scalability requirements, and regulatory compliance. A comprehensive evaluation of the advantages and limitations of each technology will help identify the optimal combination of IoT solutions to enhance visibility, efficiency, and security in the drug supply chain (Nguyen et al., 2022).

DATA-DRIVEN DEMAND FORECASTING

IoT devices and smart sensors provide valuable data on product usage, patient demand, and inventory levels. By analyzing this data, pharmaceutical companies can optimize inventory management, enhance demand forecasting accuracy, and mitigate stockouts.

IoT-enabled drug supply chain management offers several potential benefits, including increased transparency, efficiency, and security, which can be leveraged to ensure a rapid and effective response to a pandemic. IoT devices, such as sensors and RFID tags, provide real-time tracking and monitoring of pharmaceutical products throughout the supply chain. This enhanced transparency enables stakeholders to have visibility into the location, condition, and status of medications. By minimizing information gaps and improving visibility, IoT enhances supply chain coordination, reduces delays, and facilitates proactive decision-making.

During a pandemic, this transparency is crucial for identifying bottlenecks, optimizing logistics, and ensuring a steady flow of essential drugs to affected areas. IoT enables data-driven decision-making by capturing real-time data on product usage, patient demand, and inventory levels. This data helps pharmaceutical companies optimize inventory management, streamline distribution routes, and reduce waste. By improving demand forecasting based on real-time insights, companies can minimize stockouts and excess inventory, ensuring the right medications are available when and where they are needed the most. The efficiency and optimization facilitated by IoT-enabled drug supply chain management lead to faster response times and better resource allocation during a pandemic, improving the overall effectiveness of the response effort (Sharma et al., 2020).

Security and safety are critical in drug supply chain management, particularly during a pandemic. IoT technologies, such as blockchain-based product authentication, can create a tamperproof digital record of a drug's journey, ensuring the integrity and authenticity of pharmaceutical products (Hussain et al., 2022). This helps prevent the entry of counterfeit drugs into the supply chain, protecting public health. Furthermore, IoT-enabled temperature sensors and monitoring systems ensure proper storage and transportation conditions for temperature-sensitive medications. This minimizes the risk of compromised product efficacy and patient safety. By enhancing security and safety measures, IoT enables pharmaceutical companies to maintain the quality and effectiveness of medications throughout the supply chain, particularly during a pandemic when patient's well-being is of utmost importance (Khang et al., 2023c).

The combination of increased transparency, efficiency, and security provided by IoT-enabled drug supply chain management enables a rapid and effective response to a pandemic. Real-time tracking and monitoring enable proactive decision-making, ensuring that critical medications are available in the right quantities and locations. Improved demand forecasting based on real-time data allows for agile planning and rapid scalability of production and distribution efforts. The optimized inventory management and streamlined logistics facilitated by IoT enable healthcare organizations and authorities to respond swiftly to changing demand patterns and ensure timely delivery of essential medications to affected areas. By leveraging IoT technologies, the drug supply chain becomes more resilient, responsive, and capable of meeting the challenges posed by pandemics (Atlam and Wills, 2020).

THE ECONOMIC IMPACT OF IOT-ENABLED DRUG SUPPLY CHAIN MANAGEMENT

Analyzing the economic impact of IoT-enabled drug supply chain management during a pandemic reveals significant cost savings and benefits for healthcare organizations, pharmaceutical companies, and patients. The return on investment (ROI) of implementing IoT in drug supply chain management can be evaluated based on various factors, and recommendations can be provided to justify the investment in IoT-enabled solutions. IoT-enabled drug supply chain management offers potential cost savings for healthcare organizations in several areas. Improved inventory management through real-time tracking and monitoring helps prevent stockouts and excess inventory, reducing carrying costs and minimizing waste. Accurate demand forecasting based on real-time data minimizes overstocking and reduces the need for emergency or rush orders, resulting in cost savings (Subhashini and Khang, 2023).

Additionally, IoT-enabled temperature monitoring ensures proper storage and transportation conditions, reducing the risk of spoilage or degradation of medications, and lowering financial losses. These cost-saving measures contribute to improved financial performance for healthcare organizations during a pandemic. Pharmaceutical companies can benefit from increased operational efficiency by implementing IoT-enabled drug supply chain management. Real-time tracking and monitoring enable proactive decision-making, optimizing logistics and reducing delivery times. This improved efficiency minimizes bottlenecks and enhances resource allocation, resulting in cost savings and increased productivity. IoT-enabled demand forecasting helps streamline production and distribution efforts, preventing underutilization of resources and minimizing production costs. By leveraging IoT technologies, pharmaceutical companies can enhance their operational efficiency and overall financial performance during a pandemic.

The economic impact of IoT-enabled drug supply chain management extends to patients through improved healthcare outcomes. Timely delivery of medications to affected areas during a pandemic leads to better patient care and reduced hospitalization rates. IoT-enabled solutions help prevent stockouts and ensure the availability of critical medications, enhancing patient safety and reducing the burden on healthcare facilities. By minimizing disruptions in the supply chain, IoT contributes to improved patient outcomes and overall healthcare system efficiency, resulting in potential cost savings related to reduced treatment costs and improved public health (Khang et al., 2023b).

To justify the investment in IoT-enabled solutions for drug supply chain management, companies can evaluate the ROI based on several factors. These include the cost savings derived from improved inventory management, streamlined logistics, and optimized resource allocation. The ROI can also be assessed through enhanced operational efficiency, reduced product wastage, and improved customer satisfaction. Furthermore, the prevention of counterfeit drugs and the safeguarding of product quality through IoT-enabled security measures contribute to brand reputation and customer trust, providing additional economic benefits. By quantifying these potential cost savings, efficiency gains, and patient benefits, companies can demonstrate a positive ROI for implementing IoT in drug supply chain management (Ogbuke et al., 2022).

In the following, we show the costs related to drug supply chain considering two scenarios of using IoT and not using it. The data has been collected from internet (Figure 12.2).

ETHICAL AND PRIVACY IMPLICATIONS OF USING IOT FOR DRUG SUPPLY CHAIN MANAGEMENT

Here, we investigate the ethical and privacy implications of using IoT for drug supply chain management during a pandemic, including issues related to data ownership, security, and confidentiality.

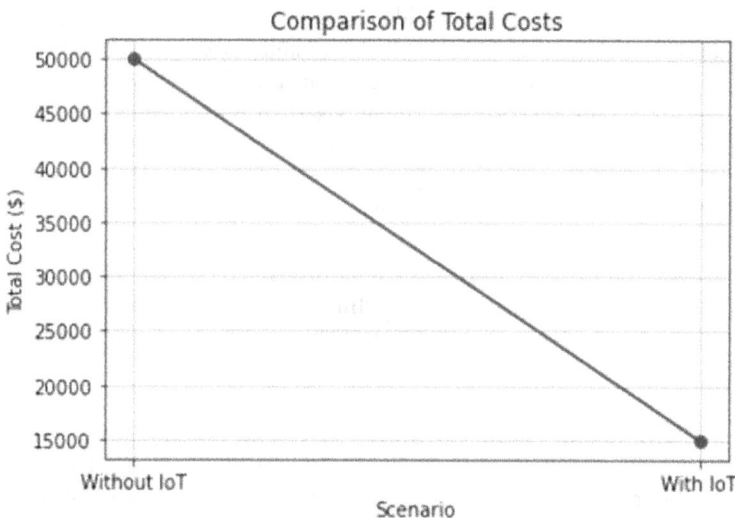

FIGURE 12.2   The difference in costs between the two scenarios represents the potential financial benefits of adopting IoT-enabled solutions

Develop guidelines and best practices for ensuring the ethical and responsible use of IoT in drug supply chain management (Michele and Furini, 2019).

The utilization of IoT in drug supply chain management during a pandemic raises significant ethical and privacy considerations. Addressing these concerns is vital to ensure the responsible and ethical implementation of IoT technologies. In this study, we explore the ethical and privacy implications and propose guidelines for mitigating potential risks.

First, it is crucial to establish clear ownership and control of the data collected through IoT devices in the supply chain. Agreements among stakeholders should be defined to determine data ownership rights, responsibilities, and permissible uses. Mechanisms should be devised to respect individual privacy rights while enabling necessary data sharing for public health purposes. Implementing robust security measures is imperative to safeguard IoT devices, networks, and data against unauthorized access, manipulation, or breaches. Encryption, access controls, and authentication protocols should be employed to ensure data confidentiality. Regular updates to security measures and vulnerability assessments are necessary to address emerging threats. Obtaining informed consent from individuals whose data is collected through IoT devices is essential (Khang et al., 2023a).

The purpose, scope, and potential risks associated with data collection and use should be clearly communicated. Individuals should be given the option to opt-out or exercise control over their data whenever feasible. Collecting only the necessary data required for drug supply chain management and avoiding the collection of excessive or sensitive personal information unrelated to the IoT system's purpose are important. Data anonymization techniques should be applied to protect individual identities whenever possible. Transparency should be fostered by providing clear information about the use of IoT devices and data collection practices (Rani et al., 2023).

Establishing mechanisms for individuals to access, review, and correct their personal data is crucial. Accountability should be maintained by assigning responsible individuals or entities to oversee data governance and compliance. Adherence to relevant data protection and privacy regulations, such as the General Data Protection Regulation (GDPR) or regional/local privacy laws, is necessary. Keeping abreast of evolving regulations and adjusting practices accordingly is important. The data collected through IoT devices should be used solely for legitimate purposes, such as improving supply chain efficiency, enhancing public health response, and mitigating risks. It is essential to avoid utilizing data for discriminatory practices, profiling, or unauthorized secondary purposes. Conducting privacy impact assessments to evaluate potential risks and privacy implications of IoT implementations is recommended (Bhambri et al., 2022).

Proactive identification and addressing of privacy concerns should be integrated into the design and implementation of IoT systems. Engaging in open dialogue and collaboration with stakeholders, including healthcare providers, regulators, patients, and advocacy groups, is crucial. Seeking their input and perspectives helps address ethical concerns and ensures alignment with societal values. Continuous monitoring and evaluation of the ethical and privacy implications of IoT-enabled drug supply chain management should be conducted. Mechanisms for promptly reporting and addressing privacy incidents or breaches should be implemented (Salehi-Amiri et al., 2022).

By following these guidelines and best practices, healthcare organizations and pharmaceutical companies can uphold ethical standards, protect individual privacy, and ensure the responsible use of IoT in drug supply chain management during a pandemic (Rani et al., 2021).

RECOMMENDATIONS FOR COMPANIES

Here are some improved sentences for companies; by implementing these recommendations, companies can overcome challenges and leverage the potential of IoT to enhance their drug supply chain management during the COVID-19 outbreak (Jia et al., 2012).

- Evaluate the company's existing infrastructure, systems, and resources to assess its readiness for IoT implementation. Identify any potential gaps and develop a comprehensive roadmap for seamless integration of IoT technologies.
- Foster collaborations with reliable IoT technology providers to carefully select appropriate devices, sensors, and platforms that align with the company's specific requirements. Consider factors such as compatibility, scalability, and data security when making these decisions.
- Integrate IoT devices, data analytics, and cloud-based platforms into the company's supply chain management system will enable real-time tracking, demand forecasting, and temperature monitoring capabilities, optimizing overall operations and efficiency.
- Cultivate strategic partnerships with stakeholders across the pharmaceutical supply chain, including manufacturers, distributors, and healthcare providers. Collaboratively establish industry standards for IoT-enabled supply chain management to promote interoperability and facilitate effective information sharing.
- Prioritize robust cybersecurity measures, including data encryption, access controls, and comprehensive data privacy protocols, to safeguard sensitive information. Compliance with regulatory requirements regarding patient information and data privacy should be a central focus (Shamsuzzoha et al., 2020).

Beside these recommendations, in the following we list some others for justifying investment in IoT-enabled solutions.

- Conduct a comprehensive cost–benefit analysis to quantify the potential financial savings and operational improvements that can be achieved through IoT-enabled drug supply chain management.
- Highlight the reduction in stockouts, waste, and emergency orders, leading to cost savings in inventory management.

- Emphasize the improved efficiency in production, distribution, and logistics, resulting in optimized resource utilization and reduced operational costs.
- Showcase the patient benefits, such as timely availability of medications, improved healthcare outcomes, and reduced treatment costs, contributing to long-term economic gains.
- Consider the potential value of brand reputation, customer trust, and regulatory compliance achieved through IoT-enabled security measures.
- Collaborate with industry partners, technology providers, and regulatory bodies to establish industry standards and foster interoperability, demonstrating the long-term sustainability and scalability of IoT-enabled solutions.
- Develop a phased implementation plan, starting with pilot projects to showcase the feasibility and benefits of IoT in drug supply chain management, allowing for gradual investment and risk mitigation (Harouni et al., 2022).

CONCLUSION

In conclusion, IoT technologies offer pharmaceutical companies the means to address key challenges in drug supply chain management during a pandemic. Real-time visibility, accurate demand forecasting, temperature monitoring, and product authentication are among the significant benefits provided by IoT-enabled solutions. Proactive integration of IoT, collaboration with technology providers, and establishment of industry partnerships empower companies to optimize drug supply chain operations, enhance patient safety, and ensure the timely delivery of critical medications even amidst challenging circumstances.

IoT-enabled drug supply chain management provides increased transparency, efficiency, and security, which can be leveraged to ensure a rapid and effective response to a pandemic. By harnessing real-time data, optimizing inventory, and ensuring the integrity of pharmaceutical products, IoT empowers pharmaceutical companies to address the demands of a crisis in a timely and efficient manner.

NOTES

1. https://www.lifeline.philips.com/.
2. https://www.ous.eversensediabetes.com/.

REFERENCES

Aich S., S. Chakraborty, M. Sain, H.-i. Lee, and H.-C. Kim. A review on benefits of IoT integrated blockchain based supply chain management implementations across different sectors with case study, pages 138–141, 2019. https://ieeexplore.ieee.org/abstract/document/8528677/

Atlam H. F. and G. B. Wills. IoT security, privacy, safety and ethics. In *Digital Twin Technologies and Smart Cities*, pages 123–149, 2020. https://link.springer.com/chapter/10.1007/978-3-030-18732-3_8

Bhambri P., S. Rani, G. Gupta, and A. Khang. *Cloud and Fog Computing Platforms for Internet of Things.* CRC Press, 2022. https://doi.org/ 10.1201/9781003213888

Brown I. and J. Russell. Radio frequency identification technology: An exploratory study on adoption in the South African retail sector. *International Journal of Information Management*, 27(4):250–265, 2007. https://www.sciencedirect.com/science/article/pii /S0268401207000400

Michele De R. and M. Furini. IoT healthcare: Benefits, issues and challenges. In *Proceedings of the 5th EAI International Conference on Smart Objects and Technologies for Social Good*, pages 160–164, 2019. https://dl.acm.org/doi/abs/10.1145/3342428.3342693

Georgios L., S. Kerstin, and A. Theofylaktos. Internet of things in the context of industry 4.0: An overview, 2019. http://dspace.vsp.cz/handle/ijek/103

Harouni M., M. Karimi, A. Nasr, H. Mahmoudi, and Z. Arab Najafabadi. Health monitoring methods in heart diseases based on data mining approach: A directional review. In *Prognostic Models in Healthcare: Ai and Statistical Approaches*, pages 115–159. Springer, 2022. https://link.springer.com/chapter/10.1007/978-981-19-2057-8_5

Harouni M., M. Karimi, and S. Rafieipour. Precise segmentation techniques in various medical images. In *Artificial Intelligence and Internet of Things*, pages 117–166, 2021. https://www.taylorfrancis.com/chapters/edit/10.1201/9781003097204-6/precise-seg-mentation-techniques-various-medical-images-majid-harouni-mohsen-karimi-shadi -rafieipour

Hussain S. H., T. B. Sivakumar, and A. Khang (Eds.). Cryptocurrency methodologies and techniques. In *The Data-Driven Blockchain Ecosystem: Fundamentals, Applications, and Emerging Technologies* (1st ed.), 2(9), pages 20–29. CRC Press, 2022. https://doi .org/10.1201/9781003269281-2

Javaid M. and I. H. Khan. Internet of things (IoT) enabled healthcare helps to take the challenges of Covid-19 pandemic. *Journal of Oral Biology and Craniofacial Research*, 11(2):209–214, 2021. https://www.sciencedirect.com/science/article/pii/S2212426821000154

Jia X., Q. Feng, T. Fan, and Q. Lei. RFID technology and its applications in internet of things (IoT), pages 1282–1285, 2012. https://ieeexplore.ieee.org/abstract/document/6201508/

Karimi M., M. Harouni, A. Nasr, and N. Tavakoli. Automatic lung infection segmentation of Covid-19 in CT scan images. In *Intelligent Computing Applications for COVID, 19*, pages 235–253. CRC Press, 2021. https://www.taylorfrancis.com/chapters/edit/10.1201 /9781003141105-12/automatic-lung-infection-segmentation-covid-19-ct-scan-images -mohsen-karimi-majid-harouni-afrooz-nasr-nakisa-tavakoli

Karimi M., M. Harouni, E. I. Jazi, A. Nasr, and N. Azizi. Improving monitoring and controlling parameters for Alzheimer's patients based on IoMT. In *Prognostic Models in Healthcare: AI and Statistical Approaches*, pages 213–237. Springer, 2022. https://link .springer.com/chapter/10.1007/978-981-19-2057-8_8

Khang A., A. Misra, V. Abdullayev, and E. Litvinova (Eds.). *Machine Vision and Industrial Robotics in Manufacturing: Approaches, Technologies, and Applications.* CRC Press, 2024b. https://doi.org/10.1201/ 9781003438137

Khang A., S. Chowdhury, and S. Sharma. *The Data-Driven Blockchain Ecosystem: Fundamentals, Applications, and Emerging Technologies* (1st ed.). CRC Press, 2022. https://doi.org/10.1201/9781003269281

Khang A., S. K. Gupta, S. Rani, and D. A. Karras (Eds.). *Smart, Cities: IoT Technologies, Big Data Solutions, Cloud Platforms, and Cybersecurity Techniques.* CRC Press, 2023c. https://doi.org/10.1201/9781003376064

Khang A., S. K. Gupta, V. Shah, and A. Misra (Eds.). *AI-aided IoT Technologies and Applications in the Smart Business and Production.* CRC Press, 2023a. https://doi.org /10.1201/9781003392224

Khang A., S. K. Gupta, V. A. Hajimahmud, J. Babasaheb, and G. Morris. *AI-Centric Modelling and Analytics: Concepts, Designs, Technologies, and Applications* (1st ed.). CRC Press, 2023b. https://doi.org/10.1201/9781003400110

Khang A., Vugar Abdullayev, Vladimir Hahanov, and Vrushank Shah. *Advanced IoT Technologies and Applications in the Industry 4.0 Digital Economy* (1st ed.). CRC Press, 2024a. https://doi.org/10.1201/9781003434269

Khanh H. H., and A. Khang. The role of artificial intelligence in blockchain applications. In *Reinventing Manufacturing and Business Processes through Artificial Intelligence*, 2 pages 20–40. CRC Press, 2021. https://doi.org/10.1201/9781003145011-2

Kumar S. and A. K. Pundir. Blockchain—Internet of things (iot) enabled pharmaceutical supply chain for Covid-19, pages 10–14, 2020. http://www.ieomsociety.org/detroit2020/papers/375.pdf

Lu D. and T. Liu. The application of IoT in medical system. In *2011 IEEE International Symposium on IT in Medicine and Education*, 1, pages 272–275. IEEE, 2011. https://ieeexplore.ieee.org/abstract/document/6130831/

Mohammadi Dashti M. and M. Harouni. Smile and laugh expressions detection based on local minimum key points. *Signal and Data Processing*, 15(2):69–88, 2018. https://jsdp.rcisp.ac.ir/browse.php?a_id=658&sid=1&slc_lang=en

Nanda S. K., S. K. Panda, and M. Dash. Medical supply chain integrated with blockchain and iot to track the logistics of medical products. *Multimedia Tools and Applications*:1–23, 2023. https://link.springer.com/article/10.1007/s11042-023-14846-8

Nassereddine, M. and A. Khang. Applications of Internet of things (IoT) in smart cities. *Advanced IoT Technologies and Applications in the Industry 4.0 Digital Economy* (1st ed.). CRC Press, 2024. https://doi.org/10.1201/9781003434269-6

Nguyen A., S. Lamouri, R. Pellerin, S. Tamayo, and B. Lekens. Data analytics in pharmaceutical supply chains: State of the art, opportunities, and challenges. *International Journal of Production Research*, 60(22):6888–6907, 2022. https://www.tandfonline.com/doi/abs/10.1080/00207543.2021.1950937

Ogbuke N. J., Y. Y. Yusuf, K. Dharma, and B. A. Mercangoz. Big data supply chain analytics: Ethical, privacy and security challenges posed to business, industries and society. *Production Planning and Control*, 33(2–3):123–137, 2022. https://www.tandfonline.com/doi/abs/10.1080/09537287.2020.1810764

Rani S., M. Chauhan, A. Kataria, and A. Khang (Eds.). IoT equipped intelligent distributed framework for smart healthcare systems. *Networking and Internet Architecture*, 2:30, 2021. https://doi.org/10.48550/arXiv.2110.04997

Rani S., P. Bhambri, A. Kataria, A. Khang, and A. K. Sivaraman. *Big Data, Cloud Computing and IoT: Tools and Applications* (1st ed.). Chapman and Hall/CRC, 2023. https://doi.org/10.1201/9781003298335

Rath, Kali Charan, A. Khang, and Debanik Roy. The role of Internet of things (IoT) technology in Industry 4.0. *Advanced IoT Technologies and Applications in the Industry 4.0 Digital Economy* (1st ed.). CRC Press, 2024. https://doi.org/10.1201/9781003434269-1

Salehi-Amiri A., A. Jabbarzadeh, M. Hajiaghaei-Keshteli, and A. Chaabane. Utilizing the internet of things (iot) to address uncertain home health care supply chain network. *Expert Systems with Applications*, 208:118239, 2022. https://www.sciencedirect.com/science/article/pii/S0957417422013860

Shamsuzzoha A., E. Ndzibah, and K. Kettunen. Data-driven sustainable supply chain through centralized logistics network: Case study in a Finnish pharmaceutical distributor company. *Current Research in Environmental Sustainability*, 2:100013, 2020. https://www.sciencedirect.com/science/article/pii/S2666049020300268

Sharma A., J. Kaur, and I. Singh. Internet of things (iot) in pharmaceutical manufacturing, warehousing, and supply chain management. *SN Computer Science*, 1:1–10, 2020. https://link.springer.com/article/10.1007/s42979-020-00248-2

Subhashini R. and A. Khang (Eds.). The role of Internet of things (IoT) in smart city framework. In *Smart Cities: IoT Technologies, Big Data Solutions, Cloud Platforms, and Cybersecurity Techniques*. CRC Press, 2023. https://doi.org/10.1201/9781003376064-3

Tu M. An exploratory study of internet of things (iot) adoption intention in logistics and supply chain management: A mixed research approach. *The International Journal of Logistics Management*, 2018. https://www.emerald.com/insight/content/doi/10.1108/IJLM-11-2016-0274/full/html

Yan B. and G. Huang. Supply chain information transmission based on RFID and internet of things, 4:166–169, 2009. https://ieeexplore.ieee.org/abstract/document/5267755/

13 Smart Grid Protection Scheme Using Internet of Things (IoT)

Hinal Shah, Jaydeep Chakravorty,
and Nilesh G. Chothani

INTRODUCTION

An electric power grid can be transformed into a "smart grid" by utilizing cutting-edge communications and information technology with automated control. The use of a smart grid enables trustworthy, top-notch, and ecological solutions to the rising demand for electricity. The need for energy is rising quickly over the globe. A smart grid is necessary in order to provide consumers with an accessible, trustworthy, and environmentally-friendly source of power while also meeting a surge in demand. It incorporates entirely novel, advanced technology and tools from generation, transmission, and distribution right through to consumer electronics and appliances.

Smart energy networks, smart metering, advanced metering infrastructure (AMI), smart sensors, phasor measurement unit (PMU), DG resources, smart communication, smart protection control, etc., are all parts of the upgraded digital power grid known as the "smart-grid." Recent developments in the electrical power grid architecture will enable us to coordinate massive amount of power generation, transmission, and distribution, electric power markets, operators, customers, and service providers, as well as reduce the impact of greenhouse gases while simultaneously improving the quality of electrical power (Colmenar-Santos et al., 2016).

Power from centralized generating was intended to be delivered via current electrical distribution networks to established customers and predicted loads. Distribution networks become more decentralized and bilateral as a result of the integration of renewable energy sources (RES), microgrids, and technologies for the storage of energy, enabling network self-healing and system restructuring (Sarathkumar et al., 2021). In order to sense, control, schedule, dispatch, bill, and identify cyber threats, smart grids will likely be more closely integrated with the cyber infrastructure. Smart meters will also be used to order electricity demand online (Subhash & Bochu et al., 2014). The most widely used standard for substation automation is IEC 61850 (Benomar et al., 2020, IEC, 2011). It aims to organize the intelligence of the monitoring, automation, control, and protection systems. A connection between Constrained Application Protocol (CoAP) and IEC 61850-based substation automation systems in

DOI: 10.1201/9781003434269-13

a smart grid context is suggested by Urkia et al. (2018) as one of the primary strategies for facilitating interworking between REST and smart grid domains. A specialized web transmission protocol for usage with confined nodes and constrained networks, CoAP is based on REST. The article by 23Pedersen et al. (2010) presents yet another option for mapping the IEC 61850 to RESTful Web Services.

The decentralization of the electric grid and its interdependence make it more practical. Electronic signals, power quality monitoring, control of the protection, and distribution of electricity are key components of the smart grid. They offer numerous benefits for monitoring distributed energy generation using tools like remote sensors and load shedding in emergency scenarios. In addition to that, it helps to monitor and control parameters like voltage, phase sequence, frequency, load demand, total harmonics distortion, etc. One of the objectives of using internet of things (IoT) in a smart grid is to forecast the load based on the history of the demand. IoT will be used to manage, monitor, and analyze the data on local networks as well as in the cloud (Sarathkumar et al., 2021).

A smart grid affects every part of an electrical grid, particularly the distribution and microgrid levels (Muthamizh et al., 2016). A smart grid structure takes power from different power plants like thermal, nuclear, hydro, solar, wind, etc. Grid delivers power to smart homes, buildings, factories, vehicles, cities, etc. It is no longer a unidirectional or radial feeder. The two-way connection capabilities of smart meters can be utilized to create a computerized widely dispersed network. That may create smart grid protection challenges like bidirectional power flow, unnecessary or false tripping, due to uncertainty of non-conventional energy resources variation in load current, variation in fault current, and overreach and underreach issues.

Smart grid challenges also include fluctuations due to the elevation of generators, links, and separations of storage devices, loads, and operating modes like manual mode or automated mode (Hou & Hu et al., 2010). This chapter includes smart grid protection issues and its mitigation techniques using IoT (Figure 13.1). The application and needs of IoT in a smart grid have been discussed in the research article (Shu-wen (2011).

Also, IoT-based wireless sensor networks (WSNs) powered by Wi-Fi have larger bandwidth and non-linear transmission, and massive data gathering is very economical as mentioned in the paper by Li and Xiaoguang et al. (2011). It is capable of video monitoring that is not possible using Zigbee. The article by Vineetha and Babu (2014) addressed problems and issues with the idea of a smart grid and offered solutions. Researchers have discussed the technology development, renewable energy integration, power quality, customer support, and security issues of smart grids. Platforms for real-time cosimulations were built on the IoT by researchers (Estebsari & Barbierato, 2018).

Opal-RT's real-time simulator, an automated fault identification, isolation, restoration algorithm, and MQTT communication have all been used to represent the physical components of the two-feeder medium voltage network with a normally open switch. The adoption of the IoT in various smart grid components was examined in this study. The study has improved real-time communication between storage facilities and renewable energy sources and can increase the profitability of both types of generating units (Shahinzadeh et al., 2019). For the purpose of locating and

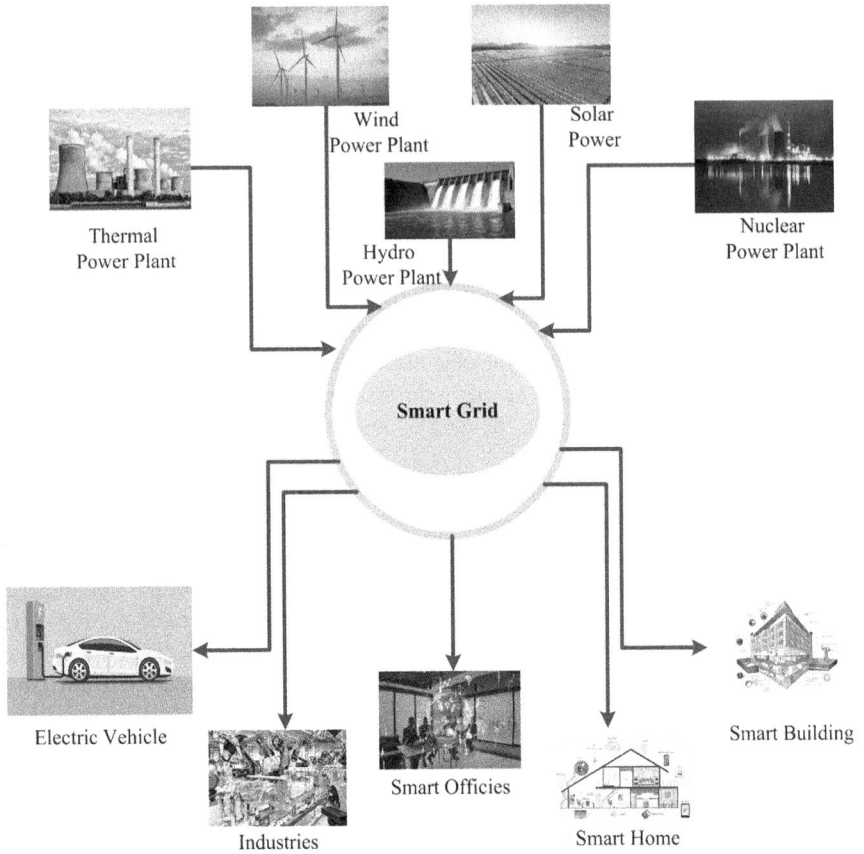

FIGURE 13.1 The architecture of the smart grid

identifying faults, a unique cloud-based IoT solution is proposed that can find one or more simultaneous faults in the presence of failing devices or sensors (Mehmood et al., 2020). The algorithm separates a distribution system into various zones using a zone-based technique. In the distribution grid, a current sensing device was placed at the edge of a zone. The purpose of the current sensor is to communicate with a cloud server via an edge device and to give time-synchronized current measurements. This chapter presents a smart grid protection overview along with IoT applications in transmission, distribution, and microgrid levels with the integration of renewable energy sources. Also, the role of IoT in the smart grid has been addressed (Khang et al., 2024a).

OVERVIEW OF SMART GRID PROTECTION

This section includes IoT in smart grids. Major protection challenges, advantages, limitations, and mitigation techniques are addressed.

SMART GRID CHALLENGES

Cybersecurity

Smart grids are vulnerable to online risks like theft of data, security breaches, and fraud. Such risks have the ability to weaken grid security, interfere with operations, and even result in power outages or damage to vital infrastructure.

Data Privacy

As data transmission and collection rise in smart grids, it is more important than ever to protect sensitive data, including consumer information and energy use trends. Unauthorized access to this kind of data might result in data misuse and privacy violations.

Physical Safety

The sensors, communication networks, and control systems that make up the smart grid are susceptible to physical assaults, damage, and tampering. To protect these important assets, appropriate physical safety precautions are necessary.

PROTECTION TECHNIQUES IN SMART GRID

To mitigate different protection issues of smart grids, the following techniques can be implemented.

- Highly secure encryption measures are implemented to preserve the security and authenticity of data transported inside the smart grid and resist unauthorized access or modification.
- Strong authentication techniques, such as passwords, digital identification numbers, and biometrics, can authenticate individuals and devices using the smart grid, lowering the possibility of unauthorized access.
- Intrusion detection systems (IDSs) keep an eye on system behavior and traffic on the network to detect any abnormal activity or potential cyberattacks. It improves the overall safety record by enabling early identification and reaction to security problems.
- Physical security measures by installing security cameras, entry restrictions, and tangible obstacles around vital smart grid infrastructure may prevent physical intrusions and attacks.

BENEFITS AND LIMITATIONS OF SMART GRID PROTECTIONS

Smart grid technology has many advantages such as smart grids allowing for real-time monitoring, fault detection, and self-healing capabilities. As a result, the electrical system is more trustworthy, encountering less downtime, and providing consumers with higher-quality power. Due to real-time monitoring of load patterns, better management of energy supply and demand, hence better optimization, as well as the overall efficiency of the smart grid, should be enhanced. Smart grids provide consumers with more control over how they monitor and manage their energy use. Eventually, that helps reduce energy consumption. Integration of renewable energy sources helps prevent greenhouse gas emissions. By including extensive monitoring and self-healing tools, smart grids can quickly spot and isolate problems, reducing the effects of disruptions brought on by calamities, accidents, or cyber threats.

Along with the benefits, there are certain limitations to the smart grid as well. It is difficult to convert a smart grid system to be fully protected due to the complexity of the system. It is necessary to continuously analyze and update the infrastructure to address security flaws. It might be difficult to integrate smart grid technologies into legacy infrastructure since older systems may not have the required security features or be compatible with contemporary security solutions. Also, strong security measures for smart grids can be expensive. It calls for expenditures in cybersecurity technologies, training, and regular updates. Some utilities or organizations may find it difficult to balance their security requirements with their financial limitations (Khang et al., 2022).

IOT IN RENEWABLE ENERGY SOURCES

Renewable energy plays a very important role in smart grid implementation. The use of green energies is promoted in recent areas to reduce carbon emissions. Many researchers are working on power generation using solar and wind energy with improved efficiency and minimizing losses of generation. IoT technology can be implemented in power generation using solar and wind energy sources as discussed below.

WIND ENERGY GENERATION WITH IoT

In terms of efficiency and dimensions, wind technologies are developing quickly. The wind energy sector has ambitious, far-reaching deployment goals, according to energy experts. The primary barrier to the development of wind energy is related to the resources' inherent erratic nature. As a result, if wind power is heavily utilized to fulfill demand, extreme instabilities could endanger the system's security. Due to that, the rest of the electrical grid can be ready for real-time operation so that it can compensate for fluctuations without experiencing abrupt transitions. Additionally, real-time data interchange between wind units and energy storage units is only possible when it is possible to operate jointly with energy storage centers more effectively.

Additionally, reliable predictive maintenance schedules may be carried out by wind farm operators because of IoT technologies and information communication technology (ICT) infrastructures, which prevent suffering significant losses. Machine learning (ML) and data mining techniques can be used to create such a strategy. Limitations in bandwidth for sending information to far-off areas and delays in data transfer for offshore wind farms are two issues at the present time. The use of IoT systems in the wind industry has brought attention to the need for more thorough plans for frameworks that are affordable, secure, and safe for the design and operation of wind farms as well as for the installation and maintenance of turbines (Rana et al., 2021).

A wind turbine's main parts include the tower, yaw system, blades, rotor hub, pitch system, drive shaft, high-speed shaft, brake system, gearbox, a doubly fed induction generator (DFIG), power converter, wind sensor, nacelle, transformer, and the central controller. Many sensors and actuators are included in the controller level. Each intrinsic component's performance and overall condition can be reported by the sensors. Using a group of actuators, the control system regulates and directs each component. The different types of sensors are used to measure parameters belonging to the respective sensors as mentioned in Table 13.1.

In order to connect the physical layer of the wind turbines to the cyber layer, a network architecture, supervisory control and data acquisition (SCADA), RTU (remote terminal unit), PLC (programmable logic controller), and condition monitoring systems or devices are used (Moness & Moustafa, 2016). In a wind farm, the term "network" refers to a secure link that enables data and control signals to be sent between the controller and various subsystems as well as between intelligent machines. The IoT-based controlling system is said to be more costly than existing SCADA systems, but it has a better ability to detect issues due to a large amount of information with a higher sampling rate that has been provided (Pettener, 2011).

SOLAR ENERGY GENERATION WITH IoT

A larger percentage of renewable energy sources must be used to meet the world's energy needs as a result of global warming and environmental concerns. Out of all renewable energy sources, solar energy has the greatest potential for power production. As a result, this source is taken into account as a significant contributor to future clean power systems. PVs can be implemented widely as systems for concentrated solar power, solar parks, or distributed generation. Photovoltaic arrays (solar panels), converters, inverters, and battery storage are the essential components of a photovoltaic system. In order to harvest solar energy more effectively, modern PV systems benefit from advanced technology such as the maximum power point tracker (MPPT) controlling scheme, GPS solar tracker, solar irradiance sensors, anemometer, and other task-specific devices. Because of a logically lower capital cost per kW, conventional PV systems can also be employed for end-user distributed generating, such as rooftop mounting or building integration of solar power.

TABLE 13.1
Sensors Are Used in IoT-based Wind Farm

S. No.	Type of Sensors	Parameters
1	Environmental sensors	Wind speed, humidity, and lighting
2	Mechanical sensors	Positions, speeds, angles, stresses, and strain
3	Electrical sensors	Voltages, currents, power factors, frequencies, and faults
4	Temperature sensors	Bearings, oils, windings, and electronic components
5	Fluid sensors	Pressure, levels, and flow

The majority of PV systems nowadays are grid-connected rather than stand-alone. The output power of PV systems is mostly influenced by the surrounding temperature and the strength of the sun's radiation. It is important to note that shade and dirt may significantly hinder the operation of a PV system, which can result in a sharp reduction in output power. Additionally, the PV system's efficiency declines in warmer climates. When there is partial cloud cover, the MPPT system tilts the panel toward the brightest region of the sky or straight in front of the sun. It is essential to have a storage facility because solar energy must be stored whenever it is available and delivered from the storage when it is required (Bhambri et al., 2022).

IoT stands for the coordination of similar objects with unique identities, including computational devices, mechanical equipment, and objects. By closing the gap between operational technologies and information technology without the need for machine-to-human or human-to-human transfers, this coordination helps transmit data and information across the network. IoT can support real-time data sharing of all information gathered from PV sensors and enable remote control of the operation of solar panels for problem and breakdown detection as well as proactive and preventive maintenance.

The performance of the PV arrays must be closely monitored because it has an impact on the PV unit's profitability and dependability. The weather has a significant impact on the cyclical and dynamic intensity of solar radiation. Thus, it is impossible to generate at an identical rate. The state of charge of the batteries and the voltage levels of the power converters are two more system components that are indirectly impacted by this issue. Environmental factors such as rain, dust accumulation, the presence of snow on the PV panel, and the impact of surface coating all contribute to the poor performance of the PV system. Because PV systems can be as close as a rooftop-mounted PV system or as far away as a solar park in the desert, it is difficult for humans to monitor all of the PV panels to prevent losses and failures because frequent site visits and maintaining the record of operational data are required, which is a laborious task when the PV plant is situated in a remote area (Manoj & Nallapaneni et al., 2018). In order to monitor the PV system characteristics and save the necessary data in a cloud-based platform, an uninterrupted real-time monitoring system must be installed alongside the PV panels.

IOT IN POWER GRID INTEGRATED WITH RENEWABLE ENERGY SOURCES

IoT technology is well suited for power system protection problems specifically when renewable energy sources are integrated with the grid. Transmission, distribution, and microgrid protection using IoT are discussed in the following sections.

IoT in Transmission System Protection

Transmission lines are a medium to transfer power from the generating unit to the distribution system. IoT integration in the transmission sector is crucial from two perspectives. The first is the contribution of IoT to better congestion management, and the second is its effect on preserving system security. Intelligent electronic devices (IEDs) equipped with the internet of things (IoT) can be put in the transmission sector, in order to inform the operator of the electrical condition of the transmission lines, such as losses, fault location, fault time, and type of faults. Phasor measuring units (PMU) can use the global positioning system (GPS) for time synchronization to detect the magnitude and angle of voltage and current at a specific point in the line.

A SCADA system can also be used to transmit voltage and current signals to the control room. IoT systems have a higher sampling rate compared to SCADA and PMU techniques. Also, faster operation and control are possible in wireless IoT networks. Distance relays are located at remote locations for transmission line protection. In the presence of renewable energy sources, relays are malfunctioning. Distance relay characteristics may create overreach and underreach issues due to the intermittent nature of renewable energy sources (Hinal Shah, Jaydeep Chakravorty, and Nilesh G. Chothani, 2023). To overcome these issues, different machine learning and deep learning (DL) (Khang et al., 2023a) techniques are implemented for fault classification and detection and based on that decision can be implemented through IoT at a local network or at a global network using the TCP protocol. Also, research work has been carried out on adaptive protection techniques for protection of power lines and overall protection of grid

IoT in Distribution System

Benefits of IoT integration at the distribution layer include online monitoring of consumer consumption patterns, intelligent energy generation and consumption control, and the identification of issues with low-voltage transmission lines, implementation of crisis demand response programs, deployment of self-healing schemes, management of power loss, and remote monitoring and control during unplanned disasters, among others. The distribution system no longer remains radial due to the integration of renewable energy sources. Bidirectional power flow, variation in load current, and fault current direction relays are used. To measure the power or unit consumption, advanced metering technologies are implemented. The IoT system addressed all these issues. It helps to transfer and update the data very quickly from a remote location. Additionally, in order for the distribution operator to be able to conduct rigorous monitoring and supervision of the distribution grid, the data gathered from all feeders and buses must be digitalized and shared through regional ICT-based networks.

IoT in Microgrid

A small-scale group of loads known as a microgrid localized on a particular feeder or distribution system and able to satisfy any or all of its demand via microsources, such as small-scale some energy sources: wind turbines, solar panels, micro-turbines, diesel generators, or gas turbines, may be coupled using a combined heat and power method (Markovic et al., 2013). Additionally, an energy storage facility can be used to store the extra energy produced by small-scale renewable resources when demand is less.

Microgrid designs can be connected to the main grid or isolated (off-grid or stand-alone), which is particularly useful for remote places. The isolated mode is also referred to as independent mode, islanding mode, or off-grid mode, whereas the linked mode is also known as collaborative mode. In cooperation with an upstream network, the microgrid can sell the excess generation of its own resources to the grid.

In other studies, hybrid methods that combine the cooperative operation of micro-sources and storage facilities are presented. Additionally, some academics have offered the notion of interconnected microgrids to reduce reliance on the main grid. The efficiency, power quality, and security of interconnected microgrids are three alarming problems that now plague modern microgrids. Wireless sensor networks, Zigbee devices, Ardinuo, or NodeMCU can be used at the local level (physical layer) of application. Cloud computing can be done for critical parameters as well as for higher-level IoT implementation. IoT level will increase simultaneously, and the cost and complexity of the system will also increase (Rani et al., 2023).

IOT IN FAULT IDENTIFICATION AND CLASSIFICATION

The following points are taken into consideration for IoT-based fault identification and categorized in smart grids.

- **Hardware Configuration**: The NodeMCU development board, predicated on the ESP8266 microcontroller with integrated Wi-Fi connectivity, is extensively used.
- **Sensors**: Based on the specific requirements for fault detection, connect the appropriate sensors to the NodeMCU board. To measure electrical parameters, current transformers and voltage transformers are connected to the grid. Both the transformers are working as transducers. Depending on the application, temperature sensors, accelerometers, or other pertinent sensors can also be included. Using signal conditioning components convert electrical parameters to appropriate signals to feed the microcontroller. Analog-to-digital converter (ADC) is also used to convert analog signals to appropriate digital values.
- **Operating Power**: Give the NodeMCU board and any associated sensors an appropriate power source.
- **Programming for NodeMCU**: Use an appropriate programming language, such as the Arduino IDE or the NodeMCU Lua-based firmware, to create the firmware code for the NodeMCU board. To enable data transfer and connection with a central server or cloud-based platform, configure the Wi-Fi connectivity. Put the required algorithms in place. Using the gathered sensor data, develop the required fault detection and classification algorithms. This can involve rule-based methods, ML techniques, or a combination of both (Khang et al., 2023b).
- **Data Transmission and Communication**: Construct a connection with a central server or cloud-based platform using the Wi-Fi capabilities of the NodeMCU. Establish data transmission protocols to communicate the

gathered sensor data to a server or the cloud, such as MQTT (message queuing telemetry transport) or HTTP (hypertext transfer protocol).

- **Cloud-based/Server Platform**: Obtain the NodeMCU's transmitted data and process it on a server or cloud computing platform. Utilizing the gathered sensor data, perform fault detection and classification algorithms on the network side. Create findings for identifying faults and categorization based on the review of the received data.

Figure 13.2 has represented NodeMCU and MQTT-based communication protocol application model to detect faults and apply tripping signal to isolate faulty section. Sensor is used for sensing electrical parameters. The NodeMCU Wi-Fi module has published various parameters to MQTT clients through the MQTT broker. Based on the requirement, MQTT client command subscribed by the microcontroller actuates the actuator accordingly as shown in Figure 13.2.

CONCLUSION

Some of the crucial details emphasizing the importance of IoT in fault detection in a smart grid are as follows:

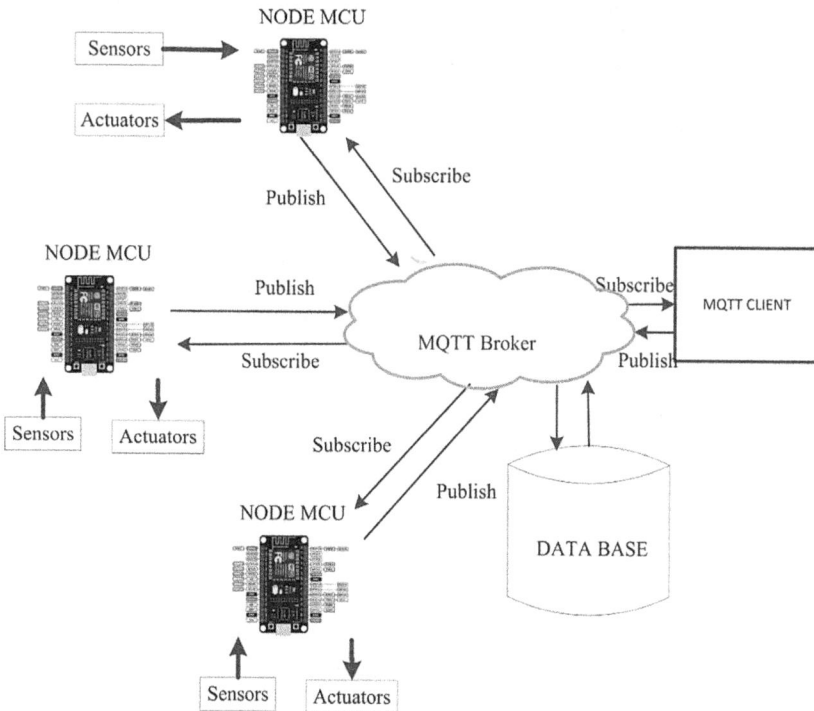

FIGURE 13.2   IoT-based fault identification in smart grid with MQTT protocol

- **Monitoring in Real Time**: IoT sensors installed throughout the grid gather and send data on a variety of factors, including voltage, current, power quality, and environmental variables. Monitoring in real time: IoT sensors installed throughout the grid gather and send data on a variety of factors, including voltage, current, power quality, and environmental variables. This ongoing monitoring allows for the quick identification of irregularities and potential problems.
- **Data Analytics**: To find patterns and discrepancies suggestive of shortcomings, IoT data is analyzed using advanced analytics techniques, including ML or DL algorithms. Smart grid systems may successfully discriminate between fault and normal operation by training algorithms on past data. This ongoing monitoring allows for the quick identification of abnormalities and potential problems (Khang, 2024b).
- **IoT**: IoT-based fault detection algorithms can quickly recognize and categorize various fault types, such as short circuits, overloads, or equipment malfunctions. These algorithms accurately identify and locate grid issues by using real-time data from IoT devices.
- **Remote Access and Control**: IoT makes smart grid systems accessible remotely, allowing administrators to keep an eye on and manage grid operations from a single place. By enabling operators to rapidly analyze the situation, identify the impacted areas, and conduct the appropriate corrective activities, this remote capability improves fault diagnosis and response.
- **Proactive Maintenance**: IoT devices provide ongoing grid component monitoring, including transformer, circuit breaker, and power line monitoring. IoT-based technologies provide proactive maintenance by identifying early warning indications of problems or equipment deterioration, reducing the risk of serious damage, and improving asset management.
- **Improved Grid Resilience**: The IoT makes smart grids more resilient by enabling real-time fault identification and short response times. It is possible to minimize the impact on the grid, cut downtime, and ensure a dependable power supply to end users by quickly identifying and isolating faults.

Overall, by providing real-time monitoring, intelligent data analysis, remote access, and control, IoT integration in smart grids improves defect diagnosis. Smart grids can become more effective, trustworthy, and robust by utilizing the power of IoT. This improves problem detection and reaction times, which ultimately boosts the performance of the electrical system as a whole (Khang et al., 2023c).

REFERENCES

Benomar Z., Longo F., Merlino G., and Puliafito A. (2020). Enabling secure RESTful web services in IoT using OpenStack. In Proceedings of the IEEE 17th International Conference on Mobile Ad Hoc and Sensor Systems (MASS), New Delhi, 10–13 December. https://ieeexplore.ieee.org/abstract/document/9356053/

Bhambri P., Rani S., Gupta G., and Khang A. (2022). *Cloud and Fog Computing Platforms for Internet of Things*. CRC Press. https://doi.org/10.1201/9781003213888

Colmenar-Santos A., Reino-Rio C., Borge-Diez D., and Collado-Fernández E. (2016). Distributed generation: A review of factors that can contribute most to achieve a scenario of DG units embedded in the new distribution networks. *Renewable* and Sustainable *Energy* Reviews, 59(C), 1130–1148. https://www.sciencedirect.com/science/article/pii/S1364032116000538

Estebsari E. Patti, and Barbierato L. (2018). Fault detection, isolation and restoration test platform based on smart grid architecture model using Internet-of-things approaches. In IEEE International Conference on Environment and Electrical Engineering and 2018 IEEE Industrial and Commercial Power Systems Europe (EEEIC / I&CPS Europe), Palermo, pp. 1–5. https://doi.org/10.1109/EEEIC.2018.8494449

Hou C., Hu X., and Hui D. (2010). Hierarchical control techniques applied in micro-grid. In Proceedings of the IEEE Conference Power System Technology (POWERCON), Hangzhou, October. https://ieeexplore.ieee.org/abstract/document/5666418/

International Electrotechnical Commission. (2011). *IEC 61850: Communication Networks and Systems for Power Utility Automation.* International Electrotechnical Commission (IEC), Geneva. IEC Standards, Parts, pp. 1–10 (Edition. 2.0). http://www.singaporestandardseshop.sg/data/ECopyFileStore/110714104958info_iec61850-8-1%7Bed2.0%7Db.pdf

Khang A. (Eds.) (2024). *AI and IoT-Based Technologies for Precision Medicine.* IGI Global Press. https://doi.org/10.4018/979-8-3693-0876-9

Khang A., Abdullayev Vugar, Hahanov Vladimir, and Shah Vrushank. (2024). *Advanced IoT Technologies and Applications in the Industry 4.0 Digital Economy* (1st ed.). CRC Press. https://doi.org/10.1201/9781003434269

Khang A., Gupta S. K., Hajimahmud V. A., Babasaheb J., and Morris G. (2023). *AI-Centric Modelling and Analytics: Concepts, Designs, Technologies, and Applications* (1st ed.). CRC Press. https://doi.org/10.1201/9781003400110

Khang A., Gupta S. K., Shah V., and Misra A. (Eds.) (2023). *AI-Aided IoT Technologies and Applications in the Smart Business and Production.* CRC Press. https://doi.org/10.1201/9781003392224

Khang A., Hahanov V., Abbas G. L., and Hajimahmud V. A. (2022). Cyber-physical-social system and incident management. *AI-Centric Smart City Ecosystems: Technologies, Design and Implementation* (1st ed.), 2(15). CRC Press. https://doi.org/10.1201/9781003252542-2

Khang A., Vrushank S., and Rani S. (2023). *AI-Based Technologies and Applications in the Era of the Metaverse* (1st ed.). IGI Global Press. https://doi.org/10.4018/978-1-6684-8851-5

Kumar Manoj, Nallapaneni, Atluri Karthik, and Palaparthi Sriteja. (2018). *Internet of Things (IoT) in Photovoltaic Systems.* https://doi.org/10.1109/NPEC.2018.8476807

Li L., Xiaoguang H., Ke C., and Ketai H. (2011). The applications of WiFi-based wireless sensor network in internet of things and smart grid. In 6th IEEE Conference on Industrial Electronics and Applications, Beijing, pp. 789–793. https://doi.org/10.1109/ICIEA.2011.5975693

Markovic D. S., Zivkovic D., Branovic I., Popovic R., and Cvetkovic D. (2013). Smart power grid and cloud computing. *Renewable and Sustainable Energy Reviews*, 24, 566–577. https://www.sciencedirect.com/science/article/pii/S136403211300227X

Mehmood M., Ulasyar A., Khattak A., Imran K., ShehZad H., and Nisar S. (2020). Cloud based IoT solution for fault detection and localization in power distribution systems. *Energies*, 13(11). https://doi.org/10.3390/en13112686

Moness M., and Moustafa A. M. (2016). A survey of cyber-physical advances and challenges of wind energy conversion systems: Prospects for internet of energy. *IEEE Internet of Things Journal*, 3(2), 134–145. https://ieeexplore.ieee.org/abstract/document/7264965/

Muthamizh Selvam M., Gnanadass R., and Padhy N. P. (2016). Initiatives and technical challenges in smart distribution grid. *Renewable and Sustainable Energy Reviews*, 58, 911–917. https://doi.org/10.1016/j.rser.2015.12.257

Pedersen A. B., Hauksson E. B., Andersen P. B., Poulsen B., Træholt C., and Gantenbein D. (2010). Facilitating a generic communication interface to distributed energy resources: Mapping IEC 61850 to RESTful services. In Proceedings of the First IEEE International Conference on Smart Grid Communications, Gaithersburg, MD, 4–6 October. https://ieeexplore.ieee.org/abstract/document/5622020/

Pettener A. L. (2011). *SCADA and Communication Networks for Large Scale Offshore Wind Power Systems*, pp. 1–6. https://doi.org/10.1049/cp.2011.0101

Rana G., Khang A., Sharma R., Goel A. K., and Dubey A. K. (Eds.) (2021). *Reinventing Manufacturing and Business Processes through Artificial Intelligence*. CRC Press. https://doi.org/10.1201/9781003145011

Rani S., Bhambri P., Kataria A., Khang A., and Sivaraman A. K. (2023). *Big Data, Cloud Computing and IoT: Tools and Applications* (1st ed.). Chapman and Hall/CRC. https://doi.org/10.1201/9781003298335

Sarathkumar D., Srinivasan Mohanambigai, Alexander Dr. S.Albert, Ravi Samikannu, Dasari Narasimha, and Antony Raj Raymon. (2021). A technical review on classification of various faults in smart grid systems. *IOP Conference Series: Materials Science and Engineering*, 1055(1), 012152. https://doi.org/10.1088/1757-899X/1055/1/012152

Shah Hinal, Chakravorty Jaydeep, and Chothani Nilesh G. (2023). Protection challenges and mitigation techniques of power grid integrated to renewable energy sources: A review. *Energy Sources, Part A: Recovery, Utilization, and Environmental Effects*, 45(2), 4195–4210. https://doi.org/10.1080/15567036.2023.2203111

Shahinzadeh H., Moradi J., Gharehpetian G. B., Nafisi H., and Abedi M. (2019). IoT architecture for smart grids. In International Conference on Protection and Automation of Power System (IPAPS), Iran, pp. 22–30. https://doi.org/10.1109/IPAPS.2019.8641944

Shu-wen W. (2011). Research on the key technologies of IOT applied on smart grid. In International Conference on Electronics, Communications and Control. (ICECC), Ningbo, pp. 2809–2812. https://doi.org/10.1109/ICECC.2011.6066418

Subhash Bochu, and Rajagopal V. (2014). Overview of smart metering system in smart grid scenario. In Power and Energy Systems Conference: Towards Sustainable Energy, PESTSE, pp. 1–6. https://doi.org/10.1109/PESTSE.2014.6805319

Urkia M. I., Casado-Mansilla D., Mayer S., Bilbao J., and Urbieta A. (2018). Integrating electrical substations within the IoT using IEC 61850, CoAP and CBOR. *IEEE Internet of Things Journal*, 14, 8. https://ieeexplore.ieee.org/abstract/document/8661513/

Vineetha C. P., and Babu C. A. (2014). Smart grid challenges, issues and solutions. In International Conference on Intelligent Green Building and Smart Grid (IGBSG), Taipei, pp. 1–4. https://doi.org/10.1109/IGBSG.2014.6835208

14 Internet of Things (IoT) as a Game Changer to the Education Sector

Kavitha Rajamohan, Sangeetha Rangasamy,
Neil Manoj C., Lalitha Mary, Raison Sabu,
and Sankar Sam Jose

INTRODUCTION

The internet of things (IoT), which seamlessly integrates linked devices and sensors, has the potential to revolutionize education (Arora & Kaushik, 2020). IoT makes it possible for innovative teaching strategies and individualized learning possibilities. Through wearable technology and smart classrooms, students may collaborate, access information, and use interactive materials. A variety of educational tools are available for teachers to use in order to creatively engage pupils. IoT also produces useful data that aids academic institutions in better understanding student behavior and curriculum customization for improved learning results. However, while using IoT in education, privacy, security, and fair access are crucial factors to take into account. For the internet of things to reach its full potential, bridging the digital gap, guaranteeing device security, and safeguarding student data are essential. The numerous IoT applications in educational settings including classrooms, online learning environments, and institutions are examined in this chapter. We can better grasp how IoT might alter education and prepare students for the digital world by looking at its benefits, drawbacks, and effects. For innovation, effectiveness, and future-ready student preparation, IoT integration in education has enormous promise.

CURRENT EDUCATION SYSTEM

The current education system, which has evolved through time, consists of a complicated network of institutions that are governed by regulations and curriculum. Students come to class, participate in lessons, and get instruction. Students' achievement is determined by subjects, exams, and assessments. In both teaching and evaluating students, teachers are essential. But there are issues with the system:

DOI: 10.1201/9781003434269-14

- **One-Size-Fits-All Strategy**: Individual learning preferences are not taken into account, which impedes personalized learning and innovation.
- **Limited Access and Inequality**: Due to regional and financial differences, some people have less access to high-quality education, which perpetuates educational inequality.
- **Rigid Curriculum**: A rigid curriculum that prioritizes memorization over critical thinking and practical abilities may not be able to adapt to the changing requirements of the world.
- **Technological Gaps**: Inequalities in the use of technology in education caused by a lack of resources, infrastructure, and personnel.

Educators and entrepreneurs investigate incorporating cutting-edge technology like IoT, artificial intelligence (AI), and virtual reality (VR) to solve these issues. These technologies can help with individualized instruction, digital access, teamwork, and the development of skills for the twenty-first century.

INTERNET OF THINGS

The internet of things (IoT) is a network of physical objects that can connect to each other and share data, such as vehicles, household appliances, furniture, and other items. Electronics, software, sensors, and communication are embedded in these items. The internet of things (IoT) enables the smooth and automated transmission of data between devices, allowing them to cooperate to accomplish a task without the need for human interaction. By establishing a connected world where devices can communicate with each other to offer consumers a seamless and effective experience, the IoT has the potential to completely transform the way we live, work, and communicate.

IoT IN THE EDUCATION SECTOR

The use of internet of things in education, or "Edutech Waves," has the potential to transform teaching and learning (Ramlowat & Pattanayak, 2019). IoT technology makes learning possibilities personalized and flexible, converting conventional classrooms into cutting-edge learning environments. IoT-enabled smart classrooms give instructors access to real-time data on student engagement and behavior. This data-driven strategy encourages student-centered learning and improves the efficacy of instruction. IoT devices also give students access to a multitude of educational materials, such as online libraries and interactive digital textbooks. Immersive technology like VR and AR help users understand and remember difficult concepts. IoT encourages cooperation and communication between students and teachers, facilitating distant cooperation and international viewpoints as shown as Figure 14.2.

It enhances building management, security, and administrative processes outside of the classroom. IoT use in education does, however, raise privacy, security, and ethical issues. To close the digital gap, equitable access to IoT infrastructure and training is crucial. Frameworks and standards must be established through collaboration between educators, politicians, technology suppliers, and business experts. Embracing IoT's potential may help students develop their critical thinking abilities and prepare them for the digital era, assuring success in a world that is changing quickly.

IOT ARCHITECTURE IN THE EDUCATION SECTOR

By allowing cutting-edge teaching and learning techniques, increasing operational effectiveness, and improving the educational experience as a whole, internet of things (IoT) may significantly improve the education industry (Figure 14.1).

DEVICES AND SENSORS

Install a variety of internet of things (IoT) devices and sensors around the educational facility, including wearable technology, interactive projectors, intelligent speakers, Radio Frequency Identification (RFID) tags, occupancy sensors, and temperature sensors. These gadgets will collect information and engage with the surroundings.

- **Connectivity**: Establishing a solid network infrastructure will enable all IoT devices to be connected. Wi-Fi, Bluetooth, and other compatible wireless communication protocols can be used. Ensure enough bandwidth and coverage to manage the data traffic the devices will produce.

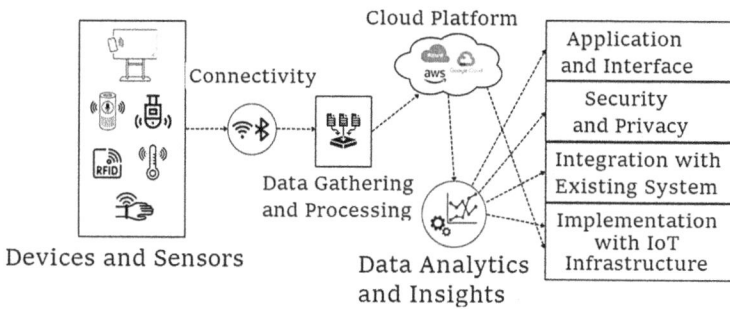

FIGURE 14.1   Architectural diagram of IoT in the education sector

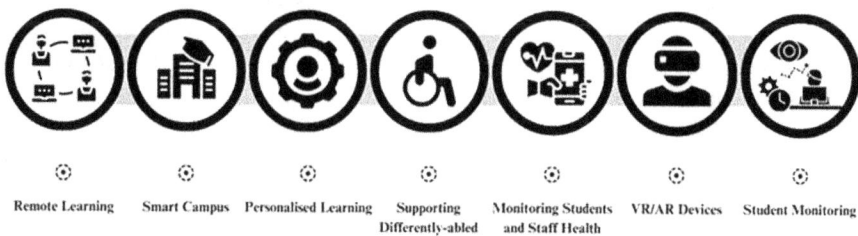

FIGURE 14.2   Applications of IoT in the education sector

- **Data Gathering and Processing**: Gather data in real time from IoT devices and sensors. Attendance records, environmental factors, classroom usage, student activities, and performance indicators are some examples of this data. Process the data locally or upload it to a cloud platform for additional analysis.
- **Cloud Platform**: To store and analyze the gathered data, use a cloud-based IoT platform. This platform should offer scalability, security, and data processing capabilities. Microsoft Azure IoT Hub, AWS IoT Core, and Google Cloud IoT Core are a few examples of well-known cloud systems.
- **Data Analytics and Insights**: Use data analytics methods to glean insightful information from the gathered data. Analyze student performance trends, pinpoint improvement opportunities, and customize the educational experience—utilize machine learning algorithms to develop prediction models to suggest tailored educational interventions or material (Khang et al., 2023a).
- **Applications and Interfaces**: Create intuitive apps and user interfaces to let administrators, instructors, and students communicate with the IoT system. This could include dashboards, web portals, and mobile apps. These user interfaces can show real-time data, deliver notifications, and make tailored suggestions.
- **Security and Privacy**: Implement strong security measures to safeguard the IoT infrastructure and data protection and privacy. Protect sensitive data by utilizing encryption, authentication techniques, and access controls.
- **Integration with Existing Systems**: Connect the IoT architecture to current learning management systems (LMS), student information systems (SIS), and administrative databases. This connection will allow for smooth data transfer and give a comprehensive overview of the educational environment.

Implementing this IoT infrastructure in the education sector allows educational institutions to improve teaching strategies, better utilize resources, monitor student progress, and offer individualized learning experiences.

IoT Devices (Smart HVAC Systems)

These gadgets include smart thermostats, environmental sensors, occupancy detectors, and HVAC management systems. The IoT Devices and Sensors layer consists of the smart HVAC systems. They record air quality, temperature, humidity, and occupancy information in various parts of the educational facility. In the Data Collection and Processing layer, the data acquired by the smart HVAC systems is analyzed alongside data from other IoT devices. The HVAC data may be subjected to real-time analytics by this layer, which can optimize temperature settings depending on occupancy or modify airflow based on external factors.

The Cloud Platform (IoT Backend) stores and manages the acquired HVAC data and other IoT data. The HVAC systems may be remotely monitored and controlled with extra capabilities, enabling facility managers to adjust and create schedules from a single platform. The Data Analytics and Insights layer may examine the HVAC data to find trends, reduce energy use, and offer information on the effectiveness of the HVAC systems. Making data-driven judgments using this information will improve the educational institution's overall energy management.

NETWORK COMMUNICATION

The Network Communication component of the IoT architecture is in charge of creating and overseeing the communication between the various levels and features. This part ensures that the IoT devices and sensors, the Data Collection and Processing layer, the Cloud Platform, the Data Analytics and Insights layer, the Applications and Interfaces layer, the Security and Privacy layer, and the Integration with Existing Systems layer are all connected securely and privately.

Various wireless and cable communication technologies may be used by the network communication component. Wi-Fi, Bluetooth, Ethernet, and other appropriate protocols may be used. Ensuring the network infrastructure has enough bandwidth, low latency, and high dependability is critical to facilitate the data flow between the various components. To safeguard communication and data, adequate network security measures should be implemented, including firewalls, intrusion detection systems, and encryption.

ANALYTICS AND STORAGE SOLUTIONS

Advanced analytics on the gathered data is carried out via the Analytics Solutions component. To extract insights, trends, and predictions from the data, it may be necessary to use machine learning algorithms, data mining techniques, and statistical analysis (Khang et al., 2023b). The data gathered must be stored and managed via the storage solutions component. It may comprise databases, data lakes, or cloud storage services to store and organize IoT data safely. The storage solutions should offer scalability, dependability, and suitable security measures to safeguard the data against loss or unauthorized access. In order to provide seamless integration and effective administration of the IoT data, the Cloud Platform acts as the primary hub for data processing, analytics, and storage (Rana et al., 2021).

APPLICATIONS OF IOT IN THE EDUCATION SECTOR

The internet of things, or IoT, has transformed several industries, including education. IoT applications in the education industry provide creative solutions to improve learning experiences, efficiency, and student engagement by merging smart devices, sensors, and connectivity.

REMOTE LEARNING

In the world of education, remote learning—also known as distance learning or online learning—has grown in popularity, especially in recent years as a result of the COVID-19 pandemic. By establishing a linked and interactive virtual classroom environment, the internet of things (IoT) integration in remote learning has the potential to improve the educational experience. Here is a detailed explanation of how IoT can be used in distance education:

- **Infrastructure**: IoT devices like routers, smartboards, and video conferencing systems provide reliable internet access and infrastructure for distance learning.
- **Interactive Learning Experiences**: Internet of things (IoT) devices improve interactivity in distance learning. IoT-enabled whiteboards or interactive screens, for instance, may be utilized to give captivating presentations, share multimedia materials, and promote in-person collaboration between students and teachers.
- **Personalized Learning**: IoT can provide remote settings for personalized learning. Connected technologies, such as smart tablets or wearable sensors, may monitor each student's progress, gather information on their learning habits, and offer recommendations and resources that are specifically suited to their needs.
- **Monitoring and Evaluation**: During distance learning sessions, IoT devices may keep an eye on students' involvement and development. For instance, eye-tracking sensors may assess levels of attention, while activity monitors can gauge alertness and physical activity.
- **Internet of Things for Education (IoT4E)**: IoT gadgets made expressly for learning can improve remote learning. Programmable robots, STEM kits, and scientific probes are some examples of the gadgets that let kids explore and experiment with different ideas remotely. Through internet platforms, IoT4E devices may be managed and watched, promoting interactive and hands-on learning.
- **Real-Time Feedback and Support**: IoT devices make it possible for distant learners to receive real-time feedback and support. Connected devices allow teachers to keep track of their students' progress, give immediate feedback on assignments, and provide counseling during virtual office hours. Students can stay on track and make required changes to their learning strategy with the aid of this quick feedback.
- **Remote Access to Materials**: IoT can offer remote access to learning materials including online databases, digital libraries, and e-books. With the help of connected devices, students may access a variety of information, multimedia materials, and interactive learning resources from any location, extending their learning possibilities outside of traditional classroom settings.

Even though the internet of things (IoT) has many advantages for remote learning, privacy, security, and ethical issues must be addressed. Data about students should be protected, and IoT devices should be used responsibly.

SMART CAMPUS

The internet of things (IoT), which powers smart campuses, is an innovative strategy for converting conventional educational establishments into technologically cutting-edge and interconnected learning environments. IoT devices, sensors, and data analytics are used by smart campuses to improve a variety of campus operations, facility management, and educational outcomes (Madyatmadja et al., 2021). An in-depth description of how IoT may be used in smart campuses is provided below:

- **Campus Infrastructure and Operations**: The management of campus infrastructure and operations may be improved with the usage of IoT devices. Connected sensors, for instance, may track how much energy, water, and garbage are used, as well as how efficiently resources are being managed. This lowers expenses, improves sustainability, and makes the campus more ecologically friendly.
- **Better Security and Safety**: IoT-based security solutions on campuses increase security and safety. Using linked security cameras, access control systems, and biometric authentication technologies, campus entrances can be watched, unauthorized entry may be found, and emergency alerts can be provided in real time.
- **Dynamic Illumination**: IoT-enabled lighting systems can dynamically modify illumination levels depending on occupancy, the availability of natural light, and the time of day. Connected sensors are able to detect motion and occupancy, which helps to reduce energy consumption and provide a welcoming, well-lit campus atmosphere.
- **Environmental Monitoring**: Within the campus, IoT sensors are able to track environmental variables including temperature, humidity, air quality, and noise levels to guarantee a safe and favorable learning environment. For instance, the IoT system can automatically alter ventilation or initiate alarms if the air quality starts to worsen in order to maintain a comfortable environment.
- **Smart Transportation and Parking**: IoT devices can help campus parking management be more effective. Connected sensors can identify available parking spaces and direct employees and students to the closest open spot, easing traffic congestion and enhancing campus mobility. IoT-based transportation systems may also optimize routes, deliver real-time updates on bus timetables, and improve the effectiveness of on-campus transportation.
- **Better Disaster Management**: Internet of things (IoT) devices can improve disaster planning and response on campuses. Panic buttons, emergency alert systems, and position tracking systems are examples of connected gadgets that may deliver real-time notifications, allow swift communication, and help effective evacuation in disaster circumstances.

When adopting IoT on smart campuses, it's critical to ensure privacy, security, and ethical issues. To preserve data privacy and guarantee responsible usage of IoT devices, the appropriate standards and protections should be in place.

PERSONALIZED LEARNING

An educational strategy known as personalized learning adapts instruction, content, and pace to each student's unique requirements, skills, and interests. Numerous uses of the internet of things (IoT) are available in the educational space that can encourage and assist individualized learning (Spyrou et al., 2018). IoT can be used in the following ways to improve personalized learning:

- **Adaptive Learning Platforms**: IoT devices can gather real-time data on student's activities, including their progress, levels of engagement, and learning preferences. Adaptive learning platforms that analyze individual learning styles and offer students individualized recommendations and content can be developed using this data. To meet the unique needs of each student, adaptive learning platforms can dynamically change the level of difficulty of activities, offer additional support resources, and offer customized assessments.

- **Smart Classrooms**: The internet of things (IoT) can make conventional classrooms smart classrooms (Burunkaya & Duraklar, 2022). For instance, networked whiteboards have the ability to record student interactions and create digital records, enabling teachers to monitor student progress and give tailored comments. A comfortable learning environment customized to the tastes of the students can be created using smart lighting and temperature control technologies, which can improve focus and engagement.

- **Wearable Devices**: Students' physiological data, including heart rate, sleep habits, and stress levels, can be tracked via IoT wearables like smartwatches and fitness trackers. By examining this data, instructors can learn more about the emotional and mental health of their pupils, enabling them to provide individualized interventions and support as necessary. For example, if a wearable device detects elevated stress levels, it can alert the instructor, who can then give the necessary advice.

- **Personalized Content Delivery**: Based on the interests, learning preferences, and skill levels of the students, IoT can enable the delivery of personalized content. For instance, Radio Frequency Identification (RFID) tags can be incorporated into books, allowing for the creation of smart shelves that can offer more resources or relevant books based on the reading habits of the students. IoT gadgets are also capable of locating pupils inside of a school and sending them notifications or content that is pertinent to their location.

- **Collaborative Learning Spaces**: By connecting pupils and teachers in various locations, IoT devices can promote collaborative learning. Remote students can engage in an immersive learning environment using video conferencing systems with IoT sensors, interacting with classmates and instructors in real time. Virtual reality (VR) and augmented reality (AR) tools that are IoT-powered can also be used to build shared learning environments where students can work together to solve problems.

By creating inclusive and accessible learning environments, the internet of things (IoT) in the education sector could significantly assist differently abled students (Mathew & Walarine, 2018). IoT can be used in the following ways to support students with disabilities:

- **Assistive Devices**: IoT can facilitate the creation of cutting-edge assistive technology for kids with disabilities. IoT-enabled mobility aids or prostheses, for instance, can gather information on movement patterns, make real-time adjustments to settings, and offer individualized support. For kids with physical limitations, these tools can improve independence, mobility, and all aspects of learning.
- **Sensory Support**: Students with sensory limitations, like visual or hearing impairments, can benefit from IoT. To deliver data and content in accessible formats, IoT devices can be coupled with braille displays, screen readers, or audio-based systems.
- **Environmental Controls**: IoT can assist in regulating learning environments to meet the unique needs of students with disabilities. For pupils who are visually challenged, networked lighting systems can change brightness levels, and temperature controls can assure comfort for those who suffer from ailments like sensory processing disorders.
- **Communication and Interaction**: For children who have trouble speaking or communicating, IoT devices can improve the prospects for interaction and communication. Alternative communication techniques can be made easier by connected devices with voice recognition and natural language processing capabilities. Students may ask questions, take part in debates, and communicate their ideas more clearly by interacting with IoT devices using voice commands or gestures.
- **Personalized Learning**: IoT can enhance individualized instruction for students with disabilities by gathering real-time information on their progress, difficulties, and preferences. IoT-enabled adaptive learning platforms are able to provide content that is specifically suited to the learning requirements and skills of impaired students, as well as interactive activities and adaptive evaluations.
- **Safety and Accessibility**: IoT gadgets can help make educational facilities safer and more accessible for students with disabilities. For students who have mobility issues, networked surveillance systems, for instance, can offer real-time monitoring and alert systems, ensuring their safety and security. IoT-based navigation tools can help blind students find their way around campus on their own and find accessible pathways.

When deploying IoT solutions in the education sector, it is imperative to take the unique needs and preferences of students with disabilities into account. To further protect the sensitive data collected by IoT devices, privacy and security measures must be put in place.

MONITORING STUDENTS AND STAFF HEALTH

In addition to encouraging wellness, safety, and early intervention, the use of IoT in educational institutions can have a substantial positive impact on staff and student health (Hong-tan et al., 2021). Here are a few instances of how IoT can be used in this situation:

- **Wearable Health Devices**: IoT-enabled wearables, like smartwatches or fitness trackers, can monitor students' and staff members' vital signs, activity levels, and sleep patterns. Real-time information on heart rate, steps taken, calories burnt, and sleep quality can be obtained from these gadgets.
- **Health Monitoring Systems**: In educational facilities, IoT can be incorporated into health monitoring systems. Environmental variables including temperature, humidity, and air quality can be monitored by sensors installed in lecture halls, libraries, or public spaces. With the aid of these sensors, a secure and comfortable learning environment may be created, spotting potential health risks like poor air quality or high temperatures (Khang et al., 2023c).
- **Access and Attendance Systems**: IoT can be utilized to automate access management and attendance tracking systems. Biometric scanners or ID cards with IoT capabilities can reliably track and record staff and student presence.
- **Alert Systems for Health Emergencies**: IoT devices can be used to quickly identify and address health emergencies. For instance, connected medical alert systems can be deployed in public spaces like classrooms or hallways, enabling users to rapidly notify administrators or medical staff in the event of an emergency like an allergic reaction, an injury, or a severe illness.
- **Monitoring of Mental Health**: IoT can be very helpful in observing and assisting with mental health in educational settings. Stress levels or emotional indicators can be detected by smart sensors built into wearable technology or even seating configurations. By identifying people who might need more support or intervention, this information can help people get timely help and access to mental health resources.

It is crucial to remember that while installing IoT systems for health monitoring, privacy and security issues should come first. IoT-enabled health surveillance systems have the potential to enhance the overall educational experience and contribute to the holistic development of individuals within the educational community (Khang et al., 2023d).

VIRTUAL AND AUGMENTED REALITY DEVICES

Virtual and augmented reality (AR/VR) technologies have a lot of potential for the educational sector. They may totally alter how students learn by creating immersive and interesting learning environments (Fetaji et al., 2020). Here are some examples of AR and VR applications that might be utilized in classrooms:

- **Virtual Field Trips**: Students may be given interactive field trips to a number of locations using augmented reality and virtual reality. Using AR or VR technology, students may explore historical sites, museums, natural wonders, and even made-up worlds. They may examine objects, engage with computer-generated characters, and discover more information about the subject.
- **Experiments and Simulations**: Virtual laboratories and simulations made with AR and VR may be quite lifelike. For instance, students can do chemical experiments in a virtual lab during science class, manipulating fictitious materials and witnessing their responses. Virtual reality (VR) can imitate operations in medical education, giving trainees the opportunity to practice procedures without the risk involved with actual patients. These simulations give students useful hands-on experience, encourage critical thought, and aid in the development of practical skills.
- **3D Visualization**: By bringing three-dimensional objects and ideas to life, AR and VR may improve comprehension and engagement. Students can investigate 3D models of cells, organs, or animals in biology classrooms, for instance, viewing them from various perspectives and engaging with their architecture. Students can visualize and modify 3D models of buildings or structures in architecture or engineering lessons. These visual aids help pupils better understand difficult ideas and increase their spatial awareness.
- **Immersive Language Learning Experiences**: It is possible with AR and VR that students may practice their language abilities by interacting with virtual native speakers in true-to-life situations. For instance, while using the target language, students can have discussions, place orders at a virtual restaurant, or go about a virtual city. By enhancing pronunciation, vocabulary, and cultural awareness, this method makes learning a language more enjoyable and useful.
- **Historical Reconstructions**: AR and VR can recreate historical events, giving students the chance to see and take part in important historical occasions. They can investigate historic civilizations from the past, engage in combat, or converse with virtual historical individuals. These reconstructions offer a more immersive and interesting way to study history, encouraging critical thinking, empathy, and a greater comprehension of historical circumstances.

While there are many advantages to using AR and VR in education, effective implementation calls for careful planning, teacher preparation, and access to the right technology and software (Bacca et al., 2014).

STUDENT MONITORING

The internet of things (IoT) education industry can improve student monitoring for attendance and exam evaluations by integrating IoT devices and technology. IoT can assist with student monitoring in the educational sector in the following ways:

- **Tracking Student Attendance**: It can be done using IoT devices like smart ID cards or wearable tech. When students enter a classroom or school building, these gadgets can be fitted with sensors or Radio Frequency Identification (RFID) tags that instantly identify their presence. Teachers and administrators can receive real-time attendance data from these devices via transmission to a centralized system.
- **Location Tracking**: IoT-enabled technologies can keep track of a student's whereabouts inside a school. To ensure that students are present in the proper places, sensors built into ID cards or smart gadgets, for instance, can track students as they move from one classroom to another. This promotes student safety and offers perceptions into their behavior and movement patterns.
- **Remote Test Proctoring**: It is possible using IoT devices and technology. IoT-enabled cameras and microphones can keep an eye on students while they take an online exam to make sure they are following the rules and to prevent cheating. These gadgets can also record audio and video, which may be replayed afterwards for evaluation.
- **Data Analytics**: The data collected from IoT devices can be analyzed to gain insights into student attendance patterns, behavior, and performance. This data can help identify trends and patterns, allowing educators to make data-driven decisions for improving student engagement and success. For instance, if students consistently have low attendance, appropriate interventions can be implemented to support their attendance and academic progress.
- **Security and Safety**: IoT devices can help ensure the safety of students on school grounds. IoT-based surveillance systems, for instance, may keep an eye on important locations and notify managers in the event of security breaches or other crises. Additionally, panic buttons and wearable gadgets with emergency alert functions can be enabled by IoT devices for students, enabling them to immediately inform authorities in the event of a safety risk.

EDUTECH-BASED IOT SOLUTIONS

It's important to note that while IoT offers significant benefits in student monitoring, privacy, and data security, considerations should be given utmost priority as shown as Figure 14.3. Proper consent and transparency should be maintained when collecting and using student data, ensuring compliance with relevant regulations and policies.

TEACHING

By delivering real-time data, interactivity, and individualized learning experiences, IoT can improve instructional techniques (Tripathi & Ahad, 2018). The following are some examples of IoT-based educational solutions:

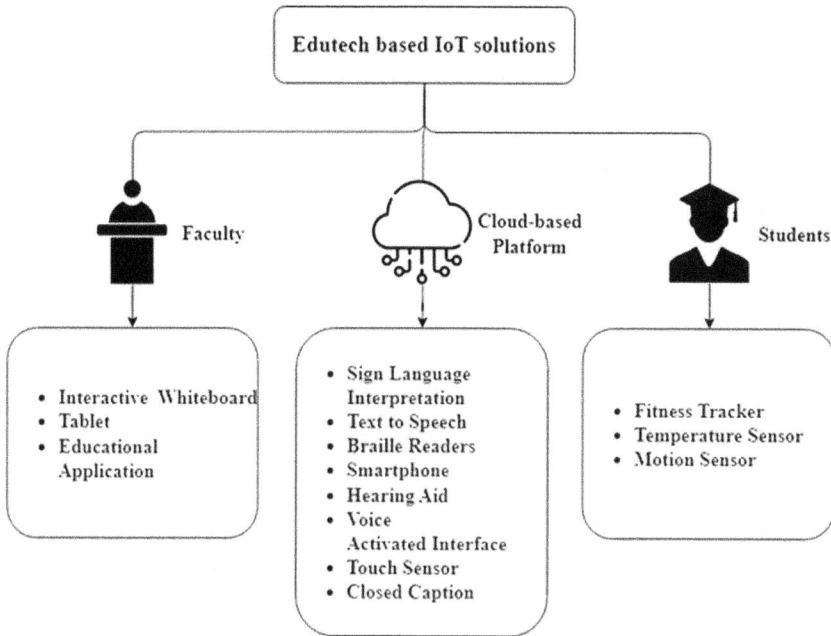

FIGURE 14.3   Edutech-based IoT solutions

- **Smart Classroom Management**: IoT sensors can be used to keep an eye on factors like temperature, humidity, and lighting in the classroom. For both students and teachers, this information can be used to create the ideal learning environment.
- **Interactive Learning Tools**: IoT-enabled tools like tablets, smart whiteboards, and projectors can support interactive learning activities. They can make it possible to distribute multimedia content, collaborate in real-time, and use adaptive learning.
- **Wearable Technology**: IoT-powered wearable technology can monitor pupils' stress levels, heart rates, and physical activity. With the use of this information, teachers can tailor their instruction and take appropriate action if necessary.

LEARNING

By offering individualized and immersive learning possibilities, IoT-based Edutech solutions can revolutionize the learning process (Liu et al., 2022). IoT sensors and beacons deployed in learning spaces can enable location-based learning, as an example:

- **Smart Learning Spaces**: Depending on where they are in the learning environment physically, students can access pertinent material, videos, or tests.

- **Integration of IoT Devices for Gamification**: IoT devices can be integrated with gamified platforms and VR simulations to produce interactive and interesting learning experiences. IoT sensors can monitor students' progress, provide incentives, and modify the course material in response to their performance.
- **Platforms for Adaptive Learning**: IoT can gather information about students' learning habits, preferences, and development. This information can be utilized to tailor learning routes, provide appropriate resources, and modify the curriculum to meet the needs of certain students.

EVALUATION

IoT-based EduTech options can improve and streamline the evaluation process while giving teachers access to real-time data on students' performance. IoT applications that can be evaluated include:

- **Automated Assessment**: By gathering and processing data, IoT devices can provide automated assessments. For instance, sensors in exam rooms might look for odd behavior or keep an eye on students' movements to ensure accuracy.
- **Learning Analytics**: Tools for learning analytics can be used to process data produced by IoT. These tools can give teachers insightful information about the learning styles, areas of strength and weakness, and need for intervention of their pupils.
- **Real-Time Feedback**: During the learning process, IoT devices can give students real-time feedback. For instance, smart pens can review handwriting, identify errors, and provide recommendations. Quizzes with IoT capabilities can offer fast feedback and explanations for wrong responses.

Overall, there are many opportunities to enhance teaching, learning, and evaluation, thanks to the integration of IoT with EduTech solutions. These approaches could increase efficacy, personalization, and engagement in educational environments.

GOVERNMENT AND MANAGEMENT INITIATIVES

GOVERNMENT AND MANAGEMENT INITIATIVES AT NATIONAL LEVEL

Assam's Online Career Guidance Portal

The government of Assam intends to implement IoT in education, notably via an online job counseling platform. This portal leverages the internet of things to offer tailored job advice based on student profiles, aptitude tests, and interests. IoT evaluations examine students' skills and reveal their strengths and limitations. Students may choose their careers more intelligently with the use of current labor market data. Remote counseling and mentorship sessions are made possible by IoT, and personalized career interventions are supported by data analysis. Information about employment roles are kept current, thanks to industry collaboration. Students in Assam may easily utilize the user-friendly site.

Jharkhand's DigiSAT

The DigiSAT (Digital Satellite) programs are a way for the administration and government of Jharkhand to make use of the possibilities of IoT in education. For efficient communication, data transmission, and remote monitoring, this entails integrating IoT technologies and deploying DigiSAT infrastructure. The dissemination of material, interactive sessions, and student participation are all made possible by IoT devices. DigiSAT employs satellite technology to bring educational content to off-the-grid locations with spotty internet connectivity.

DigiSAT's IoT implementation offers in-the-moment analytics, individualized learning, and monitoring. Remote inspections and assessments are also made possible by IoT devices. IoT hardware and software are used in teacher training programs to improve instructional methods. In educational institutions, efforts are undertaken to enhance infrastructure and internet connectivity. DigiSAT programs are part of Jharkhand's effort to close the digital gap and give everyone access to high-quality education.

Himachal Pradesh's Harghar Pathshala

To use IoT technology in education, the government of Himachal Pradesh has launched the Harghar Pathshala (School at Home in India) program. This program seeks to promote lifelong learning, particularly in remote and difficult-to-reach locations. The government prioritizes enhancing internet access, establishing hotspots, introducing IoT devices, and offering broadband connections. In order to access digital learning materials, take part in online classes, and interact with professors and classmates, students are given IoT-enabled devices including tablets, smartphones, and laptops.

IoT is used to create interactive learning platforms and applications that improve student engagement and facilitate effective distant learning. IoT devices enable real-time monitoring of student progress, engagement, and performance, enabling teachers to keep track of attendance, assess student progress, and offer prompt help. Digital textbooks, multimedia content, and other information are offered through IoT-based distribution systems. Online tests, too. Teachers are given instruction and assistance so they may use interactive platforms, IoT devices, and online teaching strategies. By giving parents up-to-date information on attendance, growth, and performance in real time, IoT technology is also utilized to engage parents in the educational process. The Harghar Pathshala program's integration of IoT seeks to close the digital divide, guarantee inclusive education, improve distant learning, encourage digital literacy, and provide top-notch instruction to all students in Himachal Pradesh.

Government and Management Initiatives in International Level

Future Schools Initiative in Ireland

The Future Schools Initiative in Ireland seeks to modernize teaching by using cutting-edge technologies. It emphasizes teacher preparation, infrastructural upgrades, and curriculum development. The project supports teachers' professional development, encourages interdisciplinary learning, and provides cutting-edge infrastructure. Smart classrooms are made possible by cutting-edge technology like AI, data analytics, and IoT that personalize education. Through partnerships with neighborhood organizations, students have access to practical applications. Inclusion, closing the digital gap, and assisting kids with special needs are given priority in the effort. The objective is to create an inclusive, cutting-edge educational system (Khang et al., 2024a).

Edutech Strategy in the United Kingdom

IoT is incorporated into education as part of the UK's Edutech Strategy to enhance teaching and learning. Using IoT devices, it gathers real-time data on student engagement and circumstances. Based on this information, educators adjust the courses and create adaptable learning paths. Monitoring of the environment, resource management, safety, and security are also improved by IoT (Department for Education, 2019). Data analytics offer information for making decisions and boosting academic standards. Inclusion, reducing the digital gap, and assisting kids with special needs are given priority by the effort. The goal is to build an educational system that is prepared for the future (Kali et al., 2024).

EXAMPLES OF IOT USED IN EDUCATION

The learning experience is transformed by integrating IoT gadgets in education. Both teachers and students can benefit from these cutting-edge technological advancements. Learning may be interactive and customized, thanks to IoT gadgets like smartboards, wearables, and sensors. For educators, real-time data analytics offer insightful information that improves student engagement and academic results (Nassereddine and Khang, 2024).

KAJEET SMART BUS

The Kajeet Smart Bus (India) system seeks to improve student safety while also reducing distractions and providing dependable internet access on school buses. It comprises a router with Wi-Fi and 5G connection as well as Kajeet's Sentinel cloud platform (India). Using safe educational websites, students may browse the internet and accomplish their assignments. The technology helps managers manage devices, track buses in real time, and create reports on speed limit infractions. It lessens disruptive student behavior, permits internet work while traveling, and enhances driver concentration. Filters that are the Children's Internet Protection Act (CIPA) compliant restrict offensive websites and social media, allowing students to concentrate on their coursework (Khang, 2024c).

C-PEN

Pen scanners are compact linked devices that are becoming increasingly popular because of how much they help streamline learning. A variety of wireless pen scanners made by C-Pen are intended to quickly scan text and transmit it through Bluetooth to a desktop program or a smartphone so that students can work with it. By scanning a text, translating it, and turning it into an audio file, students can use this tool to learn independently without the help of an instructor. Children can learn to study independently without needing assistance from their parents or teachers, thanks to this concrete internet of things example for school children. C-Pen, which is driven by end-user apps, enhances how written information is processed, making it simpler to read and absorb educational materials, learn new languages, and do a lot more.

IPEVO VZ-X WIRELESS DOCUMENT CAMERA

IPEVO has discovered a method that can enhance remote learning and assist in resolving significant issues brought on by the epidemic. In order to assist teachers in streaming lessons and lectures remotely, capturing and visualizing educational materials, connecting the camera to multiple devices, and modifying the camera's video feed, the company created the wireless document camera VZ-X (VZ-X Wireless HDMI USB 8MP Document Camera-Overview | IPEVO, 2021). Teachers can wirelessly attach the camera to their PCs, iPads, and even smartphones. The package also includes Visualizer software, which aids in controlling document cameras, in addition to the device. Teachers may handle features like zooming, stop motion, video filters, and a lot more with the Visualizer program. Additionally, the camera works with other programs like Seesaw, Camtasia, and Open Broadcaster Software (OBS) Studio. The tool supports teachers as they navigate the shift from traditional methods of teaching to modern and innovative student-relevant technologies.

IOT TOOLS USED IN THE EDUCATION SECTOR

SCHOOL LEVEL

Through the creation of interactive and individualized learning environments, the use of IoT tools in education is revolutionizing the classroom. Using IoT-enabled devices, students may access instructional content, communicate with classmates, and get real-time feedback. This innovative method improves student collaboration, information acquisition, and engagement while preparing them for a digital future. By implementing IoT solutions, educators may cater to the requirements of each student, promote autonomous learning, and promote instructional creativity. IoT solutions enable instructors and students to adopt a learner-centered approach, resulting in a dynamic and interesting educational environment.

Primary Education

IoT tools can be used in elementary education to design interactive, fun learning experiences for young students. In this context, some of the IoT tools employed include:

Smart Toys: By combining sensors and connection, IoT-enabled toys can offer instructional value. These toys can encourage learning through play by involving kids in interactive, immersive experiences (Komis et al., 2021). For instance, smart construction blocks with integrated sensors can impart fundamental programming ideas or foster problem-solving abilities.

Virtual reality (VR) and augmented reality (AR) technology provide young learners with immersive and hands-on learning possibilities. Students can explore virtual worlds, visit historical locations, or interact with 3D objects using VR headgear or AR apps, which improves their comprehension and engagement with a variety of subjects.

Secondary Education

IoT tools can be used in secondary school to help in the development of more sophisticated skills and knowledge. IoT tools in secondary education include, for instance:

Smart Classrooms: An interactive and connected learning environment can be provided by classrooms with IoT capabilities. Collaborative learning, access to digital resources, and real-time feedback are all made possible through smart boards, interactive displays, and connected gadgets. Students get access to multimedia materials, can take part in online debates, and can participate in interactive activities.

IoT devices can be utilized for data collection and analysis in fields like science and mathematics. IoT sensors can be used in experiments by students to gather real-time data that can later be analyzed and interpreted. This practical technique improves comprehension and critical thinking abilities (Khanh and Khang, 2021).

Higher Secondary Education

Personalized and dynamic learning environments are being created in the classroom, thanks to the use of IoT solutions in education. With the aid of IoT-enabled devices, students may access instructional materials, work with others, and get immediate feedback. As they become ready for a digital future, this innovative strategy improves student engagement, teamwork, and knowledge acquisition (Aldowah et al., 2017). By implementing IoT solutions, educators can accommodate each student's unique learning needs, promote autonomous learning, and promote instructional creativity. IoT technologies, in general, enable students and teachers to adopt a learner-centered approach, therefore forming a dynamic and interesting educational experience.

UNIVERSITY LEVEL

IoT has several applications in the educational sector, although its direct usage with the STEAM platform in particular may be restricted. IoT devices, on the other hand, may increase practical experiences and offer real-world applications in the context of education and STEAM (Science, Technology, Engineering, Arts, and Mathematics) learning. Here are a few illustrations:

- **IoT Experiments**: Students use IoT gadgets for data collection and analysis in various scientific disciplines like environmental monitoring.

- **Internet-Connected Devices**: Students combine IoT devices like Arduino or Raspberry Pi with STEAM projects for real-world control and programming.
- **VR/AR Integration**: IoT and virtual/augmented reality create immersive learning experiences where students interact with IoT sensors in virtual environments.
- **Data Gathering**: IoT devices help students collect data for science experiments, such as environmental sensors for temperature or air quality measurements.
- **Interactive Learning**: Teachers incorporate IoT devices into STEAM classes, allowing students to design and program projects using hardware like Arduino.
- **IoT and VR Integration**: Students use IoT-connected VR devices to explore virtual environments and interact with IoT sensors for hands-on learning.
- **IoT and Gamification**: IoT-enabled game controllers enhance engagement by allowing students to manipulate physical objects and perform virtual experiments.
- **IoT-Enabled Collaborations**: Students use IoT devices and the STEAM platform for remote collaboration, exchanging data and creating solutions together.

IoT in STEAM education offers hands-on activities, data analysis, and innovative interactions to enhance learning across STEAM disciplines (Khang, 2024b).

Special Schools and Universities

Sectors like agriculture, healthcare, and education have seen significant improvements due to the development of intelligent technologies. The use of wearable technology to track physiology and physical health has grown. Innovative classroom technologies are intended to improve interaction and immersion in educational materials. Difficulties hamper their widespread adoption in institutions in their execution. This essay will look at the elements of smart classrooms, assess their benefits and drawbacks, and gauge how well they work to get the desired results. Additionally, the future ramifications of IoT technologies and their role in influencing the construction of smart classrooms are examined.

The use of intelligent educational technologies and highly efficient and effective learning management systems is crucial due to the rapid growth in educational needs. In this article, we proposed a synergistic framework by fusing the core components of smart technology with a flow theory approach that can induce overall learning efficacy with the added benefit of maintaining the smart learning and management environment. This was done in consideration of the benefits of classical flow theory. The provided framework incorporates a wide range of technical and communication tools to energize the learning process and modify or sustain the needs of various pupils under one roof. The suggested framework can assist in preventing discrimination and give low-level/low-grade pupils more confidence in a supportive learning environment. Additionally, regular monitoring and analysis of the student's performance using information processing platforms based on the internet of things (IoT) can significantly increase learning efficacy and quality. However, building IoT applications for students of various levels in the same class is a complex problem that requires rigorous standardization. This effort seeks to support users in developing skills, adaptation, and use of technology in a learning context to achieve successful learning outcomes, which leads to big data, within this framework (Rani et al., 2021).

Connected Security Systems

Security and safety at educational institutions have been revolutionized by IoT devices. Real-time monitoring, access management, and emergency response are all made possible by connected security systems. Surveillance cameras, motion detectors, and IoT-enabled hardware allow for remote monitoring and rapid emergency response. Security is improved through centralized access control and IoT-based door locks. With the use of linked wearable technologies and panic buttons, networked security systems accelerate coordination and reaction times in emergency situations. IoT-based security solutions provide improved administration and safety in schools overall.

Access control, remote administration, emergency response capabilities, real-time monitoring, and data-driven insights are some of the ways that linked security systems powered by the internet of things (IoT) improve safety in educational institutions. These innovations make workplaces safer for visitors, staff, and students, promoting a positive learning environment.

Virtual Reality and Augmented Reality Devices

Virtual reality (VR) and augmented reality (AR) devices have made significant contributions to the education sector, transforming the way students learn and engage with educational content (Liyuan, 2020). Here are some ways in which VR and AR devices have helped:

- **Immersive Learning**: VR devices create engaging virtual worlds, offering hands-on experiences and deeper understanding of concepts.
- **Visualizing Abstract Concepts**: VR and AR help students visualize complex ideas in a tangible and interactive way.
- **Virtual Field Trips**: Students can explore distant locations and cultural sites through virtual experiences.
- **Simulation-Based Training**: VR and AR provide safe environments for practical training and skill development.
- **Personalized Learning**: Adaptive content in VR/AR can be tailored to individual students' needs and learning styles.
- **Collaboration and Interactivity**: VR/AR devices foster teamwork, problem-solving, and communication skills.

- **Accessibility and Inclusivity**: VR/AR promote equal opportunities for students with disabilities or limitations.
- **Engagement and Motivation**: Immersive experiences in VR/AR enhance student engagement and curiosity.

In summary, VR and AR devices have revolutionized education by providing immersive and interactive learning experiences, visualizing abstract concepts, facilitating virtual field trips, enabling simulation-based training, supporting personalized learning, fostering collaboration, and promoting inclusivity (Lee, 2012). As these technologies continue to advance, they have the potential to enhance education further and transform the way students acquire knowledge and skills.

STRENGTHS, WEAKNESSES, OPPORTUNITIES, AND CHALLENGES OF EDUCATION SECTOR

Stakeholders in the education sector involved in the implementation and adoption of IoT face various strengths, weaknesses, opportunities, and challenges. The integration of IoT in education offers several advantages. It enables immersive and interactive learning experiences that promote student engagement and improve information retention. IoT gadgets and analytics provide personalized education, adapting the learning process to individual student needs and enhancing academic outcomes. IoT solutions also streamline administrative tasks, such as resource management, security, and attendance tracking, leading to more efficient campus management (Figure 14.4).

RESEARCH ISSUES, SOLUTIONS, AND OPPORTUNITIES

IoT applications in the educational space offer both challenges and opportunities for research. Sensitive student data is collected and transmitted through IoT devices, raising privacy and security concerns that call for strict privacy policies, encryption mechanisms, and access control (Mohammadian, 2019). The adoption of IoT solutions is also hampered by infrastructure and connectivity issues, particularly in distant locations, necessitating cooperation with technology providers and governmental organizations to close the gap. Additionally, the cost of IoT deployment is a

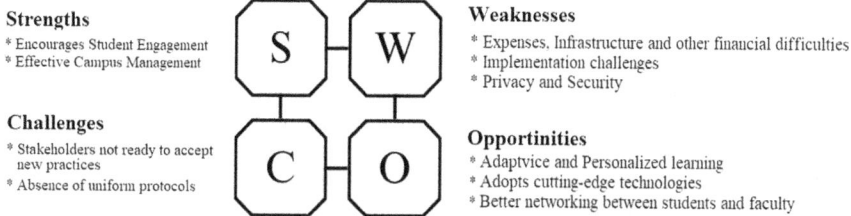

Strengths
* Encourages Student Engagement
* Effective Campus Management

Weaknesses
* Expenses, Infrastructure and other financial difficulties
* Implementation challenges
* Privacy and Security

Challenges
* Stakeholders not ready to accept new practices
* Absence of uniform protocols

Opportinities
* Adaptvice and Personalized learning
* Adopts cutting-edge technologies
* Better networking between students and faculty

FIGURE 14.4   Strengths, weaknesses, challenges, and opportunities faced by stakeholders

major obstacle, thus finding affordable solutions and funding options is essential to making IoT available to educational institutions.

Despite these difficulties, there are lots of prospects for IoT in education. By delivering personalized and immersive settings via linked devices and sensors, it improves learning experiences by enabling real-time feedback and collaborative learning. Another advantage is effective resource management, as IoT-powered solutions optimize asset tracking and energy use, cutting costs, and fostering sustainability. Additionally, IoT device data enables data-driven decision-making, giving educators the ability to modify lesson plans and student support services based on insights and trends. Overcoming infrastructure, cost, privacy, and security issues is essential for a successful IoT implementation in the education sector. Educational institutions can improve learning processes, better manage resources, and make data-driven decisions by utilizing IoT, ultimately resulting in a more effective and individualized educational system (Rani et al., 2023).

CONCLUSION

To sum up, IoT applications in the education sector have the power to transform established teaching methods and influence how people learn in the future. The opportunities provided by IoT are substantial, despite the fact that issues like privacy, security, infrastructure, and cost need to be addressed. IoT delivers personalized and immersive learning experiences, improves resource management, and enables data-driven decision-making by integrating smart devices, sensors, and connections. IoT enables educational institutions to make data-driven decisions and optimize resource consumption, while providing students with engaging, adaptive learning environments (Khang, 2024d).

It is crucial for educators, decision-makers, and stakeholders to embrace the opportunities presented by IoT in education as technology develops. We can design a more effective, interesting, and responsive educational system by overcoming the obstacles and taking advantage of the opportunities. Ultimately, teamwork, innovation, and dedication to giving students the finest learning experiences are necessary for the effective application of IoT in the education sector. IoT has the potential to alter education and open up new opportunities for both students and educators with the appropriate approach (Khang & Vrushank et al., 2023).

REFERENCES

Arora, J. B., & Kaushik, S. (2020). IoT in Education: A Future of Sustainable Learning. In R. Singh, A. Gehlot, V. Jain, & P. Malik (Eds.), *Handbook of Research on the Internet of Things Applications in Robotics and Automation* (pp. 300–317). IGIGlobal. https://doi.org/10.4018/978-1-5225-9574-8.ch015

Aldowah, H., Ul Rehman, S., Ghazal, S., & Naufal Umar, I. (2017). Internet of Things in Higher Education: A Study on Future Learning. *Journal of Physics: Conference Series*, 892, 012017. https://doi.org/10.1088/1742-6596/892/1/012017

Bacca, J., Baldiris, S., Fabregat, R., Graf, S., & Kinshuk. (2014). Augmented Reality Trends in Education: A Systematic Review of Research and Applications. *Journal of Educational Technology & Society*, 17(4), 133–149. https://www.jstor.org/stable/jeductechsoci.17.4.133

Burunkaya, M., & Duraklar, K. (2022). Design and Implementation of an IoT-Based Smart Classroom Incubator. *Applied Sciences*, 12(4), 2233. https://doi.org/10.3390/app12042233

Department for Education. (2019, April 3). Realising the Potential of Technology in Education. GOV.UK. https://www.gov.uk/government/publications/realising-the-potential-of-technology-in-ed ucation

Fetaji, B., Fetaji, M., Asilkan, O., & Ebibi, M. (2020). Examining the Role of Virtual Reality and Augmented Reality Technologies in Education. *New Trends and Issues Proceedings on Humanities and Social Sciences*, 7(3), 160–168. https://doi.org/10.18844/prosoc.v7i3.5247

Hong-tan, L., Cui-hua, K., Muthu, B., & Sivaparthipan, C. B. (2021). Big Data and Ambient Intelligence in IoT-Based Wireless Student Health Monitoring System. *Aggression and Violent Behavior*, 101601. https://doi.org/10.1016/j.avb.2021.101601

Khang, A., Gupta, S. K., Shah, V., & Misra, A. (Eds.). (2023a). *AI-aided IoT Technologies and Applications in the Smart Business and Production*. CRC Press. https://doi.org/10.1201/9781003392224

Khang, A., Gupta, S. K., Hajimahmud, V. A., Babasaheb, J., & Morris, G. (2023b). *AI-Centric Modelling and Analytics: Concepts, Designs, Technologies, and Applications* (1st Ed.). CRC Press. https://doi.org/10.1201/9781003400110

Khang, A., Rana, G., Tailor, R. K., & Hajimahmud, V. A., (Eds.). (2023). *Data-Centric AI Solutions and Emerging Technologies in the Healthcare Ecosystem*. CRC Press. https://doi.org/10.1201/9781003356189

Khang, A., Hahanov, V., Litvinova, E., Chumachenko, S., Triwiyanto, H. V. A., Ali, R. N., Alyar, A. V., & Anh, P. T. N. (2023). The Analytics of Hospitality of Hospitals in Healthcare Ecosystem. *Data-Centric AI Solutions and Emerging Technologies in the Healthcare Ecosystem*. P (4). CRC Press. https://doi.org/10.1201/9781003356189-4

Khang, A., Abdullayev, V., Hahanov, V., & Shah, V. (2024a). *Advanced IoT Technologies and Applications in the Industry 4.0 Digital Economy* (1st Ed.). CRC Press. https://doi.org/10.1201/9781003434269

Khang, A. (Eds.). (2024b). *AI and IoT-Based Technologies for Precision Medicine*. IGI Global Press. https://doi.org/10.4018/979-8-3693-0876-9

Khang, A. (Eds.). (2024c). *AI-Oriented Competency Framework for Talent Management in the Digital Economy: Models, Technologies, Applications, and Implementation*. CRC Press. https://doi.org/10.1201/9781003440901

Khang, A., Vrushank, S., & Rani, S. (2023). *AI-Based Technologies and Applications in the Era of the Metaverse* (1st Ed.). IGI Global Press. https://doi.org/10.4018/978-1-6684-8851-5

Khang, A. (Eds.). (2024d). *AI-Oriented Competency Framework for Talent Management in the Digital Economy: Models, Technologies, Applications, and Implementation*. CRC Press. https://doi.org/10.1201/9781003440901

Khanh, H. H., & Khang, A. (2021). The Role of Artificial Intelligence in Blockchain Applications. *Reinventing Manufacturing and Business Processes through Artificial Intelligence*, 20–40. CRC Press. https://doi.org/10.1201/9781003145011-2

Komis, V., Karachristos, C., Mourta, D., Sgoura, K., Misirli, A., & Jaillet, A. (2021). Smart Toys in Early Childhood and Primary Education: A Systematic Review of Technological and Educational Affordances. *Applied Sciences*, 11(18), 8653. https://doi.org/10.3390/app11188653

Lee, K. (2012). Augmented Reality in Education and Training. *TechTrends*, 56(2), 13–21. https://doi.org/10.1007/s11528-012-0559-3

Liu, Z., Ren, Y., Kong, X., & Liu, S. (2022). Learning Analytics Based on Wearable Devices: A Systematic Literature Review From 2011 to 2021. *Journal of Educational Computing Research*, 073563312110647. https://doi.org/10.1177/07356331211064780

Liyuan, L. (2020). The Application of Virtual Reality and Augmented Reality Technology in the Field of Education. *Journal of Physics: Conference Series*, 1684, 012109. https://doi .org/10.1088/1742-6596/1684/1/012109

Madyatmadja, E. D., Yulia, T. R., Sembiring, D. J. M., & Angin, S. M. B. P. (2021). IoT Usage on Smart Campus: A Systematic Literature Review. *International Journal of Emerging Technology and Advanced Engineering*, 11(5), 45–52. https://doi.org/10.46338/ije-tae0521_06

Mathew K V. B., & Walarine, M. T. (2018). Need of Technical Educational Integration in Disability Sector for Differently-abled Empowerment. *International Journal of Engineering and Management Research*, 8(4). https://doi.org/10.31033/ijemr.v8i4.13247

Mohammadian, H. D. (2019, April 1). *IoT – A Solution for Educational Management Challenges*. IEEE Xplore. https://doi.org/10.1109/EDUCON.2019.8725213

Nassereddine, M., & Khang, A. (2024). Applications of Internet of Things (IoT) in Smart Cities. *Advanced IoT Technologies and Applications in the Industry 4.0 Digital Economy* (1st Ed.). CRC Press. https://doi.org/10.1201/9781003434269-6

Ramlowat, D. D., & Pattanayak, B. K. (2019). Exploring the Internet of Things (IoT) in Education: A Review. *Advances in Intelligent Systems and Computing*, 245–255. https://doi.org/10.1007/978-981-13-3338-5_23

Rana, G., Khang, A., Sharma, R., Goel, A. K., & Dubey, A. K, (Eds.). (2021). *Reinventing Manufacturing and Business Processes through Artificial Intelligence*. CRC Press. https://doi.org/10.1201/9781003145011

Rani, S., Chauhan, M., Kataria, A., & Khang A. (Eds.). (2021). IoT Equipped Intelligent Distributed Framework for Smart Healthcare Systems. *Networking and Internet Architecture*, 2, 30. https://doi.org/10.48550/arXiv.2110.04997

Rani, S., Bhambri, P., Kataria, A., Khang, A., & Sivaraman, A. K. (2023). *Big Data, Cloud Computing and IoT: Tools and Applications* (1st Ed.). Chapman and Hall/CRC. https://doi.org/10.1201/9781003298335

Rath, K. C., Khang, A., & Roy, D. (2024). The Role of Internet of Things (IoT) Technology in Industry 4.0. *Advanced IoT Technologies and Applications in the Industry 4.0 Digital Economy* (1st Ed.). CRC Press. https://doi.org/10.1201/9781003434269-1

Spyrou, E., Vretos, N., Pomazanskyi, A., Stylianos, A., & Leligou, H. C. (2018). Exploiting IoT Technologies for Personalized Learning. https://doi.org/10.1109/cig.2018.8490454

Tripathi, G., & Ahad, M. A. (2018). IoT in Education: An Integration of Educator Community to Promote Holistic Teaching and Learning. *Soft Computing in Data Analytics*, 675–683. https://doi.org/10.1007/978-981-13-0514-6_64

VZ-X Wireless HDMI USB 8MP Document Camera- Overview | IPEVO. (2021). IPEVO (United States). https://www.ipevo.com/products/vz-x

15 Internet of Forestry Things (IoFT) Technologies and Applications in Forest Management

Rufai Yusuf Zakari, Wasswa Shafik, Kassim Kalinaki, and Chima Jude Iheaturu

INTRODUCTION

Forests are invaluable resources that are crucial in maintaining our planet's and human societies' well-being. They offer many benefits, including climate regulation, water cycle regulation, biodiversity conservation, and providing essential resources such as food, fiber, and fuel (Sahal et al., 2021). However, forests face numerous challenges, such as deforestation, forest degradation, and the adverse impacts of climate change, which pose significant threats to their long-term sustainability (Bo and Wang, 2011). Addressing these challenges requires innovative solutions, and one such solution is the application of internet of things (IoT) technologies in forest management, also known as the internet of forestry things (IoFT) (Wang et al., 2010).

The IoT refers to a network of interconnected physical devices embedded with sensors, software, and other technologies that enable them to collect and exchange data. This network facilitates the seamless integration of the physical and digital worlds, allowing for real-time monitoring, control, and analysis of various systems and environments. In the context of forests, the IoFT harnesses IoT technologies to revolutionize how these complex ecosystems are managed.

The importance of IoFT in forest management cannot be overstated (Lal et al., 2020). By deploying IoT devices throughout forested areas, real-time data on crucial environmental parameters, such as temperature, humidity, and soil moisture, can be collected and transmitted for analysis (Fang, 2022). This data empowers forest managers and conservationists with valuable insights into the health and status of forest ecosystems, enabling them to make informed decisions and take timely actions (Hazards, 2022). The IoFT also provides information on the location and movement

DOI: 10.1201/9781003434269-15

of forest assets, including wildlife, carbon stock, personnel, and equipment, which further enhances the understanding and management of forests (Lakhwani et al., 2019).

In addition to improving the efficiency and effectiveness of forest management, the IoFT offers significant benefits regarding the overall significance and sustainability of forest ecosystems. By continuously monitoring and analyzing forest conditions, the IoFT can contribute to the early detection and prevention of environmental threats, such as forest fires and illegal logging activities (Forestry Fire, 2022). This proactive approach enables prompt interventions, mitigating potential damages and protecting forest resources. Furthermore, the IoFT can optimize forest operations, such as harvesting and transportation, leading to resource utilization and economic benefits while minimizing negative environmental impacts (Cao et al., 2022).

Despite the immense potential of IoFT technologies, some challenges and limitations need to be addressed. Data privacy and security are paramount concerns, as the collection and transmission of sensitive information require robust safeguards to prevent unauthorized access or misuse (Shafik et al., 2023a). Interoperability, the ability of different devices and systems to work seamlessly together, is another challenge that needs to be overcome to ensure the effective integration and utilization of IoFT technologies in forest management (Lakhwani et al., 2019).

This study aims to provide an overview of the current state of IoFT technologies and their applications in forest management. By examining various studies, reports, and articles, this review sheds light on the diverse applications of IoFT technologies, including forest health monitoring, forest fire prediction and detection, and detecting illegal logging activities. Furthermore, this study explores how IoFT technologies can optimize forest operations and improve the safety of forest workers. However, it also critically discusses the challenges and limitations of IoFT technologies, such as data privacy, security, and interoperability (Khang et al., 2023c).

The remainder of this chapter is arranged into seven sections as follows. The second section presents the IoFT technologies and sensors, including remote sensing, wireless sensor networks (WSNs), mobile devices, drones and unmanned aerial vehicles (UAVs), and satellite imagery. The third section presents forest monitoring, fire detection and prevention, wildlife tracking, forest inventory and resource management, forest health monitoring, forest operations optimization, ecosystem conservation, and sustainable forest certification. The fourth section presents the challenges and limitations of IoFT in forest management: connectivity and infrastructure, data management and analysis, security and privacy, diverse forest ecosystem adaptation, energy efficiency and power supply, and cost. The fifth section demonstrates real-world implementations and case studies of IoFT, IoT sensors in Alice Holt Forest, the Forest Inventory and Analysis (FIA) program in the USA, forest inventory and management in British Columbia, Canada, and soil moisture monitoring in the Amazon Rainforest, Brazil. Therefore, case studies include harnessing integrated communication technologies for Amazonian environmental monitoring, enhancing surveillance of unauthorized tree felling with energy-efficient smart IoT devices, forest environmental monitoring systems using IoT, and solar-powered IoT-based peat swamp forest environmental monitoring systems in Brunei. The sixth section avails future developments in IoFT technologies and their applications, including enhanced forest monitoring and ecosystem modeling. The seventh section presents the integration of AI, enhanced data privacy and security, interoperability, standardization, real-time decision support systems, stakeholder engagement and capacity building, long-term monitoring and adaptive management, and collaboration and knowledge sharing. Finally, the last section avails to the conclusion.

IOFT TECHNOLOGIES AND SENSORS

IoFT technologies encompass a range of advanced tools and sensors that enable comprehensive monitoring and management of forest ecosystems. Some of the key technologies and sensors used in IoFT applications are as follows.

REMOTE SENSING

Remote sensing involves using satellites or aircraft-mounted sensors to capture data about the Earth's surface without direct physical contact. In forest management, remote sensing allows for collecting valuable information about forest cover, vegetation health, land-use changes, and other relevant parameters (Forestry Fire, 2022). This data provides a broader perspective on forest conditions and dynamics over large areas, facilitating effective decision-making and resource allocation.

WIRELESS SENSOR NETWORKS (WSNs)

WSNs consist of numerous interconnected sensors deployed throughout forested areas. These sensors can collect data on various environmental parameters, such as temperature, humidity, soil moisture, and air quality (Zakari et al., 2022). WSNs enable real-time monitoring and data transmission, allowing forest managers to obtain accurate and up-to-date information on forest conditions (Fang, 2022). This data is vital for understanding ecosystem health, identifying potential risks, and implementing timely interventions (Vrushank et al., 2023).

MOBILE DEVICES

Mobile devices, such as smartphones and tablets, equipped with specialized applications and sensors, play a significant role in IoFT (Bayne et al., 2017). Forest managers and workers can use these devices to collect and transmit data in the field, enhancing data collection efficiency and reducing manual effort. Mobile devices enable real-time communication and collaboration among forest management teams (Chen et al., 2022), facilitating better coordination and decision-making.

DRONES AND UNMANNED AERIAL VEHICLES (UAVs)

Drones and UAVs offer a versatile and efficient means of collecting high-resolution imagery and data over forested areas (Bayne et al., 2017). With advanced cameras,

Light Detection and Ranging (LiDAR) sensors, and other specialized instruments, these aerial platforms can capture detailed information about forest structure, canopy cover, topography, and vegetation density. Drones and UAVs enable rapid data acquisition, making them valuable tools for forest inventory, monitoring, and mapping applications (Shafik et al., 2023b).

SATELLITE IMAGERY

Satellite imagery provides a wide-scale and consistent view of forested regions. High-resolution satellite images can offer valuable insights into forest cover changes, deforestation patterns, and landscape dynamics over time (Cao et al., 2022). Satellite imagery is beneficial for monitoring large-scale forest ecosystems, identifying land-use changes, and detecting forest fires or illegal activities. Combining satellite imagery with other IoFT technologies enhances the accuracy and efficiency of forest management practices.

APPLICATIONS OF IOFT IN FOREST MANAGEMENT

IoT has revolutionized various industries, and forest management is no exception. Within this section, this integration of IoT in forest management is demonstrated in several applications presented below.

FOREST MONITORING

IoT aids in forest monitoring by utilizing sensors, drones, and satellite systems to collect real-time data on temperature, humidity, air quality, and more. This data enables forest managers to detect forest fires, disease outbreaks, or illegal logging activities promptly, allowing for a quick response and minimizing damage (Li et al., 2022). The positive impact of IoT in forest monitoring is evident in protecting valuable forest resources, preserving biodiversity, and promoting ecosystem health (Hirawan and Mahendra, 2020). By enabling timely intervention, the IoT contributes to reducing the loss of forest cover, ensuring the safety of surrounding communities, and fostering sustainable forestry practices that benefit society.

FIRE DETECTION AND PREVENTION

These devices support fire detection and prevention by employing connected devices and systems to detect forest smoke or temperature changes. This enables early identification of forest fires, triggering rapid response and intervention. The positive impact of IoT on fire detection and prevention is significant for society as it helps prevent the spread of fires by reducing the loss of forest cover, minimizing ecological damage, and safeguarding wildlife (Gaitan and Hojbota, 2020). By preserving valuable forest resources and ensuring the safety of surrounding communities, IoT contributes to the resilience of ecosystems, protects property and lives, and supports sustainable forest management practices.

WILDLIFE TRACKING

IoFTs utilize global positioning system-assisted collars and tags attached to animals, allowing researchers and forest managers to monitor their movements, migration patterns, and behavior (Oliveira et al., 2021). This data provides valuable insights into wildlife habitats, supports conservation efforts, and helps mitigate human–wildlife conflicts. The positive impact of IoT in wildlife tracking is evident in preserving biodiversity and promoting ecological balance (Zhang et al., 2021). By understanding wildlife dynamics, IoT enables evidence-based conservation strategies, protects endangered species, and fosters harmonious coexistence between humans and wildlife. This benefits society by maintaining healthy ecosystems, supporting ecotourism, and ensuring the long-term sustainability of wildlife populations.

FOREST INVENTORY AND RESOURCE MANAGEMENT

IoT helps forest inventory and resource management by using connected devices and sensors to accurately measure tree dimensions, identify species, and estimate biomass. This data enables forest managers to make informed decisions regarding timber harvesting, ensure sustainable practices, and prevent overexploitation (Sihombing et al., 2021). The positive impact of IoT on forest inventory and resource management is evident in optimized timber harvesting planning, increased operational efficiency, and reduced environmental impact. By promoting responsible resource management, IoT contributes to the long-term sustainability of forest ecosystems, preserves biodiversity, and supports the fair and efficient use of forest resources for societal benefits (Liu et al., 2021).

FOREST HEALTH MONITORING

IoFT devices and sensors play a crucial role in assessing and maintaining the health of forest ecosystems through forest health monitoring. By providing real-time data on factors such as insect infestations, disease outbreaks, and environmental stressors, IoFT aids in the early detection of forest health issues (Hu et al., 2022). This enables forest managers to implement timely interventions, prevent widespread damage, and maintain forests' long-term productivity and resilience. Furthermore, IoFT-enabled monitoring supports evidence-based decision-making, leading to targeted management strategies that preserve biodiversity, ecosystem services, and the vital role of forests in climate change mitigation (Wong and Chu, 2021). Ultimately, by ensuring the health of forests, IoT-driven forest health monitoring positively impacts society by safeguarding valuable natural resources, supporting ecological balance, and providing essential ecosystem services (Kavita, 2023).

FOREST OPERATIONS OPTIMIZATION

IoT technologies facilitate forest operations optimization, which entails the integration of connected devices and data analytics to improve the effectiveness, productivity, and environmental sustainability of forest management activities (Liu et al.,

2020). By providing real-time monitoring and analysis of crucial parameters, for example, timber movement, equipment performance, and environmental conditions, the IoT enabled timely decision-making and optimization of logistics, leading to cost savings and reduced environmental impacts. Furthermore, IoT allows predictive maintenance of equipment, automation of tasks, and remote-control capabilities, improving worker safety, productivity, and resource allocation (Liu et al., 2021). Therefore, IoT-driven forest operations optimization positively impacts society by maximizing operational efficiency, minimizing costs, and promoting responsible and sustainable forest management practices.

Ecosystem Conservation

Ecosystem conservation is a critical endeavor that focuses on the preservation and sustainable management of natural environments, biodiversity, and ecological processes. It involves protecting and restoring ecosystems to maintain their integrity, functionality, and resilience (Kurniabudi et al., 2022). Ecosystem conservation aims to safeguard the myriad of species and habitats that depend on these ecosystems, sustain essential ecosystem services such as clean air and water, regulate climate patterns, and provide recreational and cultural values to society (Qi and Gao, 2022). By conserving biodiversity, conserving habitats, and mitigating human impacts, ecosystem conservation ensures the long-term survival of diverse species, promotes ecological balance, supports sustainable resource use, and safeguards the well-being of both present and future generations.

Sustainable Forest Certification

IoFTs further play a significant role in supporting sustainable forest certification by providing real-time data and monitoring capabilities that ensure compliance with certification standards (Vikash and Varma, 2022). This data enhances the certification process's transparency, accuracy, and traceability, providing stakeholders with reliable information on sustainable forest management practices. The positive impact of IoT on sustainable forest certification includes improved forest governance, increased market demand for certified products, and enhanced conservation efforts (Kaur et al., 2019). By bolstering credibility and accountability, the IoT contributes to preserving forests, biodiversity, and ecosystem services while empowering consumers to make informed choices and promoting responsible forest management practices that benefit society and the environment (Figure 15.1).

CHALLENGES AND LIMITATIONS OF IOFT IN FOREST MANAGEMENT

While IoFTs have demonstrated the capacity to enhance forest management, several challenges and limitations must be considered. Within this section, these notable challenges of IoFTs are explained.

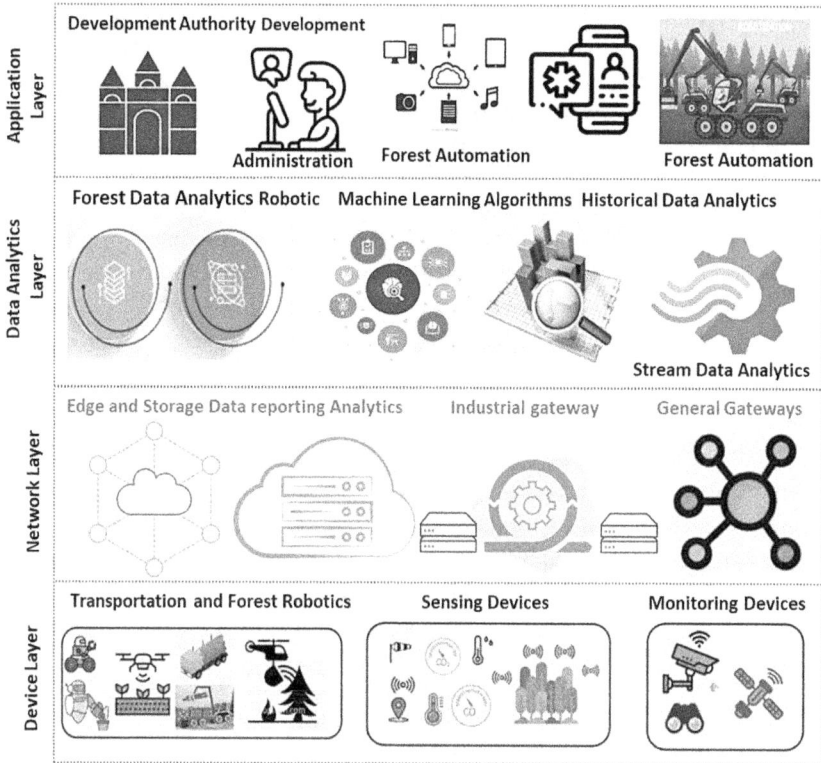

FIGURE 15.1 Application of IoFTs according to the layers

Connectivity and Infrastructure

Connectivity and infrastructure pose significant challenges and limitations for IoFTs in forest management. Forest areas often need more reliable internet connectivity, making it difficult to establish seamless communication and data transfer between IoT devices (Sahal et al., 2021). Remote locations with limited network coverage hinder the deployment and operation of IoT systems. Additionally, installing necessary infrastructure, such as network connectivity and power sources, can be challenging and costly in remote forest environments (Singh et al., 2022). Overcoming these limitations requires innovative solutions, such as satellite-based connectivity or developing low-power, long-range communication technologies, to ensure that IoFT can effectively operate in diverse forest ecosystems and provide real-time data for informed decision-making (Deng et al., 2023).

Data Management and Analysis

The vast amount of data generated by IoT devices, such as sensors and remote sensing technologies, requires robust systems and resources to collect, store, and process it effectively. Forest managers must possess the necessary skills and infrastructure to handle and interpret large datasets promptly (Chauhan and Atulkar, 2023). Data management challenges include organizing and structuring the data, ensuring data quality and accuracy, and integrating data from different sources. Moreover, data analysis poses challenges for deriving meaningful insights and actionable information from the collected data (Elhaloui et al., 2023). It requires advanced analytics techniques, including data modeling, machine learning, and artificial intelligence, to extract valuable knowledge from the data. Addressing these challenges requires investments in data infrastructure, data governance frameworks, and the development of data analytics capabilities to unlock the full potential of IoFT in forest management (Rosmiati et al., 2023).

SECURITY AND PRIVACY

Deploying IoT devices and sensors increases the risk of unauthorized access, data breaches, and privacy violations (Zakari and Abdulraza, 2016; Alli et al., 2021). Forest management data, including sensitive information on forest resources, biodiversity, and ecosystem health, must be protected from cyber threats and potential misuse (Jun et al., 2021). Robust security measures, such as encryption, authentication, and access controls, must be implemented to safeguard the integrity and confidentiality of the data. Privacy concerns arise from collecting personal or sensitive information from individuals involved in forest management activities (Zhao et al., 2022). It is crucial to establish clear policies and guidelines to address privacy issues, ensure data anonymization when necessary, and obtain informed consent for data collection and usage. Building trust and maintaining data privacy and security are essential for implementing IoFT in forest management (Kalinaki & Thilakarathne et al., 2023).

DIVERSE FOREST ECOSYSTEM ADAPTATION

Forest ecosystems vary globally, encompassing diverse environmental conditions, species compositions, and ecological dynamics. IoFT solutions must be adaptable and customizable for different forest types, including tropical rainforests, temperate forests, and arid woodlands (Shafik et al., 2021). This requires considering factors such as varying climate patterns, soil conditions, and vegetation characteristics (Kalinaki & Shafik et al., 2023). Additionally, different ecosystems may require specific sensor types, data collection frequencies, and analysis methods (Yang et al., 2021). Developing IoT technologies that can accommodate diverse forest ecosystems' unique characteristics and requirements is essential to ensuring the effectiveness and applicability of IoT-based solutions across different regions. Collaboration between forest management experts, researchers, and technology providers is crucial to understanding the specific needs of each ecosystem and tailoring IoFT applications accordingly (Hu and Chen et al., 2022).

ENERGY EFFICIENCY AND POWER SUPPLY

IoT devices in remote forest areas may face challenges accessing a stable power supply. Deploying and maintaining traditional power infrastructure can be impractical and costly in such locations. Therefore, alternative power sources like solar panels, energy harvesting, or battery-powered solutions are often utilized (Ananthi et al., 2022). However, ensuring uninterrupted power supply in all weather conditions and geographic locations remains a challenge. Additionally, energy efficiency is crucial to maximizing the lifespan of battery-powered devices and minimizing their environmental impact (Kianoush et al., 2023). Optimizing the energy consumption of IoT devices, employing low-power communication protocols, and implementing efficient data processing algorithms are essential for the sustainable and long-term deployment of IoFTs in forest management (Avazov et al., 2023). Continuous advancements in energy storage technology and power management strategies can help overcome these limitations and improve the power supply and energy efficiency of IoFT systems.

COST

The deployment and operation of IoFT devices, sensors, networks, and data analytics platforms can require a substantial initial investment. The costs include acquiring and installing the necessary hardware, developing custom software solutions, establishing network connectivity, and maintaining the infrastructure over time (Avazov et al., 2023). Additionally, ongoing maintenance, upgrades, and monitoring expenses should be considered. Limited financial resources may pose a challenge, especially for smaller-scale or community-based forest management initiatives needing help to afford the upfront and recurring costs associated with IoFT implementation (Avazov et al., 2023). Finding cost-effective solutions, leveraging economies of scale, and exploring partnerships and funding opportunities can help address this challenge and make IoFTs more accessible and feasible for forest management organizations of all sizes. Addressing these challenges and limitations requires collaboration between IoFT experts, forest management professionals, and technology providers. Overcoming these limitations can unlock the full potential of IoFTs, enabling more effective and sustainable forest management practices and promoting the conservation of valuable forest resources. There have been successful deployments of IoFTs, as illustrated in the following section.

REAL-WORLD IMPLEMENTATIONS AND CASE STUDIES OF IOFT

IoFT represents a real-world implementation of interconnected sensors, devices, and data analytics in forest ecosystems, providing valuable insights into the health of forests, facilitating proactive conservation measures, and supporting sustainable resource management. From monitoring biodiversity and detecting wildfires to optimizing timber harvesting and mitigating illegal logging, applying IoFT solutions is expanding globally, empowering stakeholders to protect and preserve our vital forest ecosystems for future generations. This section delves into the real-life implementations of IoFT and their purpose in forestry conservation.

IoT Sensors in Alice Holt Forest

Cutting-edge sensors have been meticulously affixed to the towering trees thriving in Hampshire's Alice Holt Forest, their seamless connectivity facilitated by the renowned narrow band-internet of things (NB-IoT) network spearheaded by Vodafone. This intelligent system ensures that a continuous stream of valuable data is effortlessly relayed to two eminent entities: the Department for Environment, Food and Rural Affairs (Defra) of the United Kingdom and Forest Research, an esteemed organization specializing in forestry and groundbreaking tree-related research. These invaluable datasets are subjected to advanced analytics, meticulously scrutinizing the intricate interplay between temperature, humidity, and moisture levels in the soil, elucidating their profound impact on the growth and functionality of these magnificent arboreal denizens. Precisely gauging tree growth enables scientists to glean vital insights into the potential role of trees in mitigating the severe effects of climate change, attributable to their extraordinary capacity to absorb and sequester carbon from the air we breathe. Armed with these indispensable findings, Defra and Forest Research stand poised to illuminate policymakers and the general public regarding the profound influence of our ever-evolving environment on tree development, underscoring the enormous advantages trees bestow as formidable guardians against carbon accumulation.

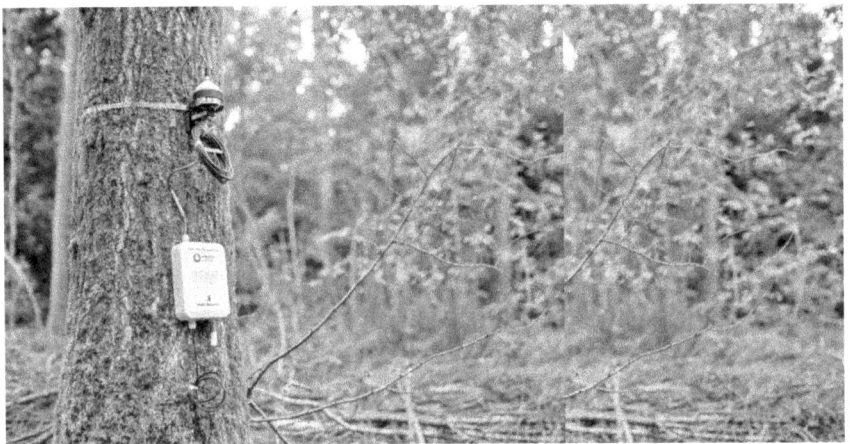

FIGURE 15.2 Vodafone-supplied sensor affixed on a tree in Alice Holt Forest (Climate change, 2023)

FOREST INVENTORY AND ANALYSIS (FIA) PROGRAM IN THE USA

The United States Forest Service utilizes IoT technology for forest inventory and monitoring. The primary objectives of the Forest Inventory and Analysis program are to assess forest health, monitor changes in forest resources over time, provide data for land management decisions, and support research on forests and their ecosystems (Khang et al., 2024a). The FIA program incorporates a combination of remote sensing, aerial imagery, and ground-based field surveys. IoT technology, such as GPS devices and mobile data collection tools, is used by field crews to accurately locate and record information about tree species, size, carbon sequestration, vegetation, soils, and other forest attributes. Forest inventory data supports a wide range of management applications, including timber harvesting planning, forest health assessments, wildlife habitat analysis, biodiversity conservation, land-use planning, and wildfire management (National Program, 2023). The information helps guide sustainable forest management practices and informs decision-making at various levels.

FOREST INVENTORY AND MANAGEMENT, BRITISH COLUMBIA, CANADA

This implementation uses IoT technologies for efficient forest inventory and management practices, enabling data-driven decision-making. Sensors used include drones equipped with light detection and ranging (LiDAR) as Figure 15.3 details. The deployed sensors to collect highly detailed three-dimensional information about the forest structure. These sensors emit laser pulses and measure the time it takes for the light to reflect, creating a detailed point cloud representation of the forest. This

FIGURE 15.3 Using LiDAR for forestry inventory management in British Columbia (BC et al., 2023)

data can be used to assess forest health, estimate tree volume, monitor vegetation changes, and plan forestry operations effectively (British Columbia, 2023).

SOIL MOISTURE MONITORING IN THE AMAZON RAINFOREST, BRAZIL

This implementation focuses on monitoring soil moisture levels in the Amazon Rainforest to enhance the management of forest ecosystems and support sustainable land-use practices. Sensors used include soil moisture sensors that are embedded in the ground at various locations within the forest. These sensors measure the moisture content of the soil, providing valuable data for understanding water availability and supporting decisions related to irrigation, planting, and overall forest health (Cama et al., 2013).

The sensors are strategically placed at different locations within the Amazon Rainforest. These sensors are often buried at various depths to capture soil moisture data at different layers. They are connected to a data acquisition system that collects the measurements at regular intervals (Maiwada et al., 2020). The collected data is then transmitted to a central database or cloud platform via wireless networks or satellite communication. Researchers, scientists, and forest managers can access this data remotely for analysis and decision-making. By studying the soil moisture patterns, insights into the water availability in different regions of the rainforest can be gained. Identifying areas prone to drought or excessive moisture is made possible and allows making informed decisions regarding irrigation, forest health management, and land-use planning.

CASE STUDIES

With its ability to connect devices and collect data from remote areas, IoFT has the potential to revolutionize how forests are monitored, protected, and managed. This section provides an overview of case studies that demonstrate the implementation of IoFT in forestry management, showcasing its benefits and highlighting real-world applications.

Harnessing Integrated Communication Technologies for Amazonian Environmental Monitoring

In this case study, the authors proposed deploying a wireless sensor network (WSN) in the middle of the Peruvian Amazon Rainforest. This network shall harness the mighty power of the existing ICT infrastructure to keep track of mystical parameters such as humidity, temperature, total solar radiation (TSR), photosynthetically active radiation (PAR), and soil moisture (BC, 2023). Within their innovative scheme, the WSN harmoniously works with the far-reaching wireless network, orchestrating a mechanism that effortlessly carries data across vast expanses spanning tens of kilometers, ultimately delivering it to a magnificent data center adorned with a splendid ensemble comprising a database, a web server, and an ever-connected portal to the boundless realms of the Internet.

Enhancing Surveillance of Unauthorized Tree Felling with Energy-Efficient Smart IoT Devices

The authors present an innovative framework aiming to automate detecting unlawful tree-cutting activities in forested areas by leveraging audio event classification (Andreadis et al., 2021). Their primary objective was to enable efficient and extensive surveillance across vast forest regions, employing ultra-low-power miniature devices with edge-computing microcontrollers and long-range wireless communication capabilities. In response to wireless edge devices' energy and resource constraints, the authors proposed a highly accurate and efficient audio classification mechanism based on convolutional neural networks.

Notably, their groundbreaking contribution lies in the system's capacity to identify a broader spectrum of deforestation-related threats through a distributed and pervasive edge-computing approach, surpassing previous research endeavors in this field. The authors comprehensively evaluated various preprocessing techniques, meticulously assessing the trade-off between classification accuracy and computational resources, memory utilization, and energy footprint. Empirical results unequivocally demonstrated the efficacy of the proposed framework in detecting and promptly alerting authorities about tree-cutting incidents, thereby facilitating a cost-effective and efficient forest monitoring paradigm using cutting-edge IoT technology (Kali et al., 2024). With an impressive accuracy rate of 85%, their solution exhibits immense potential in combatting illegal tree-cutting activities through the IoFT paradigm.

Forest Environmental Monitoring System Using IoT

This case study presents a pioneering methodology for monitoring forest ecosystems, employing the Raspberry Pi Model 3, an array of analog and digital sensors, and cutting-edge algorithms for signal analysis (Andreadis et al., 2021). Using these tools enables continuous tracking of crucial parameters such as soil moisture, gas concentrations, and temperature. Furthermore, a comprehensive examination of the ambient soundscape is conducted using a classification algorithm, enabling the precise identification of notable events such as forest background noise and chainsaw. The collected data is made readily accessible to users through an internet-based platform and a mobile application, facilitating instantaneous notifications upon detecting fire incidents, pollution sources, or instances of illegal deforestation. This IoFT-based solution for environmental monitoring is tailored to meet the specific requirements of public and private forest owners, as well as national authorities responsible for environmental protection and disaster response.

Solar-Powered IoT-based Peat Swamp Forest Environmental Monitoring System in Brunei

The innovative surveillance case study system under consideration is an eco-friendly solution, harnessing solar power for its operation and incorporating an integrated long-range (LoRa) network for seamless data collection through a centralized gateway (Essa et al., 2020). The system exhibits a comprehensive design catering to the measurement of vital environmental parameters, including soil and water temperature, soil moisture, water level, wind direction, atmospheric humidity, and pressure. The system intelligently acquires these crucial data points and seamlessly transmits them to a remote dashboard, ensuring remote accessibility for the pertinent agencies involved. Thorough laboratory testing on diverse peat soil conditions has verified the system's efficacy, yielding reliable measurement outcomes that serve as invaluable data analysis and interpretation references.

FUTURE DEVELOPMENTS IN IOFT TECHNOLOGIES AND THEIR APPLICATIONS

ENHANCED FOREST MONITORING

Enhanced forest monitoring is one of the critical areas where IoFT technologies have the potential to make a significant impact. Traditional forest monitoring methods rely on manual surveys and systematic data collection, which can be time-consuming and limited in real-time insights. However, IoFT can significantly enhance the monitoring process by deploying interconnected sensors and advanced data analytics. Aspects such as real-time data collection using IoFT technologies will enable networked sensors throughout the forest, including trees, soil, and the atmosphere (Zweifel et al., 2023). These sensors can collect data on various environmental parameters such as humidity, temperature, air quality, soil moisture, and light intensity. Moreover, through biodiversity monitoring, IoFT technologies also offer the potential to monitor wildlife populations and biodiversity within forests (Upinder et al., 2023).

For instance, the OI Pejeta Conservancy project in Kenya utilizes smart collars, tags, or embedded sensors on animals, which allows researchers to gather data on their movements, behavior, and habitat preferences (Shafik et al., 2023b). This information can aid in understanding the impact of human activities, climate change, and other factors on wildlife populations. Additionally, IoFT can help track endangered species, monitor their reproductive patterns, and facilitate conservation efforts. Integration with existing monitoring systems allows for the consolidation and analysis of data from multiple sources, including satellite imagery, weather stations, and ground-level sensors (Bo and Wang, 2011). Combining these diverse datasets can provide a more comprehensive and holistic view of the forest ecosystem, enabling better-informed decision-making and management strategies.

FOREST ECOSYSTEM MODELING

Forest ecosystem modeling is a field of study that utilizes advanced technologies, including the IoFT, to simulate and understand the behavior and dynamics of forest ecosystems (Singh et al., 2022). A vital aspect of this future trend is the integration of diverse datasets through IoFT technologies. By assimilating information from satellite imagery, LiDAR data, weather stations, and ground-based sensors, these models capture the complexity and heterogeneity of forests, providing more accurate and detailed insights into their functioning. IoFT-powered forest ecosystem modeling significantly advances our ability to monitor and simulate forest ecosystems (Khang et al., 2023d).

These models have the potential to simulate ecological processes such as tree growth, mortality, carbon sequestration, nutrient cycling, and water dynamics, enabling us to predict how forests will respond to changing environmental conditions. In conclusion, future trends in IoFT technologies will revolutionize forest management by enabling enhanced monitoring, advanced analytics, precision forestry, wildlife conservation, forest ecosystem modeling, integration with existing systems, and continuous improvement. These trends will contribute to sustainable forest management practices, conservation of biodiversity, mitigation of climate change effects, and the effective stewardship of forest resources for future generations (Lakshmi et al., 2023).

FUTURE DIRECTIONS AND RECOMMENDATIONS

INTEGRATION OF ARTIFICIAL INTELLIGENCE (AI)

Integrating AI techniques, such as machine learning and data analytics, with IoFT technologies can further enhance forest management practices (Khang et al., 2023a). AI algorithms can analyze large volumes of data collected from sensors, satellites, and other sources to identify patterns, predict forest disturbances, and provide valuable insights for decision-making. Further research should explore the potential of AI in IoFT applications, including improved forest health monitoring, early detection of pests and diseases, and optimization of forest operations (Rana et al., 2021).

ENHANCED DATA PRIVACY AND SECURITY

As IoFT technologies involve collecting and transmitting sensitive data, ensuring robust data privacy and security measures is essential. Further research and development should focus on implementing advanced encryption techniques, secure data transmission protocols, and authentication mechanisms to safeguard the confidentiality and integrity of forest data. Additionally, developing privacy-preserving algorithms that allow for data analysis while protecting individual privacy should be a priority (Kamran et al., 2023).

INTEROPERABILITY AND STANDARDIZATION

The interoperability of IoFT technologies is crucial for their seamless integration into forest management systems. Efforts should be made to establish common standards and protocols that enable different devices, platforms, and systems to exchange data and work together effectively. Research should focus on developing open-source frameworks and protocols that promote interoperability, facilitating data sharing and collaboration among various stakeholders in forest management.

REAL-TIME DECISION SUPPORT SYSTEMS

Developing real-time decision support systems powered by IoFT technologies can significantly enhance forest management practices. These systems can integrate real-time data from sensors, drones, satellites, and other sources to provide forest managers with actionable insights and recommendations. Future research should focus on developing user-friendly decision support systems that integrate diverse data sources, employ advanced analytics, and provide intuitive visualizations to aid decision-making processes.

STAKEHOLDER ENGAGEMENT AND CAPACITY BUILDING

Implementing IoFT technologies in forest management requires stakeholder engagement and capacity-building efforts. Research should focus on developing training programs, workshops, and educational materials to enhance the understanding and skills of forest managers, policymakers, and local communities in utilizing IoFT technologies effectively. Engaging stakeholders throughout the design, implementation, and evaluation phases of IoFT projects is crucial for their success and sustainability.

LONG-TERM MONITORING AND ADAPTIVE MANAGEMENT

IoFT technologies can provide valuable insights into short-term forest conditions and dynamics. However, long-term monitoring and adaptive management approaches are equally important for sustainable forest management. Future research should explore how IoFT technologies can be integrated into long-term monitoring frameworks and adaptive management strategies, allowing for continuous forest health assessment, identification of emerging risks, and adaptation to changing environmental conditions.

COLLABORATION AND KNOWLEDGE SHARING

Collaboration among researchers, practitioners, and policymakers is essential for advancing IoFT technologies in forest management. Research institutions, governmental agencies, and industry stakeholders should foster collaborative networks and platforms for knowledge sharing, data sharing, and best practice exchange. This collaborative approach can accelerate the development and implementation of IoFT technologies and ensure their wide-scale adoption and impact on forest management.

CONCLUSION

In summary, the internet of forestry things (IoFT) has revolutionized forest management by leveraging advanced technologies such as remote sensing, wireless sensor networks, drones, and satellite imagery. These IoFT technologies provide real-time data and valuable insights, empowering forest managers with informed decision-making capabilities and proactive management strategies. By enhancing the efficiency, effectiveness, significance, and sustainability of forest ecosystems, IoFT plays a pivotal role in addressing challenges and ensuring the long-term well-being of forests. From the onset, we recognized the critical importance of forests as invaluable resources for our planet and human societies and the challenges such as deforestation, forest degradation, and climate change. To tackle these challenges, we explored the potential of IoFT technologies in forest management. However, our exploration of IoFT's transformative power should not conclude here. What lies beyond the current state? How can we unlock the full potential of IoFT in advancing forest management practices? To fully grasp the implications of IoFT in forest management, we must confront potential limitations and challenges that lie ahead. By critically evaluating these concerns, we can navigate the path toward effective and responsible implementation of IoFT. Among the foremost challenges are data privacy, security, and interoperability.

To fully realize the benefits of IoFT, robust measures must safeguard sensitive information and ensure seamless integration among different IoFT systems. Gazing into the future, the potential of IoFT in forest management extends far beyond our current understanding. As technology evolves, we can anticipate even more sophisticated sensors, improved connectivity, and advanced data analytics. Integrating artificial intelligence and machine learning algorithms holds immense promise, revolutionizing forest management with predictive models and automated decision-making processes (Khang et al., 2023b).

Collaboration and standardization become paramount as we journey deeper into IoFT technologies. How can stakeholders unite to foster seamless integration and compatibility of IoFT systems? Moreover, how can capacity-building initiatives equip forest managers and workers with the necessary skills to effectively harness IoFT technologies and maximize their potential for sustainable forest management? In conclusion, IoFT technologies have reshaped the landscape of forest management by providing real-time data and insights that enhance decision-making and promote sustainable practices. Although challenges and limitations persist, addressing them through collaboration and innovation will pave the way for a future where IoFT plays a central role in safeguarding and managing our precious forest ecosystems (Khang et al., 2024b).

REFERENCES

Alli A. A., K. Kassim, N. Mutwalibi, H. Hamid, and L. Ibrahim, "Secure Fog-Cloud of Things: Architectures, Opportunities and Challenges," in *Secure Edge Computing*, M. Ahmed and P. Haskell-Dowland, Eds. (1st ed.). CRC Press, 2021, pp. 3–20, doi: 10.1201/9781003028635-2.

Ananthi J., N. Sengottaiyan, S. Anbukaruppusamy, K. Upreti, and A. K. Dubey, "Forest Fire Prediction Using IoT and Deep Learning," *Int. J. Adv. Technol. Eng. Explor.*, vol. 9, no. 87, 2022, doi: 10.19101/IJATEE.2021.87464.

Andreadis A., G. Giambene, and R. Zambon, "Monitoring Illegal Tree Cutting through Ultra-Low-Power Smart IoT Devices," *Sensors (Basel)*, vol. 21, no. 22, 2021, doi: 10.3390/s21227593.

Avazov K., A. E. Hyun, A. A. Sami, S. A. Khaitov, A. B. Abdusalomov, and Y. I. Cho, "Forest Fire Detection and Notification Method Based on AI and IoT Approaches," *Futur. Internet*, vol. 15, no. 2, 2023, doi: 10.3390/fi15020061.

Bayne K., S. Damesin, and M. Evans, "The Internet of Things - Wireless Sensor Networks and Their Application to Forestry," *New Zeal. J. For.*, vol. 61, no. 4, 2017.

BC, "How Technology Is Transforming BC's Forest Sector | Video + Resources | Naturally: Wood," https://www.naturallywood.com/resource/how-technology-is-transforming-bcs-forest-sector-video-overview/ (accessed Jun. 24, 2023).

Bo Y. and H. Wang, "The Application of Cloud Computing and the Internet of Things in Agriculture and Forestry," Proceedings - 2011 International Joint Conference on Service Sciences, IJCSS 2011, 2011, doi: 10.1109/IJCSS.2011.40.

British Columbia, "Forest Health - Province of British Columbia," https://www2.gov.bc.ca/gov/content/industry/forestry/managing-our-forest-resources/forest-health (accessed Jun. 24, 2023).

Cama A., F. G. Montoya, J. Gómez, J. L. De La Cruz, and F. Manzano-Agugliaro, "Integration of Communication Technologies in Sensor Networks to Monitor the Amazon Environment," *J. Clean. Prod.*, vol. 59, 2013, doi: 10.1016/j.jclepro.2013.06.041.

Cao L., K. Zhou, X. Shen, X. Yang, F. Cao, and G. Wang, "The Status and Prospects of Smart Forestry," *J. Nanjing For. Univ. (Nat. Sci. Ed.)*, vol. 46, no. 6, 2022, doi: 10.12302/j.issn.1000-2006.202209052.

Chauhan P. and M. Atulkar, "An Efficient Centralized DDoS Attack Detection Approach for Software Defined Internet of Things," *J. Supercomput.*, 2023, doi: 10.1007/s11227-023-05072-y.

Chen J., M. Hoppen, D. Boken, J. Reitz, M. Schluse, and J. Rosmann, "Identity, Authentication and Authorization in Forestry 4.0 Using OAuth 2.0," 3rd International Informatics and Software Engineering Conference, IISEC 2022, 2022, doi: 10.1109/IISEC56263.2022.9998287.

Climate Change, "How IoT Is Helping Tree Scientists Learn More about Climate Change," https://www.vodafone.co.uk/newscentre/features/how-iot-is-helping-tree-scientists-learn-more-about-climate-change/ (accessed Jun. 24, 2023).

Deng F., X. Chen, S. N. Shavkatovich, and M. Shabaz, "Homestead Engineering Planning Based on CAD Internet of Things Technology," *Comput. Aid.. Des. Appl.*, vol. 20, 2023, doi: 10.14733/cadaps.2023.S3.121-134.

Elhaloui L., S. El Filali, E. H. Benlahmer, M. Tabaa, Y. Tace, and N. Rida, "Machine Learning for Internet of Things Classification Using Network Traffic Parameters," *Int. J. Electr. Comput. Eng.*, vol. 13, no. 3, 2023, doi: 10.11591/ijece.v13i3.pp3449-3463.

Essa S., R. Petra, M. R. Uddin, W. S. H. Suhaili, and N. I. Ilmi, "IoT-Based Environmental Monitoring System for Brunei Peat Swamp Forest," 2020 International Conference on Computer Science and Its Application in Agriculture, ICOSICA 2020, 2020, doi: 10.1109/ICOSICA49951.2020.9243279.

Fang M., "Forestry English Corpus Construction and Application in Foreign Language Teaching under the Background of Big Data and Internet of Things," *J. Comput. Methods Sci. Eng.*, vol. 23, no. 1, 2022, doi: 10.3233/JCM-226562.

Forestry Fire. "Forestry Fire Prevention Detection and Emergency Command System Based on Internet of Things and Infrared Light Sensing Technology," *Int. J. Front., Eng. Technol.*, vol. 4, no. 6, 2022, doi: 10.25236/ijfet.2022.040606.

Gaitan N. C. and P. Hojbota, "Forest Fire Detection System Using Lora Technology," *Int. J. Adv. Comput. Sci. Appl.*, vol. 11, no. 5, 2020, doi: 10.14569/IJACSA.2020.0110503.

Hazards, *Implementation of Sensors and Artificial Intelligence for Environmental Hazards Assessment in Urban, Agriculture and Forestry Systems*, 2022, doi: 10.3390/books978-3-0365-2905-9.

Hirawan D. and D. Mahendra, "Optimization of Forest Plant Seeding Based on the Internet of Things," *IOP Conference Series: Mater. Sci. Eng.*, vol. 879, no. 1, 2020, doi: 10.1088/1757-899X/879/1/012052.

Hu X., "Analysis and Research on the Integrated English Teaching Effectiveness of Internet of Things Based on Stochastic Forest Algorithm," *Int. J. Contin. Eng. Educ. Life Long Learn.*, vol. 32, no. 1, 2022, doi: 10.1504/ijceell.2022.121222.

Hu Z., F. Chen, R. Shen, and S. Li, "Design and Implementation of Forest Fire Monitoring System Based on Internet of Things Technology," *ITM Web Conf.*, vol. 45, 2022, doi: 10.1051/itmconf/20224501063.

Ikram, K., N. A. Buttar, M. M. Waqas, M. M. O. Muthmainnah, Y. Niaz, and A. Khang, "Robotic Innovations in Agriculture: Maximizing Production and Sustainability," in *Advanced Technologies and AI-Equipped IoT Applications in High-Tech Agriculture* (1st Ed.). IGI Global Press, 2023, doi: 10.4018/978-1-6684-9231-4.ch007.

Jun Y., A. Craig, W. Shafik, and L. Sharif, "Artificial Intelligence Application in Cybersecurity and Cyberdefense," *Wirel. Commun. Mob. Comput.*, vol. 2021, 2021, doi: 10.1155/2021/3329581.

Kalinaki K., N. N. Thilakarathne, H. R. Mubarak, O. A. Malik, and M. Abdullatif, "Cybersafe Capabilities and Utilities for Smart Cities," in *Cybersecurity for Smart Cities*. Springer, 2023, pp. 71–86, doi: 10.1007/978-3-031-24946-4_6.

Kalinaki K., W. Shafik, T. J. L. Gutu, and O. A. Malik, "Computer Vision and Machine Learning for Smart Farming and Agriculture Practices," in *Artificial Intelligence Tools and Technologies for Smart Farming and Agriculture Practices*, 2023, pp. 79–100, doi: 10.4018/978-1-6684-8516-3.ch005.

Kaur P., R. Kumar, and M. Kumar, "A Healthcare Monitoring System Using Random Forest and Internet of Things (IoT)," *Multimed Tool. Appl.*, vol. 78, no. 14, 2019, doi: 10.1007/s11042-019-7327-8.

Kaur, U. and K. A. Manjit Kaur, "Futuristic Technologies in Agriculture Challenges and Future Prospects," *Advanced Technologies and AI-Equipped IoT Applications in High-Tech Agriculture* (1st Ed.). IGI Global Press, 2023, doi: 10.4018/978-1-6684-9231-4.ch020.

Kavita Mathad, K. A., "Hospital 4.0: Capitalization of Health and Healthcare in Industry 4.0 Economy," in *Data-Centric AI Solutions and Emerging Technologies in the Healthcare Ecosystem*. P (14) (1st Ed.). CRC Press, 2023, doi: 10.1201/9781003356189-19.

Khang A., *Advanced Technologies and AI-Equipped IoT Applications in High-Tech Agriculture* (1st Ed.). IGI Global Press, 2023, doi: 10.4018/978-1-6684-9231-4.

Khang A., G. Rana, R. K. Tailor, V. A. Hajimahmud (Eds.), *Data-Centric AI Solutions and Emerging Technologies in the Healthcare Ecosystem*. CRC Press, 2023c, doi: 10.1201/9781003356189.

Khang A., R. Gujrati, H. Uygun, R. K. Tailor, and S. S. Gaur, *Data-Driven Modelling and Predictive Analytics in Business and Finance*. CRC Press, 2024b. doi: 10.1201/9781032600628.

Khang A., S. K. Gupta, V. Shah, and A. Misra (Eds.), *AI-Aided IoT Technologies and Applications in the Smart Business and Production*. CRC Press, 2023b, doi: 10.1201/9781003392224.

Khang A., S. K. Gupta, V. A. Hajimahmud, J. Babasaheb, and G. Morris, *AI-Centric Modelling and Analytics: Concepts, Designs, Technologies, and Applications* (1st Ed.). CRC Press, 2023, doi: 10.1201/9781003400110.

Khang A., V. Abdullayev, V. Hahanov, and V. Shah, *Advanced IoT Technologies and Applications in the Industry 4.0 Digital Economy* (1st Ed.). CRC Press, 2024a, doi: 10.1201/9781003434269.

Kianoush S., S. Savazzi, V. Rampa, L. Costa, and D. Tolochenko, "A Random Forest Approach to Body Motion Detection: Multisensory Fusion and Edge Processing," *IEEE Sens. J.*, vol. 23, no. 4, 2023, doi: 10.1109/JSEN.2022.3232085.

Kurniabudi K. et al., "Improvement of Attack Detection Performance on the Internet of Things with PSO-Search and Random Forest," *J. Comput. Sci.*, vol. 64, 2022, doi: 10.1016/j.jocs.2022.101833.

Lakhwani K., H. Gianey, N. Agarwal, and S. Gupta, "Development of IoT for Smart Agriculture a Review," *Adv. Intell. Syst. Comput.*, vol. 841, 2019, doi: 10.1007/978-981-13-2285-3_50.

Lal S. B., A. Sharma, K. K. Chaturvedi, M. S. Farooqi, and A. Rai, "Internet of Things in Forestry and Environmental Sciences," in *Forum for Interdisciplinary Mathematics*, 2020.

Li D., X. Fu, and Y. Zan, "MPTEM: A Reliable Trust Evaluation Model for Forest IoT System," 2022 7th International Conference on Computer and Communication Systems, ICCCS 2022, 2022, doi: 10.1109/ICCCS55155.2022.9845937.

Liu D. et al., "Sensors Anomaly Detection of Industrial Internet of Things Based on Isolated Forest Algorithm and Data Compression," *Sci. Program*, vol. 2021, 2021, doi: 10.1155/2021/6699313.

Liu P., Y. Zhang, H. Wu, and T. Fu, "Optimization of Edge-PLC-Based Fault Diagnosis with Random Forest in Industrial Internet of Things," *IEEE Internet Things J.*, vol. 7, no. 10, 2020, doi: 10.1109/JIOT.2020.2994200.

Maiwada U. D., A. A. Muazu, I. K. Yakasai, and R. Y. Zakari, "Identifying Actual Users in a Web Surfing Session Using Tracing and Tracking," *JINAV J, Inf. Vis.*, vol. 1, no. 1, 2020, doi: 10.35877/454ri.jinav174.

Medida L. H., G. L. N. V. S. Kumar, and A. Khang, "Predictive Analytics for High-Tech Agriculture," in *Advanced Technologies and AI-Equipped IoT Applications in High-Tech Agriculture* (1st Ed.). IGI Global Press, 2023, doi: 10.4018/978-1-6684-9231-4.ch019.

National Program, "Forest Inventory and Analysis National Program," https://www.fia.fs.usda.gov/ (accessed Jun. 24, 2023).

Oliveira L. F. P., A. P. Moreira, and M. F. Silva, "Advances in Forest Robotics: A State-of-the-Art Survey," *Robotics*, vol. 10, no. 2, 2021, doi: 10.3390/robotics10020053.

Qi Y. and X. Gao, "Equivalence Assessment Method of Forest Tourism Safety Based on Internet of Things Application," *Comput. Intell. Neurosci.*, vol. 2022, 2022, doi: 10.1155/2022/1578005.

Rana G., A. Khang, R. Sharma, A. K. Goel, and A. K. Dubey (Eds.), *Reinventing Manufacturing and Business Processes Through Artificial Intelligence*. CRC Press, 2021, doi: 10.1201/9781003145011.

Rath, K. C., A. Khang, and D. Roy, "The Role of Internet of Things (IoT) Technology in Industry 4.0," *Advanced IoT Technologies and Applications in the Industry 4.0 Digital Economy* (1st Ed.). CRC Press, 2024, doi: 10.1201/9781003434269-1.

Rosmiati M., G. I. H., and M. Mahardika, *The Simulation of Monitoring System of Elephant Location in the Forest Using Internet of Things*, 2023, doi: 10.1109/icitisee57756.2022.10057608.

Sahal R., S. H. Alsamhi, J. G. Breslin, and M. I. Ali, "Industry 4.0 Towards Forestry 4.0: Fire Detection Use Case," *Sensors (Switzerland)*, vol. 21, no. 3, 2021, doi: 10.3390/s21030694.

Shafik W., "Cyber Security Perspectives in Public Spaces: Drone Case Study," in *Handbook of Research on Cybersecurity Risk in Contemporary Business Systems*. IGI Global, 2023a, pp. 79–97, doi: 10.4018/978-1-6684-7207-1.ch004.

Shafik W., "Making Cities Smarter: IoT and SDN Applications, Challenges, and Future Trends," in *Opportunities and Challenges of Industrial IoT in 5G and 6G Networks*. IGI Global, 2023b, pp. 73–94, doi: 10.4018/978-1-7998-9266-3.ch004.

Shafik W., S. Mojtaba Matinkhah, F. Shokoor, and M. N. Sanda, "Internet of Things-Based Energy Efficiency Optimization Model in Fog Smart Cities," *Int. J. Inform. Vis.*, vol. 5, no. 2, 2021, doi: 10.30630/joiv.5.2.373.

Sihombing P., H. Herriyance, and M. R. Syaputra, "Smart System to Prevent Forest Fire Based on Internet of Things," *Internetworking Indones. J.*, vol. 13, no. 1, 2021.

Singh R., A. Gehlot, S. Vaseem Akram, A. Kumar Thakur, D. Buddhi, and P. K. Das, "Forest 4.0: Digitalization of Forest Using the Internet of Things (IoT)," *J. King Saud Univ. Comput. Inf. Sci.*, vol. 34, no. 8, 2022, doi: 10.1016/j.jksuci.2021.02.009.

Vikash and S. Varma, "Trust-Based Forest Monitoring System Using Internet of Things," *Int. J. Commun. Syst.*, vol. 35, no. 12, 2022, doi: 10.1002/dac.4163.

Vrushank S., T. Vidhi, and A. Khang, "Electronic Health Records Security and Privacy Enhancement Using Blockchain Technology," *Data-Centric AI Solutions and Emerging Technologies in the Healthcare Ecosystem*. P (1) (1st Ed.). CRC Press, 2023, doi: 10.1201/9781003356189-1.

Wang Y., G. Zhang, and S. Q. Tan, "Application Prospect of Internet of Things Technology in Info-Forestry," Asia-Pacific Youth Conference on Communication (APYCC) 2010, 2010, pp. 95–99.

Wong Q. Y. and Y. B. Chu, "A Mobile Production Monitoring System Based on Internet of Thing (IoT) and Random Forest Classification," *Int. J. Electr. Electron. Eng. Telecommun.*, vol. 10, no. 4, 2021, doi: 10.18178/ijeetc.10.4.243-250.

Yang Z. et al., "Green Internet of Things and Big Data Application in Smart Cities Development," *Complexity*, vol. 2021, 2021, doi: 10.1155/2021/4922697.

Zakari R. Y., J. W. Owusu, H. Wang, K. Qin, Z. K. Lawal, and Y. Dong, *VQA and Visual Reasoning: An Overview of Recent Datasets, Methods and Challenges*. Computer Vision and Pattern Recognition, 2022, pp. 1–39. http://arxiv.org/abs/2212.13296.

Zakari R. Y. and N. Abdulraza, "Computer Security: A Literature Review and Classification," *Int. J. Comput. Sci. Control. Eng.*, vol. 4, no. 2, 2016.

Zhang S., Y. Li, D. Gao, and Z. Guan, "Research on Early Forest Fire Detection Based on the Internet of Things and Spectrum Analysis," *J. For. Eng.*, vol. 6, no. 3, 2021, doi: 10.13360/j.issn.2096-1359.202006045.

Zhao L. et al., "Artificial Intelligence Analysis in Cyber Domain: A Review," *Int. J. Distrib. Sens. Netw.*, vol. 18, no. 4, 2022, doi: 10.1177/15501329221084882.

Zweifel R. et al., "Networking the Forest Infrastructure Towards near Real-Time Monitoring – A White Paper," *Sci. Total Environ.*, vol. 872, 2023, doi: 10.1016/j.scitotenv.2023.162167.

16 Artificial Intelligence (AI) Equipped Edge Internet of Things (IoT) Devices in Security

Nikita Agrawal and Aakansha Saxena

INTRODUCTION

The military has a lot of promise with artificial intelligence (AI) since it can be used to gather, analyze, and present data for surveillance tasks as well as to recognize individuals and objects utilizing smart sensors. Algorithms based on machine learning are better able to foresee and predict the expenses and resourcing needed for missions and training exercises.

AI-enabled solutions in field operations can offer real-time data for on-the-spot assessments to enhance decision-making abilities. Furthermore, it can be used for HR tasks, intelligent budgeting, and support services to speed up the purchase process. A real Revolution in Military Affairs (RMA) could be sparked by artificial intelligence. RMA is a theory on how war will develop in the future that is frequently linked to military organizational and technological advice. Therefore, it is crucial for military and national security strategists to include AI's potential in their plans and strategies. As a result, it becomes more important to have information about the use of similar technology by adversaries.

A major advantage for every military force has always been having access to greater technology than its rivals. In battle, the first-mover advantage is usually a highly prized resource. Amazing ideas have come up as a result of the pressing need to end the impasse and defeat your adversary. In order to win competitions like space races or to accurately track approaching enemy aircraft, the military-industrial complex that emerged after World War II accelerated the development of computers. Therefore, it shouldn't be surprising that the most cutting-edge research and development in this subject is driven by the need for national security.

DOI: 10.1201/9781003434269-16

BACKGROUND

AI has only recently been incorporated into traditional warfare. Although USAF AC-130 gunships were somewhat capable of differentiating between different Iraqi targets during the first Gulf War, mistakes nonetheless resulted in a significant number of civilian deaths. Tomahawk cruise missiles developed in the USA were able to map the topography, including hills and houses, using a database that had already been constructed.

Drones have never before been used for surveillance or strikes halfway around the world, thanks to the War on Terror. In 2018, a Turkish micro- Unmanned aerial vehicle (UAV) killed a militant active in the Libyan Civil War without the involvement of any humans, marking the achievement of this milestone. AI-capable military equipment can efficiently manage enormous data quantities. Because they are so adept at calculating and decision-making, such technologies have also improved self-regulation, self-control, and self-action. With more powerful computers and more sophisticated AI, this function will only get better.

MOTIVATION

Defense and national security have always been our concerns as Rashtriya Raksha University students. The majority of the field's obvious advancements may be attributed to artificial intelligence and machine learning, which are now fundamental components of computer science. We investigated the subject in-depth to comprehend its historical significance, current application, and potential in the future as computer science students driven by a desire to raise awareness. Both traditional state actors and unconventional non-state actors pose equal risks to our country.

The development of one of the most technologically advanced military forces in the world is being spearheaded by the Indian Defense sector. The Indian Military will undergo a change as a result of the use of AI-based technology. It also establishes India firmly in the enormous market for defense products. Government backing and aspirations to improve the military with AI are the result of years of planning. A variety of breaking AI-based technological products in the fields of data, logistics, surveillance, weapons, and many others have been created as the outcome of this collaborative effort between the public and private sectors of industry, research organizations, academic institutions, start-ups, and innovators.

RESEARCH METHODOLOGY

The topic's study technique was based on an examination of the pertinent scientific publications and interactions with defense and military representatives. Because the technologies are still relatively new and cognizant of issues of confidentiality, access to publicly accessible information is constrained in terms of specifics, especially when it comes with regard to the scientific and functional implementation of the necessary features and capabilities. The approach only uses resources that are available with open access.

Examples pertain to areas of the army that are critical to executing contemporary combat operations on battlegrounds and ensuring the correct operation and security of the state and all people. The categorization provides an overview of the literature ranging from the various artificial intelligence (AI) techniques used to solve this problem to the variety of military applications or areas of applications. In order to better understand the evolution of AI in the military, all fields and individual algorithms are briefly presented. In the framework of ethics and social conduct, the society's primary worries about the advancement of AI were examined. On the opinions expressed by individuals from different categories of society about the development of artificial intelligence algorithms in various fields and applications, polls taken between 2011 and 2019 were compared.

LITERATURE SURVEY

A variety of breaking AI-based technological products in the fields of data, logistics, surveillance, weapons, and many others have been created as the outcome in Table 16.1.

TECHNOLOGIES USED

Big data, AI, autonomous vehicles, space, hypersonic aerial vehicles, quantum technology, biotechnology, and novel materials are examples of Extended data types (EDTs). A greater level of regulation is required to safeguard communities against malicious individuals who are skilled in creating extremely hazardous tools and solutions using readily accessible information. Despite the fact that technology related to chemicals, bacteria, radiation, and the like Chemical, Biological, Radiological and Nuclear (CBRN) are typically subject to strict control, these laws may not be sufficient. Internet of things (IoT) tools can be used as shown in Figure 16.1 (a–d).

APPLICATIONS OF AI

OPERATIONAL LEVEL

Applications for artificial intelligence (AI) in operational warfare are concerned with attaining tactical goals through military power. Utilizing big data-driven simulations and wargames, AI-enabled autonomous vehicles, as well as intelligent data collecting and analysis on the battlefield, it augments the traditional deterrent mechanism (Figure 16.2).

STRATEGIC WARFARE LEVEL

Applications of AI to warfare at the strategic level examine the extent of the conflict and its implications on sustaining strategic stability and deterrence as it escalates or de-escalates. It is propelled by AI-enabled Intelligence, Surveillance and Reconnaissance (ISR) operations, as well as by its application to missile defense

TABLE 16.1

Outcome of the Collaborative Effort between the Public and Private Sectors of Industry, Research Organizations, Academic Institutions, Start-ups, and Innovators

S. No.	Authors	Title	Source
1.	Fraga-Lamas, Fernandez-Carames	Tactical Edge IoT in Defense and National Security	IEEE, January 2023

Reported Outcomes: Edge computing strategies can be used to overcome various difficulties in the implementation with internet of things (IoT) systems in the national security and defense industries. Together, the edge computing and the internet of things (IoT) concepts have the potential to be useful since they push back against the constraints of traditional centralized cloud computing.

Challenges: While centralized cloud computing allows for easy scalability, real-time applications, and mobility support, it also comes with certain risks, including cybersecurity.

2.	Hui Wu; Haiting Han; Xiao Wang; Shengli Sun	Research on Artificial Intelligence Enhancing Internet of Things Security: A Survey	IEEE, August 20, 2020

Reported Outcomes: The specificity and difficulty of IoT security protection are systematically reviewed in this work, and we discover that two AI methodologies, machine learning (ML) and deep learning (DL), can offer new, potent capabilities to handle IoT security requirements. It lists four key IoT security issues, describes typical AI solutions, and contrasts the technological advancements and algorithms employed by those solutions.

Challenges: In addition, AI adds not only new potential issues and negative effects that could harm IoT in terms of data, algorithms, and architecture but also new possibilities for IoT security prevention.

3.	Murat Kuzlu, Corinne Fair, and Ozgur Guler	Role of Artificial Intelligence in the Internet of Things (IoT) Cybersecurity	Springer, February 24, 2021

(Continued)

TABLE 16.1 (CONTINUED)

Outcome of the Collaborative Effort between the Public and Private Sectors of Industry, Research Organizations, Academic Institutions, Start-ups, and Innovators

S. No.	Authors	Title	Source
4.	Elahe Fazel Dehkordi and Tor-Morten Grønli	A Survey of Security Architectures for Edge Computing-Based IoT	MDPI, June 30, 2022

Reported Outcomes: Cybercriminals have learnt how to use AI to their advantage during cybersecurity attacks, and some have even begun to use antagonistic AI. This review study aims to comprehensively summarize and then include the pertinent academic literature for the fields. It achieves this by fusing data from a number of recent academic publications and surveys on IoT, AI, and attacks on and against AI.

Challenges: The paper describes how IoT and AI have been used for illegal activities or had security flaws uncovered as an example in order to help readers understand present threats and develop understanding so that these vulnerabilities are taken into account in the future to prevent cyberattacks (Khang et al., 2024).

S. No.	Authors	Title	Source
5.	Eric Gyamfi and Anca Jurcut	Intrusion Detection in Internet of Things Systems: A Review on Design Approaches Leveraging Multi-Access Edge Computing, Machine Learning, and Datasets	MDPI, May 14, 2022

Reported Outcomes: Edge computing (EC) is a useful strategy to deal with these problems since it places processing and keeping information close to IoT devices and end users. The primary security and privacy concerns and attacks in the context of EC-based IoT are then covered in great detail, along with viable solutions. Second, we suggest an EC-based architecture that is secure for IoT applications. Also, designers analyze the benefits and drawbacks of a scenario based on edge computing and cloud computing, and another use case for the internet of things' edge computing is offered (Kali et al., 2024).

Challenges: The main threats to the safety and confidentiality of EC-based IoT that have been documented in the literature, as well as the associated defenses and solutions that may be used to prevent these attacks, have also been extracted and summarized. Then release a safe infrastructure architecture based on EC with reference to edge computing in the IoT. Furthermore, they presented two prospects of edge computing for the selected IoT scenario and discussed the advantages and disadvantages of each. Finally, we defined an IoT scenario based on cloud computing for transport mode detection.

Reported Outcomes: In order to overcome these limitations, the multi-access edge computing (MEC) platform was developed. The use of MEC for distributed solutions needs to be given more attention. The methods rely on MEC platforms and make use of ML techniques. Also, the research conducts a comparative examination of the publicly accessible datasets, assessment criteria, and deployment tactics used in the NIDS design.

(Continued)

TABLE 16.1 (CONTINUED)

Outcome of the Collaborative Effort between the Public and Private Sectors of Industry, Research Organizations, Academic Institutions, Start-ups, and Innovators

S. No.	Authors	Title	Source
	Challenges: The evolution of NIDS in the real world has not yet been fully confirmed. There aren't any established standards for certain detection strategies or deployment approaches to protect IoT systems. Designing a workable NIDS solution that successfully identifies cyberattacks in actual IoT systems still needs more work. Additionally, the majority of the linked studies do not take important IoT system evaluation criteria like energy consumption, processing, and storage efficiency into account.		
6.	Marta Bistron and Zbigniew Piotrowski	Artificial Intelligence Applications in Military Systems and Their Influence on Sense of Security of Citizens	MDPI, April 6, 2021
	Reported Outcomes: To conducting research on an overview of existing and future prospects for the development of artificial intelligence algorithms, with a focus on military applications, is provided in this article. The majority of the focus was on using robotics, object detection, military logistics, and cybersecurity all using AI algorithms. Also, it briefly describes the mathematical underpinnings of the Expectation-Maximization and Gaussian Mixture Model methods, as well as the ART, CNN, and SVM networks that are utilized to solve the problems under discussion.		
	Challenges: The popularity of programs like AIE demonstrates the significance of this area of study. People continue to be concerned about the potential consequences of these technologies, according to a study. This makes sense given that not even specialists are certain of the direction that artificial intelligence will go in the future.		
7.	Yao Jun, Alisa Craig, Wasswa Shafik, and Lule Sharif	Artificial Intelligence Application in Cybersecurity and Cyber Defense	Hindawi, October 27, 2021
	Reported Outcomes: Businesses in the public and private sectors may use IoT applications that are effective to help companies manage their assets, improve operational effectiveness, and create new business models. Studies assess the IoT development as a strategy for technological advancement by considering qualities, designs, applications, supporting technologies, and upcoming difficulties. IoT architecture, comprising the network management layer, the application layer, the transmission layer, and the perception layer facilitates an aging society, optimizes various means of mobility and transportation, and helps to boost power consumption, coupled with the definition and features of IoT devices (Nassereddine and Khang, 2024).		
	Challenges: There may be some difficulties, especially in the legal and ethical realms. Compliance with these rules may be difficult because ethical laws and regulations are connected to privacy, access, and information integrity, while legal laws are tied to cybersecurity.		

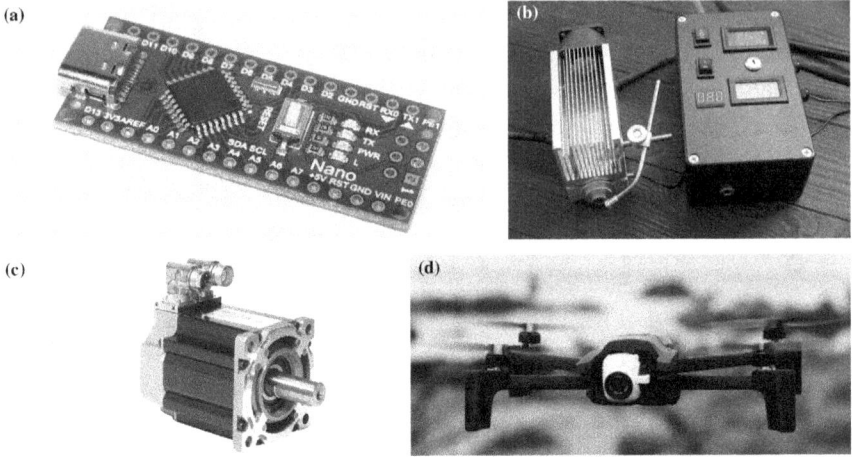

FIGURE 16.1 (a) Arduino Nano ATMEGA processor. (b) 2× laser module. (c) 4× servo motor. (d) Drone and camera

FIGURE 16.2 AI-enabled autonomous vehicles

and precise strikes (Figure 16.3). Because of AI-driven cyberspace operations that include both offensive and defensive measures, the diplomatic operations where social media has been a crucial component in disseminating false information and conducting information warfare have taken on a new dimension. Social media has proved to be a major weapon with deep fakes generated by AI having adverse effects on public opinions about certain individuals.

LAND SYSTEMS

Unmanned ground vehicles (UGV) will form the future spearheads of any blitzkrieg style push across enemy's territory (Figure 16.4). These systems need processing of

FIGURE 16.3 Propelled by AI-enabled ISR operations

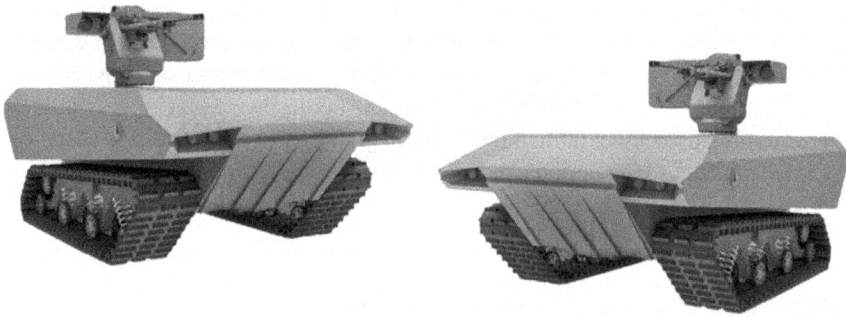

FIGURE 16.4 Unmanned ground vehicles (UGV)

peripheral response intelligence, big data knowledge in a cloud infrastructure, seamless network connections, on demand information resource extraction, and flexible organizational support in order to carry out autonomous combat operations. One such UGV was first successfully deployed in active combat by the Russian army in the Syrian Civil War in 2018.

AIR POWER

An effective method for enabling trapping systems that can operate in hazardous environments and in a variety of domains by adhering to complex patterns is the use of swarm drones and AI-informed guidance systems (Figure 16.5). Examples include the Combat Air Teaming System (CATS) program from Hindustan Aeronautics Limited (HAL), which pairs unmanned vehicles with a manned HAL Tejas aircraft. Swarm drones have already proved to be game-changer weapons in

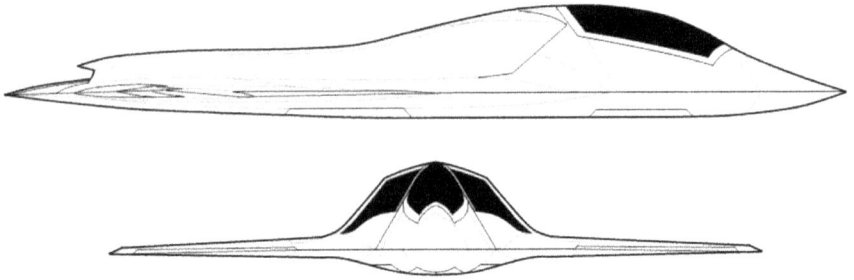

FIGURE 16.5   Next-generation air dominance (NGAD) program

the Russia–Ukraine war by successfully overwhelming air defenses with their sheer numbers.

OVER AND UNDER THE WAVES

The autonomous systems on the maritime domain consist of unmanned surface vehicles (USVs) and unmanned underwater vehicles (UUVs). An area of research into autonomous systems called "swarm technology" is less expensive than conventional systems and has the power to overcome enemies' protection mechanisms, which is where autonomy is headed. These systems can stay in the theater of operation for extended periods without human intervention giving them the advantage of collecting intelligence for long periods and attacking enemy assets at opportune time.

CYBER SECURITY

The strength of AI combined with cyberspace has given the future war, which easily integrates into the cyber framework, a new dimension. Both offensive and defensive uses of the AI-enabled system are possible in cyberspace. Applications including intelligence exploring, screening, network jamming, locating security holes, recognizing anomalies, and intrusion detection have already found application in electronic warfare for unconventional and asymmetric uses, a crucial part of discouragement. DARPA, for instance, has partnered with developing software solutions like CHASE, which utilizes AI to forecast and identify assaults, alongside BAE Systems as part of its 2016 Grand Challenge (Figure 16.6). Both sides of the conflict between Russia and Ukraine have recently deployed AI-enabled technologies and software to take down the adversary's vital infrastructure.

MILITARY LOGISTICS

Transporting personnel, equipment, and ammunition effectively is a component of military logistics. Predictive maintenance capabilities for private airplanes have been effectively deployed in the commercial sector by adding AI to their existing systems. Reports state that the US Air Force used maintenance schedules that

FIGURE 16.6 AI-enabled technologies and software in Russia and Ukraine war

were AI-enabled and customized for each aircraft. The F35 Autonomic Logistic Information System now makes use of it; a technician is informed when to inspect and repair aircraft parts using a predictive algorithm. The Army Logistic Support Activity (LOGSA) is the other that employs IBM Watson's predictive software to create maintenance schedules for the fleet of Stryker vehicles.

C2 (COMMAND AND CONTROL)

Operations in the air, sea, land, space, and cyberspace will be centralized and integrated via all-domain command and control. In order for the algorithms to establish correlations and predictions for improved decisions, the usage of AI will be beneficial for compiling data from several sources into a single database. Additionally, it will produce a unified operating picture that will consolidate information, address communication issues, and give forces access to alternative systems for significant actionable strategies (Khang et al., 2022).

LAWS

AI weaponization and intelligent warfare have changed national defense plans and contributed to the global arms race. Deadly autonomous weapon systems are one of the emerging, disruptive technologies that have caused some serious worries. Lethal autonomous weapon systems (LAWSs) are those that autonomously identify an object, engage the target, and operate the weapon system. The dynamic AI system, which is built on the system's prior knowledge, makes decisions.

COMBAT SIMULATIONS, TRAINING, AND WAR GAMES

AI-enabled systems and software are used in training and simulations to build models that replicate various military scenarios. Hence, AI helps military officers and soldiers comprehend the concepts of battle and get familiar with various fighting technologies. This software is used by the US Army to undertake warfare assessments. They do deal with companies including Orbital ATK, Leidos, SAIC, and AECOM who are supporting US army initiatives.

CASE STUDY

CASE STUDY: THE RUSSO–UKRAINIAN WAR

The conflict between Russia and Ukraine now serves as a battlefield for AI-enabled combat. In order to destroy each other's crucial infrastructure, it is claimed that both sides used AI-enabled software, hacks, the spread of deep fakes, and disinformation operations. Several drone footages have been used as part of the US Project MAVEN to collect and categorize data in order to find objects of interest.

According to reports, the USA is gathering and analyzing battlefield data from the Russia–Ukraine conflict using AI and ML algorithms to produce intelligence regarding Russian Command and Control Tactics. Also, geospatial activity has been detected using AI-based technologies like Space Know, and the Russian village of Yelna saw military activity in December 2021, soon before the war. Also, a US-based private company called Clear View AI gave Ukraine free access to their software in order to track the Russian military. This software's database, which contains 2 million images, was culled from different Russian social networking sites. Russian intelligence is likely responsible for the compromise of hundreds of systems at Ukrainian government institutions using destructive AI-enabled software.

INDIAN CAPABILITIES AND ROADMAP IN AI

India with its wide uses in a variety of industries, including healthcare (Khang et al., 2023c), agriculture (Khang et al., 2023a), education, smart cities (Khang et al., 2023b) and infrastructures, and smart mobility and transportation, the commercial establishment is growing. In 2018, NITI Aayog published the national AI strategy, which placed a strong emphasis on these topics. The development, acceptance, and support of the AI ecosystem in India are covered in the responsible AI Parts 1 and 2 report that was released in 2021.

The International Centers for Transformational AI (ICTAIs), which are focused on developing AI-based applications for societal domains, will get technology from the Centers for Research Excellence in AI (COREs) of NITI Aayog. The Defense Research and Development Organization (DRDO), which is a part of the Centre for AI and Robotics, is in charge of India's AI military research (CAIR). A high-level Defense AI Council (DAIC) was created by the United Kingdom Ministry of Defence (MoD) in 2019 with the goal of giving defense AI adoption strategic direction. A multi-stakeholder task group was established to investigate participants from the government, services, academia, and industry professionals, and begin to discuss the tactical implications of AI from a national security and global perspective.

The chapter discussed Indian AI developments in relation to the country's defense requirements and offered suggestions for future defense AI developments in fields such as aviation, maritime, land, cyber, nuclear, and biological warfare. In light of these advancements, a budget of $1 billion was allocated annually to AI-enabled projects by the DAIC and the Defense AI Project Agency (DAIPA).

The 12 AI fields specified by the national task group include the Indian army's usage of AI for C4ISR. RoboSen, an intelligent wheelchairs, micro-UAVs like Black Hornet, the AI-enabled robot for ISR operations, the small, man-portable walking robot for logistic assistance, cognitive robots for component maintenance and service-ability, and CAIR's Network Traffic Analysis (NETRA) system for real-time internet traffic monitoring are a few examples of India's AI military capabilities (C4ISR is Command, Control, Communications, Computers, Intelligence, Surveillance and Reconnaissance).

India has some drone capabilities, including swarm drones from Botlab Dynamics, Air Launched Flexible Asset Swarm (ALFA-S) from HAL and NRT, swarm drones, and Rustom 1 from DRDO. The International Accreditation Forum (IAF) has inaugurated the UDAAN-sponsored Artificial Intelligence (AI) Center of Excellence at the Air Force Station in New Delhi on July 10, 2022. A Technological Innovation Center on Autonomous Navigation and Data Collecting Systems has been established at IIT-Hyderabad (TiHAN). The unmanned ground vehicles' development by DRDO include the Mobile Autonomous Robotic System (MARS) UGV and the Arjun MK 1A battle tank-based UGV, each equipped with an indigenous geographic information system and a 120 mm cannon (INDIGIS). The defense start-up challenge's 6th version was launched this year in an effort to promote growth and development in the defense industry.

CONCLUSION

There are already some prototypes of swarm UAVs guided by AI coming up. Systems like these in the hands of terror organizations can wreak havoc on civilian lives. All these weapons produce a grim picture of future wars which may be an order of magnitude bloodier than those fought previously. In the future, we may see soldiers carrying ballistic computers on their rifles allowing them to shoot with near-perfect accuracy.

The next generation of fighter jets, i.e., the sixth generation, promise to become optionally unmanned and partially autonomous in their operations. These are going to be followed into the battlefield by various advanced drones like wingman, swarm drones, and suicide drones keeping these expensive aircraft safely away from the enemy while these drones will do the kinetic fighting. We will also see bigger and more capable unmanned ships and unmanned underwater vehicles (UUVs) soon. Some of these upcoming systems will likely be groundbreaking and revolutionary courtesy of computers and AI advancements and will forever change the way we perceive warfighting. Hence, all nations and militaries are investing heavily in it (Rana et al., 2021).

REFERENCES

Bistron, M., Piotrowski, Z., Artificial Intelligence Applications in Military Systems and Their Influence on the Sense of Security of Citizens. *Electronics*, vol. 10, 2021, p. 871. https://www.mdpi.com/2079-9292/10/7/871

Fazel Dehkordi, E., Grønli, T.-M., A Survey of Security Architectures for Edge Computing-Based IoT. *IoT*, vol. 3, no. 3, Sept. 2022, pp. 332–365. www.mdpi.com, https://doi.org/10.3390/iot3030019

Fraga-Lamas, P., Fernández-Caramés, T. M., Tactical Edge IoT in Defense and National Security. *IoT for Defense and National Security, IEEE*, 2023, pp. 377–396, doi: 10.1002/9781119892199.ch20

Gyamfi, E., Jurcut, A., Intrusion Detection in Internet of Things Systems: A Review on Design Approaches Leveraging Multi-Access Edge Computing, Machine Learning, and Datasets. *Sensors*, vol. 22, no. 10, Jan. 2022, p. 3744. www.mdpi.com, https://doi.org/10.3390/s22103744

Hui, W., Han, H., Wang, X., Sun, S., Research on Artificial Intelligence Enhancing Internet of Things Security: A Survey. *IEEE Access*, 2020. 10.1109/ACCESS.2020.3018170

Khang, A., Hahanov, V., Abbas, G. L., Hajimahmud, V. A., Cyber-Physical-Social System and Incident Management. *AI-Centric Smart City Ecosystems: Technologies, Design and Implementation* (1st Ed.), 2 (15), (2022). CRC Press. https://doi.org/10.1201/9781003252542-2

Khang, A., Gupta, S. K., Shah, V., Misra, A. (Eds.), *AI-aided IoT Technologies and Applications in the Smart Business and Production* (2023a). CRC Press. https://doi.org/10.1201/9781003392224

Khang, A., Gupta, S. K., Hajimahmud, V. A., Babasaheb, J., Morris G., *AI-Centric Modelling and Analytics: Concepts, Designs, Technologies, and Applications* (1st Ed.) (2023b). CRC Press. https://doi.org/10.1201/9781003400110

Khang, A., Rana, G., Tailor, R. K., Hajimahmud, V. A. (Eds.), *Data-Centric AI Solutions and Emerging Technologies in the Healthcare Ecosystem* (2023c). CRC Press. https://doi.org/10.1201/9781003356189

Khang, A., Abdullayev, V., Hahanov, V., & Shah, V., *Advanced IoT Technologies and Applications in the Industry 4.0 Digital Economy* (1st Ed.) (2024). CRC Press. https://doi.org/10.1201/9781003434269

Kuzlu, M., et al., "Role of Artificial Intelligence in the Internet of Things (IoT) Cybersecurity." *Discover Internet of Things*, vol. 1, no. 1, Feb. 2021, p. 7. Springer Link, https://doi.org/10.1007/s43926-020-00001-4

Nassereddine, M., Khang, A., Applications of Internet of Things (IoT) in Smart Cities. *Advanced IoT Technologies and Applications in the Industry 4.0 Digital Economy* (1st Ed.) (2024). CRC Press. https://doi.org/10.1201/9781003434269-6

Rana, G., Khang, A., Sharma, R., Goel, A. K., Dubey, A. K. (Eds.), *Reinventing Manufacturing and Business Processes Through Artificial Intelligence* (2021). CRC Press. https://doi.org/10.1201/9781003145011

Rath, K. C., Khang, A., Roy, D., The Role of Internet of Things (IoT) Technology in Industry 4.0. *Advanced IoT Technologies and Applications in the Industry 4.0 Digital Economy* (1st Ed.) (2024). CRC Press. https://doi.org/10.1201/9781003434269-1

17 Internet of Things (IoT) Integration with 5G and 6G Wireless Technologies

Shahzada Asif, Rishabh Pal, Vandana Dubey,
Priti Kumari, and Sarika Shrivastava

INTRODUCTION

Nowadays, people are connected to the internet to process information worldwide. People used their devices such as computers, mobile phones, and laptops for accessing the internet for search information, reading online documents, sending emails and messages, listening to online music or watching videos or download them, and for many things (Figure 17.1) (Linux Hint, 2023). Moreover, internet of things (IoT) is the technology used to connect devices and reduce human interaction, make the devices fully automatic, reduce human effort, etc.

In technical terms, the IoT is one of the most important and promising innovations today. The IoT is the collective network of interconnected objects and the technology that enables interaction among them as well as with the public internet. IoT is the interconnection of embedded devices, and these devices can be computers, various types of sensors, and other electronic devices too. The connection between devices will take place by using the internet, which can be either 5G or 6G (Shafique et al., 2020; Laghari et al., 2021).

Moreover, the IoT integrates everyday "things" with the internet. These days, the user uses everything connected with technology whether we use cell phones, smart watches, smart home appliances, and so on. The IoT has evolved because of the conjunction of numerous technologies, such as pervasive computing, commodity sensors, and connection with artificial intelligence and machine learning. Many modern products, including vacuums, vehicles, and robots, employ sensors to gather data and react to users in a sophisticated manner (Swamy and Kota, 2020). IoT is highly used in many fields such as smart car parking, autonomous car, smart waste management, smart traffic signals, smart houses, and home security system, smart car parking, autonomous car, smart waste management, and smart traffic signals (Al-Masri et al., 2020; Stoyanova et al., 2020).

DOI: 10.1201/9781003434269-17

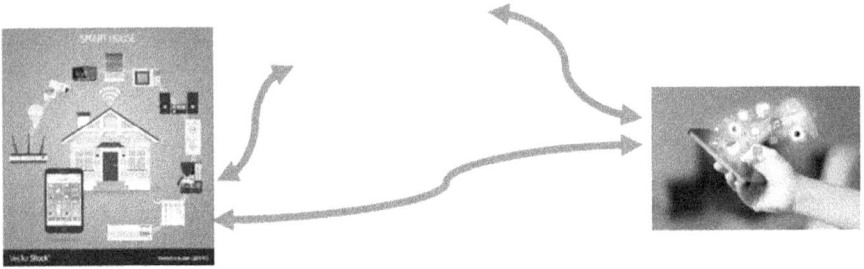

FIGURE 17.1   Connectivity of things and people with the internet

In summary, the internet of things is a type of technology that is all about electronic devices which were connected with the other devices using Bluetooth, or the internet and use sensors to operate and control them.

REQUIREMENTS OF IoT

Connectivity

The ability to join the internet via a wired/wireless link, such as 4G/5G/6G, is required for IoT devices. The internet of things (IoT) depends on connectivity because it enables communication between IoT devices and a centralized server or cloud platform. IoT devices can connect to the internet via a variety of technologies and protocols, such as Wi-Fi, Bluetooth, and cellular networks. The location, power needs, and data transfer requirements of the device will all have an impact on the technology that is chosen. For instance, if an automated irrigation system is installed in a remote field without connecting to a Wi-Fi signal, it can link to the internet via a cell phone network.

Sensors

IoT gadgets need detectors that can gather information about their environment, i.e., moisture, humidity, temperature, movement, or noise. The actions or user information are triggered by this data. Connection between sensors and devices can be made internally or externally via a sensor module or smart hub, for example. Temperature sensors, humidity and motion sensors, and sound sensors are a few of the typical sensor types utilized in the internet of things. They might be incorporated into the gadget or connected by means of an outside component, like a sensor component or a smart hub (Hahanov et al., 2022). Sensor data is frequently utilized to start actions or give users information. For example, an automated irrigation system can utilize environmental sensor data to decide the appropriate amount of fluid to supply to grass, or an intelligent surveillance system can employ facial recognition to assess whether an individual accessing a property is an authorized user.

Intelligence

The data that IoT gadgets collect must be analyzed, examined, and used for making decisions. This can be accomplished using internal software or by transferring data to a centralized computer for processing. For instance, a smart surveillance camera might use face recognition to evaluate whether someone entering a home is an approved user, while a sophisticated irrigation system might utilize climate sensor data to decide how much water to apply to a lawn. Intelligence is a required element for the IoT. It makes it possible for IoT devices to study, interpret, and take decisions using the data they collect. Lack of intelligence results in an IoT device's collapse to complete its intended duty or provide value to the consumer (Hajimahmud et al., 2022).

INPUT AND OUTPUT DEVICES FOR IoT

Input Device

Input devices give input to the cloud to react on the problem occurs. The devices are LDR, push button, temperature sensor, fingerprint sensor, smoke sensor, keypad, sound detection sensor, etc. (IoT, 2023; Twi-Global, 2023).

LDR

LDR stands for light-dependent resistor (LDR), which is a light sensor. This device is a common input device used in various devices and further upcoming projects. These photo resistors are delicate to sunlit which can alliterate their resistance reliant on the deteriorating light on them.

Pushbutton

Pushbutton is another type of input device that is used to permit to flow of information in the form of electricity (Figure 17.3).

Temperature Sensor

It is also a type of input device, which can interface through Arduino to give the value of the temperature of the surrounding temperature (Figure 17.4). The temperature sensor is made up of the resister so it senses the temperature from the outer surrounding and when the outer temperature gets increase the resister also gets increase.

FIGURE 17.2   LDR architecture (GoogleUserContent, 2023)

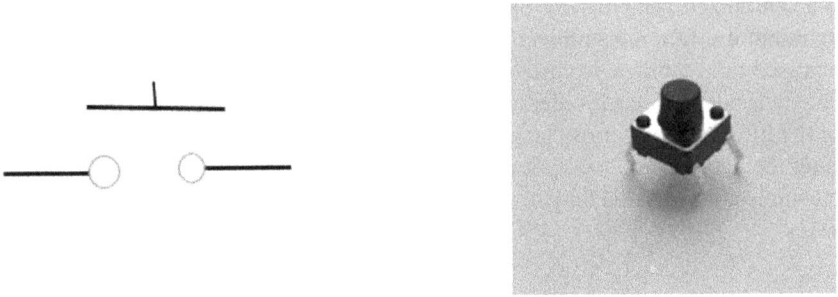

FIGURE 17.3   Symbol and circuit of the pushbutton

FIGURE 17.4   Temperature sensor

Fingerprint Sensor

This is used to take the input of fingers imprint for safety purposes. There are diverse types of sensors, and the most used sensor is r503 which can be connected to the Arduino (Figure 17.5).

Output Devices

The output devices used in Arduino are the output device that takes instruction from the Arduino and provide result in a user-readable form. Some output devices that are used are LEDs, motors, LCDs, and many more devices (IoT, 2023; Twi-global, 2023).

Light-Emitting Diode (LEDs)

With Arduino, LEDs are employed as output tools, and they may be used for a variety of applications. For instance, various kinds of indications are connected to dissimilar LED colors (Figure 17.6).

FIGURE 17.5   Fingerprint sensor

FIGURE 17.6   LED

Motors

DC motors are used to run or control another device. For instance, if water is to be pumped using DC motor present in Arduino, then it is assumed to be an output device.

Liquid Crystal Display (LCDs)

To show the productivity of the Arduino program, these displays are linked to Arduino, which comes in different sizes like 16×2; these display dimensions are commonly used for projects.

IMPLEMENTATION OF IoT WITH ARDUINO

A tiny circuit board that serves as a microcontroller, i.e., used for the creation of electronic gadgets, is designed and manufactured by Arduino. The Arduino is a freely available software and hardware company, projects, and user community. Both the hardware and the applications require licenses in order to be manufactured (Khang et al., 2024b); the hardware is licensed under a CC BY-SA license, and the software is licensed through the GNU LESSER General Public License (LGPL) or the GNU LESSER General Public License (GPL). The C and C++ coding languages, as well as the conventional API, formerly known as the Arduino language, are utilized to program this microcontroller. Arduino board is used to communicate with the system to provide instruction to the microcontroller with the help of USB (Universal Serial Bus) that is used for loading the program in the microcontroller (Linux Hint, 2023; Arduino, 2023).

FIGURE 17.7   Motors

FIGURE 17.8 LCDs

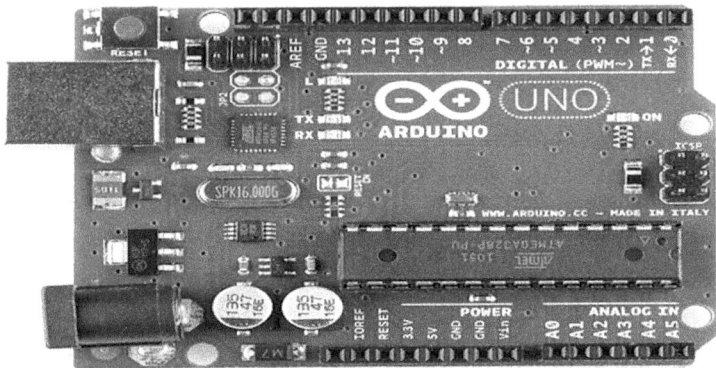

FIGURE 17.9 Arduino UNO

Arduino Pins

Arduino is comprised of different kinds of pins and each of these pins performs specific tasks. The list of Arduino pin is demonstrated in Figure 17.10.

USB Plug

Arduino board gets powered by the computer devices with the help of a USB cable. USB means Universal Serial Bus, i.e., used to load the program into the microcontroller.

- **External Power Supply**: Arduino board gets a direct power supply by connecting AC mains to the external power supply which is called Barrel Jack.
- **Reset Pin**: There are two different ways to reset the programs of microcontroller. Primarily, the reset button present on the board is used. Second, the external reset button can be connected through the reset pin present in the Arduino.
- **Volt Power Pin**: It is used to supply 3.3 outcome voltage.
- **Volt Power Pin**: This pin was used to provide the power of five output voltage devices which are connected to the Arduino.

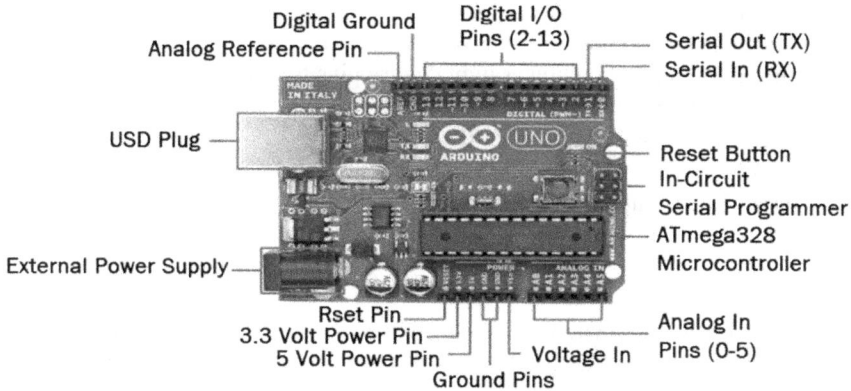

FIGURE 17.10 Arduino pins

- **Ground Pin**: The ground pin was indicated by the GND in the Arduino board which is used to close the electrical circuit and provide a common reference level throughout the circuit that you use.
- **Voltage In**: It is used to provide an external power supply, which is AC mains. Voltage In indicated by Vin in on Arduino board.
- **Analog Pins**: The Arduino board contains five analog pins which were indicated by A0–A5. These pins are used to sense or read signals which were sent by sensors like moisture sensors, and sensors for temperature and then translate them into digital form, which can be understandable to the microcontroller.
- **Microcontroller**: In human, there is a brain that instructs the parts of the body that Arduino board also consists of the brain which was called a microcontroller. The integrated circuit of the Arduino board is slightly different from each other. The microcontroller loads a new program from the Arduino IDE.
- **In-Circuit Serial Programmer (ICSP) Pin**: ICSP is an AVR, a minute programming header comprising MOIS, MISO, SCK, RESET, VCC, and GND. ICSP is also known as the Serial Peripheral Interface (SPI), which would help in the expansion of the output.
- **Serial In (RX)**: Serial in (RX) was used to transfer characters over serial from your computer to the Arduino; we have to just open the serial monitor and type something in the field next to the send button.
- **Serial Out (TX)**: These pins are used for communication with computers. The serial communication pins use TTL logic levels that don't connect directly to the serial ports; these operate at the ±12 V and will damage the Arduino board. TX was more complex than the RX.
- **TX and RX LEDs**: The TX LED sparks during the transmission of the data, the blinking of the LED depended on the speed of the sending of the serial data, and the RX LED bling when any process or data gets received.

- **Digital I/O Pins**: There are 14 digital I/O pins on the Arduino board, of which 6 are used for the Pulse Width Modulation (PWM) outcome. These pins are set up to function as digital entry pins for reading logical numbers (0 or 1) and as digital outcome pins to drive various modules, including LEDs, servo motors, and other devices.
- **AREF**: AREF, which stands for Analogue Reference Pin, has been used for a while to supply external voltage for the input pins' maximum limits.

APPLICATIONS OF THE IoT

Removing the requirement for a from person to person or human-to-computer communication system, the IoT enables the interconnection of computer devices, mechanical equipment, items, animals, and unique identifiers/individuals. The IoT provides objects, computing devices, or unique identifiers and people's ability to transfer data across a network without human-to-human or human-to-computer interaction (IoT, 2023; Twi-global, 2023). The major application area of IoT is demonstrated in Figure 17.11.

Figure 17.11 depicts the IoT applications significantly improve today's lives. They are:

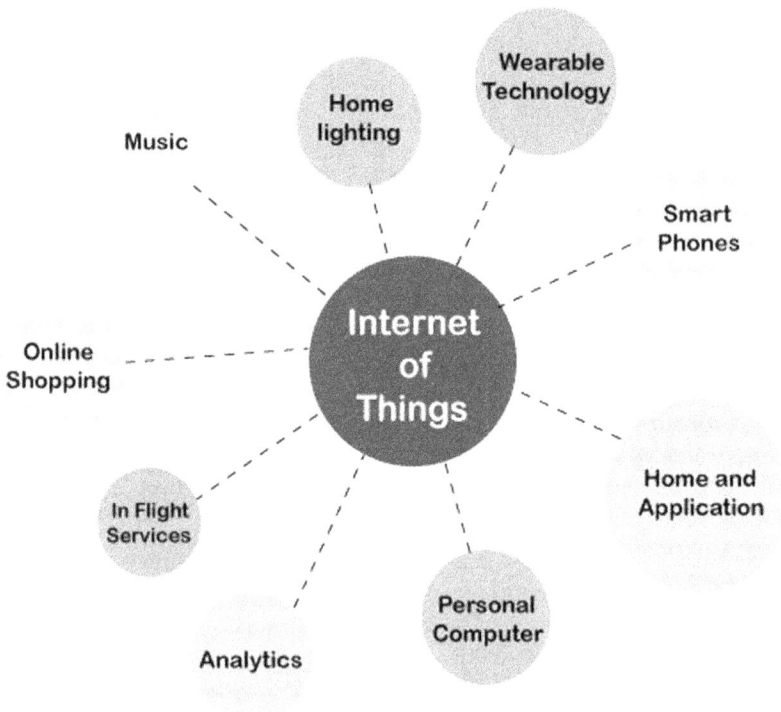

FIGURE 17.11   Application of IoT (JavaPoint, 2023)

- **Wearables**: One of the first businesses to adopt IoT was the wearable technology sector, which is the defining feature of IoT applications. These days, we have smart watches, fit bits, and heart rate monitors. To assist diabetics, the Guardian glucose monitoring device has been created. It connects a radiofrequency monitoring device to a tiny electrode known as the glucose sensor to measure the levels of glucose in our bodies.
- **Smart Home Applications**: When discussing IoT applications, the smart home is likely the first thing that comes to mind. As an example, Mark Zuckerberg uses AI to automate his home. The home automation system invented by Alan Pan employs built-in capabilities to play a series of musical notes (Rana et al., 2021).
- **Health Care**: Responsive medical-based schemes can become active wellness-based with the help of IoT technologies. Resources used in current medical research lack crucial information from the real world. Clinical studies make use of controlled surroundings, old data, and volunteers. The power, accuracy, and availability of the gadget are all improved through the internet of things. IoT concentrates on creating systems as opposed to mere tools. This is how IoT-enabled medical equipment functions.
- **Smart Cities**: Most people are familiar with the term "smart city." Technology is used in smart cities to deliver services. The smart city comprises enhancing social services and transportation, encouraging stability, and providing inhabitants with a voice.
- **Agriculture**: By the year 2050, it is expected that there will be around 10 billion individuals on the planet. For agriculture to adequately sustain such a large population, technology must be incorporated. There are numerous options in this regard. The smart greenhouse is one of them.
- **Industrial Automation**: It is one of the zones where a higher investment return is dependent on the quality of the items. Anyone can redesign products and their packaging to offer IoT applications that deliver improved performance in terms of cost and user experience.
- **Hacked Car**: A linked automobile is a technologically advanced vehicle with WAN and internet connectivity. The user may take advantage of features like enhanced navigation, in-car entertainment, and fuel efficiency, thanks to technology.
- **Healthcare**: Healthcare uses smart technologies to perform real-time monitoring. It collects and transmits health information like ECG, blood pressure, blood sugar, and oxygen levels. In the event of an emergency, the patient can get in touch with the doctor via a smart smartphone application.
- **Smart Retail**: Retail IoT solutions provide customers with a novel experience. Customers don't need to wait in queue for long because the checkout system uses IoT applications to read product tags and subtract the full from the client's payment app.
- **Smart Supply Chain**: Customers use a smart supply chain to automate shipment and delivery. Additionally, it provides information on current conditions and supply chains.

- **Smart Farming**: Farmers may cut back on leftovers and increase output. The technology permits the usage of sensors to monitor grounds. Farmers can keep an eye on the area's condition.

5G WIRELESS TECHNOLOGY

The fifth generation of wireless technology is recognized as 5G. It is the most recent mobile technology and will, among other things, speed up wireless connections. There are an infinite number of users using mobile phones worldwide, and there is a requirement for high-speed internet to transfer data in the last two decades. Today's 5G wireless technology provides relatively fast data speed along with low latency. The development of the 5G mobile network meets with the latest applications, and the bandwidth of the system increased widely. Moreover, IoT and communication with machines need high speed of internet and high speed of data transfer. 5G network has developed in different directions like unlimited data transfer, a huge amount of active connection, and new types of electronic devices powered by sustainable sources (Mahdi et al., 2021; Gustavsson et al., 2021).

In India, mobile network companies have undergone a tectonic shift. Now network markets have become highly competitive with invent of new network technology that offers inexpensive and better voice and data plans. This leads to a cold war between the network companies and affects all the customers. The National Digital Communications Policy of 2018 attracted investment worth US$100 billion in the broadcastings sector in 2022.

In the future, 4G/LTE may not reach expected speeds due to the congestion of networks, and the growing demands for new technologies, such as unmanned aerial vehicles (UAVs) or autonomous vehicles as well as many new technologies are being introduced, these all-new technologies need high speeds to operate while 4G has not yet met requirement.. The 5G technology (5G offers a maximum speed of 20 Gbps) provides network administration features, i.e., network slicing to help mobile operators to generate numerous virtual networks using a solitary physical 5G network (Chowdhury et al., 2019; Kim, 2021) as shown in Figure 17.12.

HOW 5G TECHNOLOGY WORKS?

There are two key components of the 5G technology, i.e., the Radio Access Network (RAN) and Core Network (Chowdhury et al., 2019; Salameh and El Tarhuni, 2022).

Radio Access Network (RAN): The core of 5G wireless connectivity is composed of 5G small and macro cells, which make up the RAN. Since the millimeter wave spectrum can only travel a limited distance, the 5G minor cells are situated in the large cluster, which offers long-range coverage. Multiple input, multiple output (MIMO) antennae with multiple connections are used by macro cells to transmit and receive enormous volumes of data at once (Pan et al., 2022).

Core Network: All information and internet connections for the 5G wireless network will be handled by the Core Network. The 5G network's major benefits include improved internet integration and the availability of new services.

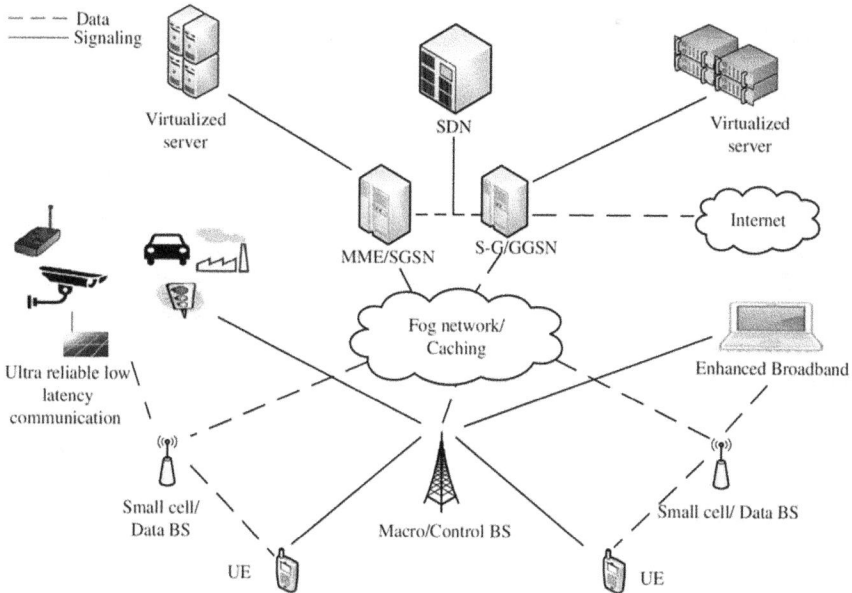

FIGURE 17.12   5G network architecture

EVOLUTION OF 5G TECHNOLOGY

The fifth wireless technology was first launched in South Korea and the USA in April 2019. Japan plans to launch the 5G wireless technology at the time of the Tokyo Summer Olympics (Figure 17.13).

The Central Government in India has also set a board to launch 5G wireless technology; the government has already set the program of launching a three-year program in March 2018 (Shukurillaevich et al., 2019). A comparison of 5G with other wireless technologies is demonstrated in Table 17.1.

PARAMETER OF 5G WIRELESS TECHNOLOGY

There are various parameters for 5G wireless technology, which are discussed in Table 17.2.

INTEGRATION OF 5G AND IoT

When IoT is connected to the devices, then there is a requirement of the internet to access the sensors. In the case of huge traffic, the response time of the sensor decreases, which further increases the latency for the users. For instance, if a self-driving car connected to the Cloud and the internet and the wireless network suffers from the data traffic, then the instruction that comes from the sensor will work slowly and it may face an accident. So, to solve these problems, the integration of a 5G

The Evolution of **5G**

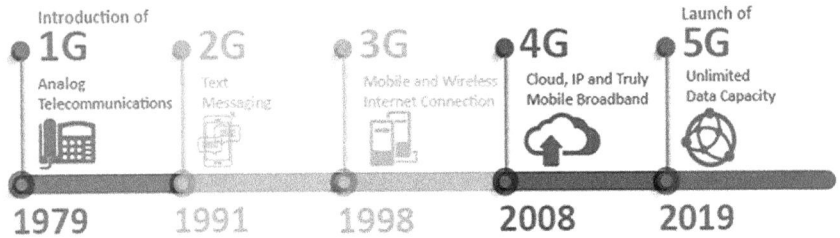

FIGURE 17.13   Evolution of 5G wireless technology

wireless network with IoT provides better speed to the device or sensors (Chowdhury et al., 2019; Kim, 2021).

Advantages of the 5G Wireless Technology

- The 5G wireless technology provides high speed.
- This technology provides low latency and high capacity.
- 5G technology has high bandwidth which provides seamless transfer of the data.

Disadvantage of the 5G Wireless Technology

- 5G technology has increased the risk of hacking, which means that the new technology has less security and the hacker can easily access the data (Khang et al., 2022).
- This technology has a high speed of network running because of which the battery of the device gets drained fast.

To overcome the above-mentioned issues, the 6G wireless technology comes into existence.

6G WIRELESS TECHNOLOGY

6G means sixth-generation wireless technology, which is the advanced version of 5G technology. This technology needs high frequency than the 5G network, i.e., 1000 times faster or reduce the latency (1/1000th). In other words, one of the major goals of 6G internet would be to support 1-microsecond latency communication that is about 1000 times faster than the 1-millisecond throughput. 6G will probably be a broadband cellular network like its predecessors (5G, 4G, 3G, etc.) in which the services are divided into small geographical areas termed cells (Chowdhury et al., 2019) (Salameh and El Tarhuni, 2022; Kim, 2021).

TABLE 17.1

Comparison between 1G, 2G, 3G, 4G, and 5G

Technology/Feature	1G	2G	3G	4G	5G
Evolution	1970	1980	1990	2000	2010
Development	1984	1999	2002	2010	2015
Data rate	2 Kbps	14.4–64 Kbps	2 Mbps	200 Mbps to 1 Gbps for low mobility	10–100 Gbps
Famous standard	AMPS	2G: GSM, CDMA 2.5G:GPRS EDGE, 1×RTT	WCDMA, CDMA-2000	LTA, WiMAX	Not defined to date
Technology behind	Analog cellular technology	Digital cellular technology	Broad bandwidth CDMA, IP technology	Undefined IP and seamless combination of broadband. LAN/WAN/PAN/WLAN	Undefined IP and seamless combination of broadband. LAN/WAN/PAN/WLAN
Service	Voice	2G: Digital voice, SMS	Integrated high-quality audio, video, and data	Dynamic information access, wearable device	Dynamic information access, wearable device with AI capabilities
Multiplexing	FDMA	FDMA, CDMA 2G:	CDMA packets	CDMA packets	CDMA packets
Type of switching	Circuit	Circuit			
Handoff	Horizontal	Horizontal	Horizontal	Horizontal	Horizontal and vertical
Core network	PSTN	PSTN	Packet network	Internet	Internet

TABLE 17.2
Parameters of 5G Technology

Parameter	Description
Peak data rate	10 Gbps uplink and 20 Gbps downlink minimum per mobile base station
Real-world data rate	Download speed = 100 Mbps and upload speed = 50 Mbps
Spectral efficiency	30 bits/Hz downlink, 15 bits/Hz uplink
Latency	Extreme latency for 4 ms compared to 20 ms for LTE
Connection density	Minimum 1 million linked devices per square km to permit IoT support

This 6G technology provides an improvement over existing technology and provides conjunction with artificial intelligence (AI) and machine learning. This technology supports the applications that are used in mobile, i.e., virtual/augmented reality (VR/AR). Many companies like Airtel, Apple, LG, Samsung, Xiaomi, and Nokia, research institutes like "Technology Innovation Institute," some countries like the USA, India, China, Singapore, and South Korea, and many other countries have shown their interest in 6G networks (Electronics Notes, 2023). 6G technology may start from or appear in the technical community in 2030 to 2035; if there is a fall in the 5G then we all have to switch to the 6G proposal.

IoT becomes more accelerated in the launch of 6G wireless networks because it has become faster and connects more people with the sensor and technical networks. Countless sensors are embedded in the physical world to transfer many data with several devices (Khanh and Khang, 2021). This technology made businesses more powerful and connected people from all over the world with each other (Electronics Notes, 2023). 6G technology is expected to bring a huge change in telecommunication as shown in Figure 17.14.

However, there is no universally accepted standard for the 6G network. It is expected that 6G will be using terahertz and millimeter waves as part of its technology. A systematic architecture of future 6G technology is shown in Figure 17.14 and the evolution of 6G wireless technology is demonstrated in Figure 17.15.

Terahertz and Millimeter Wave Progress

It is believed that 6G will be using millimeter waves and terahertz radiations that range from 30 GHz to 300 GHz and from 300 GHz to 3000 GHz respectively. The use of these waves will make 6G more sensitive to obstacles. In the terahertz frequency spectrum, Purple Mountain Laboratories of China announced in January 2022 that its research team had for the first time in a lab setting reached a speed of 206.25 gigabits per second. Chinese researchers claim to have transmitted 1 TB of data over a distance of 1 km (3300 feet) in only seconds utilizing vortex millimeter waves in February 2022, setting a new record for information transfer speed.

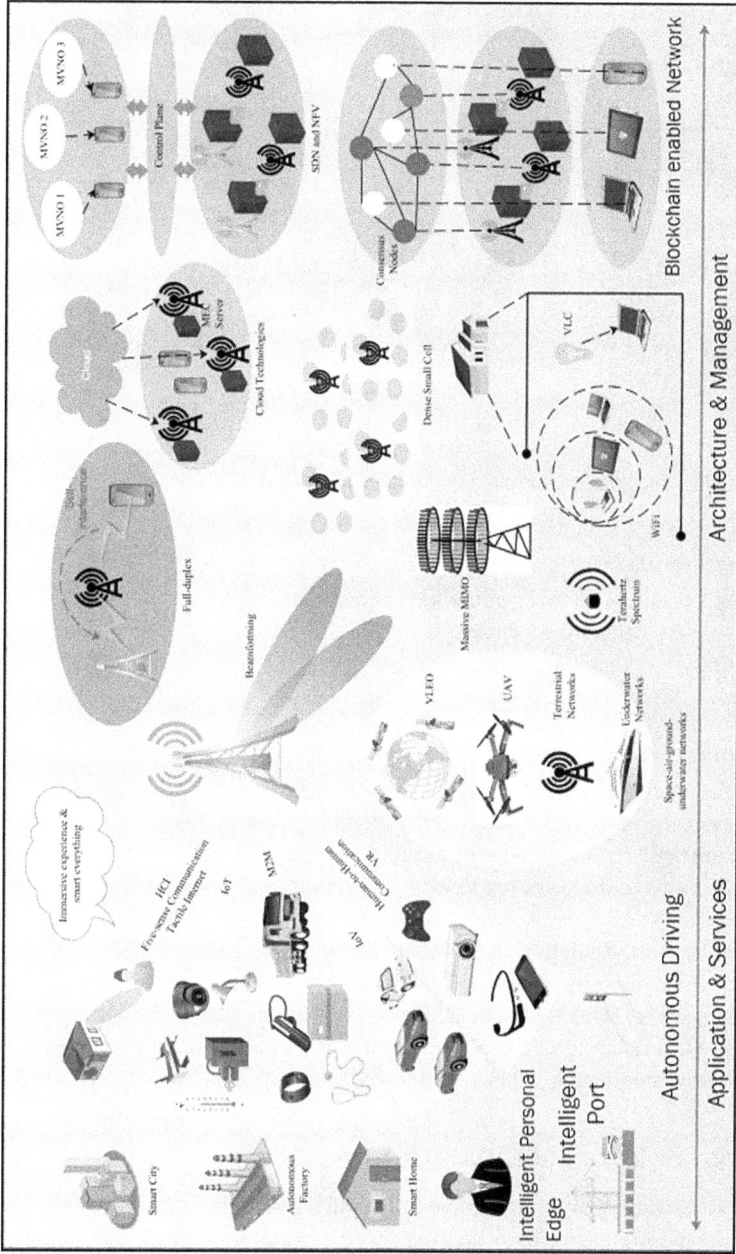

FIGURE 17.14 6G wireless technology (Guo et al., 2021)

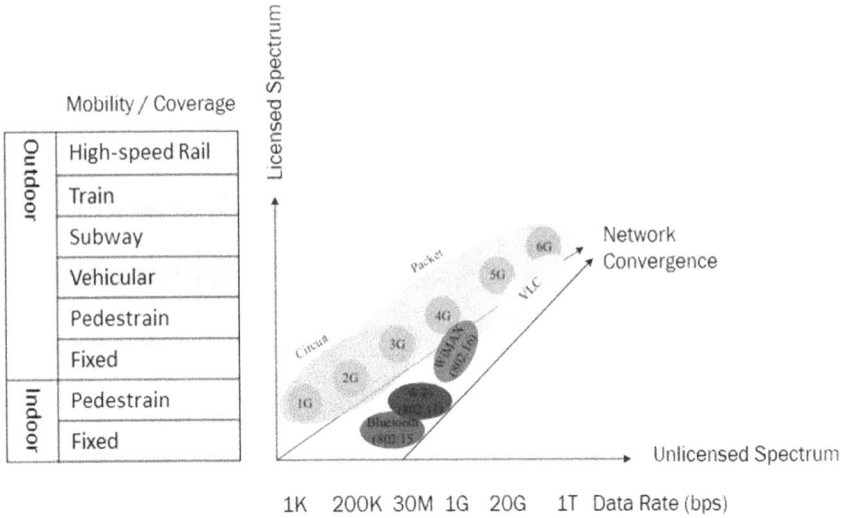

FIGURE 17.15   6G wireless technology evolution (Guo et al., 2021)

Advantages of 6G
- Supports a huge number of mobile connections.
- Have independent frequencies.
- Have high data rates.
- Have a large area of coverage.

Disadvantages of 6G
- Higher radiations harm health.
- It is expensive as compared to other technology available.
- It is difficult to use because of its complexity.
- As it is new to the outer world, compatibility issues are more likely to arise.

According to the speculation in 6G technology, there will be 30–300 GHz millimeters and THz radiation 300–3000 GHz will be used, and the wave frequency or wave propagation will be more sensitive to the obstacle than the frequency of the microwave that was used in the 5G network and Wi-Fi that was more delicate than the radio wave used in 1G, 2G, 3G, and 4G.

INTEGRATION OF IoT WITH 6G WIRELESS TECHNOLOGY

Nowadays, there are a lot of disruptive IoT applications that are big in number, data- and computation-intensive, and delay-sensitive. Examples include augmented and virtual reality online gaming, autonomous vehicles, and smart things. There is a need for technological breakthroughs and evolutions for wireless communications and networking toward the sixth-generation (6G) networks due to the mismatch between the requirements of such enormous IoT-enabled applications and the fifth-generation (5G) networks (Khang et al., 2024a).

The highly anticipated expanded 5G capabilities of 6G, including Tbps data rate, less latency, cm-level localization, and others, will be crucial in enabling enormous IoT devices to run without a hitch with a wide range of service requirements (Kali et al., 2024).

CONCLUSION

IoT brings people to connect to smart technologies to save time. Nowadays technology is at its peak; all devices are connected to the internet and solve many problems in less period. A huge number of devices are associated with the sensors and need no or less human effort to work, and the devices run automatically by connecting to the internet. To put it another way, the internet of things (IoT) is becoming increasingly popular as a means of interacting with smart gadgets as a result of the rising demands placed on wireless networks (Nassereddine and Khang, 2024). This necessitates evaluating the improved wireless networks as next-generation communication systems, such as 5G and 6G wireless communication.

The 5G networks can serve a wide range of internet of everything (IoE)-based services; however, they are unable to fully satisfy the needs of the newest smart applications. 6G wireless communications are necessary to accommodate enormous data-driven applications and the growing user base to solve this problem. Moreover, it can be said that the integration of IoT with 5G and 6G will make it possible to connect the physical world with the digital world. This integration possesses some advantages such as enabling faster data transfer speed, higher bandwidth, better security and privacy, and so on (Khang et al., 2024c).

REFERENCES

Al-Masri, E., Kalyanam, K. R., Batts, J., Kim, J., Singh, S., Vo, T., Yan, C. (2020). Investigating messaging protocols for the Internet of things (IoT). *IEEE Access*, 8, 94880–94911. https://ieeexplore.ieee.org/abstract/document/9090208/

Arduino. (2023). Available online 2023. https://linuxhint.com/arduino-output-devices/

Chowdhury, M. Z., Shahjalal, M., Hasan, M. K., Jang, Y. M. (2019). The role of optical wireless communication technologies in 5G/6G and IoT solutions: Prospects, directions, and challenges. *Applied Sciences*, 9(20), 4367. https://www.mdpi.com/2076-3417/9/20/4367

Electronics Notes. (2023). Available online 2023. https://www.electronics-notes.com/articles/connectivity/6g-mobile-wireless-cellular-technology-basics.php

GoogleUserContent. (2023). https://lh3.googleusercontent.com/tRwXyd-Jg2xC-WXZSo5hRtYLZJ8b_kOc_nk-uyC-sV6YaYKTuwX3nr0aWL7cLabraEVLGfZymrx bc9LvoUkv23T5HuLW-PiagQW78bw and https://lh3.googleusercontent.com/W0l nNBe69sUnwcG3nbxZsDztQo_gcqIvw4JwPa1uAWYysKEz2DzMgEFwO45zDe CvMTAEXyQl30iga4nQ7ePabj6V8147qptQIpDS3i48w

Guo, F., Yu, F. R., Zhang, H., Li, X., Ji, H., Leung, V. C. (2021 March 4). Enabling massive IoT toward 6G: A comprehensive survey. *IEEE Internet of Things Journal*, 8(15), 11891–11915. https://ieeexplore.ieee.org/abstract/document/9369324/

Gustavsson, U., Frenger, P., Fager, C., Eriksson, T., Zirath, H., Dielacher, F., Carvalho, N. B. (2021). Implementation challenges and opportunities in beyond-5G and 6G communication. *IEEE Journal of Microwaves*, 1(1), 86–100. https://ieeexplore.ieee.org/abstract/document/9318749/

Hahanov, V., Khang, A., Litvinova, E., Chumachenko, S., Hajimahmud, V. A., Alyar, V. A. (2022). The key assistant of smart city - sensors and tools. In *AI-Centric Smart City Ecosystems: Technologies, Design and Implementation* (1st Ed.), 17(10). CRC Press. https://doi.org/10.1201/9781003252542-17

Hajimahmud, V. A., Khang, A., Hahanov, V., Litvinova, E., Chumachenko, S., Alyar, V. A. (2022). Autonomous robots for smart city: Closer to augmented humanity. In *AI-Centric Smart City Ecosystems: Technologies, Design and Implementation* (1st Ed.), 7(12). CRC Press. https://doi.org/10.1201/9781003252542-7

JavaPoint. (2023). https://static.javatpoint.com/blog/images/internet-of-things-applications.png

IoT. (2023). Available Online 2023. https://www.researchgate.net/publication/ 320532203_Internet_of_Things_IoT_Definitions_Challenges_and_Recent_Research_Directions

Kim, J. H. (2021). 6G and Internet of Things: A survey. *Journal of Management Analytics*, 8(2), 316–332. https://www.tandfonline.com/doi/abs/10.1080/23270012.2021.1882350

Khang, A., Hahanov, V., Abbas, G. L., Hajimahmud, V. A. (2022). Cyber-physical-social system and incident management. In *AI-Centric Smart City Ecosystems: Technologies, Design and Implementation* (1st Ed.), 2(15). CRC Press. https://doi.org/10.1201/9781003252542-2

Khang, A., Abdullayev, V., Hahanov, V., Shah, V. (2024a). *Advanced IoT Technologies and Applications in the Industry 4.0 Digital Economy* (1st Ed.). CRC Press. https://doi.org/10.1201/9781003434269

Khang, A., Misra, A., Abdullayev, V., Litvinova, E. (Eds.). (2024b). *Machine Vision and Industrial Robotics in Manufacturing: Approaches, Technologies, and Applications.* CRC Press. https://doi.org/10.1201/ 9781003438137

Khang, A., Gujrati, R., Uygun, H., Tailor, R. K., Gaur, S. S. (2024c). *Data-Driven Modelling and Predictive Analytics in Business and Finance.* CRC Press. https://doi.org/10.1201/9781032600628

Khanh, H. H., Khang, A. (2021). The role of artificial intelligence in blockchain applications. *Reinventing Manufacturing and Business Processes through Artificial Intelligence*, 2, 20–40. CRC Press. https://doi.org/10.1201/9781003145011-2

Laghari, A. A., Wu, K., Laghari, R. A., Ali, M., Khan, A. A. (2021). A review and state of art of Internet of Things (IoT). *Archives of Computational Methods in Engineering*, 1–19. https://link.springer.com/article/10.1007/s11831-021-09622-6

Linux Hint. (2023). Available online 2023. https://linuxhint.com/list-of-arduino-input-devices/

Mahdi, M. N., Ahmad, A. R., Qassim, Q. S., Natiq, H., Subhi, M. A., Mahmoud, M. (2021). From 5G to 6G technology: Meets energy, internet-of-things and machine learning: A survey. *Applied Sciences*, 11(17), 8117. https://www.mdpi.com/2076-3417/11/17/8117

Nassereddine, M., Khang, A. (2024). Applications of Internet of things (IoT) in smart cities. In *Advanced IoT Technologies and Applications in the Industry 4.0 Digital Economy* (1st Ed.). CRC Press. https://doi.org/10.1201/9781003434269-6

Pan, T., Zhu, Y., Pan, T., Zhu, Y. (2022). *Getting started with Arduino.* Maker Media, Inc. https://link.springer.com/chapter/10.1007/978-981-10-4418-2_1

Rana, G., Khang, A., Sharma, R., Goel, A. K., Dubey, A. K. (Eds.). (2021). *Reinventing Manufacturing and Business Processes through Artificial Intelligence.* CRC Press. https://doi.org/10.1201/9781003145011

Rath, K. C., Khang, A., Roy, D. (2024). The role of Internet of Things (IoT) technology in Industry 4.0. In *Advanced IoT Technologies and Applications in the Industry 4.0 Digital Economy* (1st Ed.). CRC Press. https://doi.org/10.1201/9781003434269-1

Salameh, A. I., El Tarhuni, M. (2022). From 5G to 6G—Challenges, technologies, and applications. *Future Internet*, 14(4), 117. https://www.mdpi.com/1999-5903/14/4/117

Shafique, K., Khawaja, B. A., Sabir, F., Qazi, S., Mustaqim, M. (2020). Internet of things (IoT) for next-generation smart systems: A review of current challenges, future trends and prospects for emerging 5G-IoT scenarios. *IEEE Access*, 8, 23022–23040. https://ieeexplore.ieee.org/abstract/document/8972389/

Shukurillaevich, U. B., Sattorivich, R. O., Amrillojonovich, R. U. (2019, November). 5G technology evolution. In *2019 International Conference on Information Science and Communications Technologies (ICISCT)* (pp. 1–5). IEEE. https://ieeexplore.ieee.org/abstract/document/9011957/

Stoyanova, M., Nikoloudakis, Y., Panagiotakis, S., Pallis, E., Markakis, E. K. (2020). A survey on the internet of things (IoT) forensics: Challenges, approaches, and open issues. *IEEE Communications Surveys and Tutorials*, 22(2), 1191–1221. https://ieeexplore.ieee.org/abstract/document/8950109/

Swamy, S. N., Kota, S. R. (2020). An empirical study on system level aspects of Internet of Things (IoT). *IEEE Access*, 8, 188082–188134. https://ieeexplore.ieee.org/abstract/document/9218916/

Twi-Global. (2023). Available online 2023. https://www.twi-global.com/technical-knowledge/faqs/what-is-the-internet-of-things-iot

18 Blockchain Technologies and Internet of Things (IoT)-based Blockchain Applications

Krishna Kanodia, Priyanka Kujur, and Sanjeev Patel

INTRODUCTION

Blockchain supports direct peer-to-peer transactions of digital assets, contrary to traditional methods. In the early stages, blockchain was created to promote the renowned digital currency Bitcoin. Nakamoto first suggested Bitcoin in 2008 (Nakamoto, 2008), and it was formally launched in 2009. Following that, the economic sector had tremendous development, reaching a total of 10 billion dollars in 2016. The term "blockchain" refers to a simple grouping of blocks, which utilize a public ledger to keep each committed transaction. Blockchain operates in a decentralized environment by utilizing a variety of crucial techniques, including digital signatures, cryptographic hashes, and distributed consensus methods. Due to the decentralized nature of all the transactions, no middlemen are required to validate or check any of them.

Blockchain may be employed for a variety of purposes well beyond cryptocurrencies, even if Bitcoin is its most well-known use. Blockchain enables transactions to be completed without a bank or other middleman. It may be utilized for many different financial services, including online payments, remittances, and digital assets. Utilizing blockchain itself has developed an existence of its own and is present in many different applications in a number of different sectors, including manufacturing, distribution, healthcare, finance, and the government (Rani et al., 2021).

The blockchain is ready to innovate and revolutionize numerous different applications, including remote service delivery, distributed credentialing, and moving computing to data sources, the movement of products (supply chain), and the sale of digital art. Additional usages for blockchain technology include distributed assets (like the generation and dissemination of power), crowdfunding, e-voting, identification management, and the administration of public publications. This chapter focuses on the utilization of blockchain technology in mitigation of distributed denial-of-service (DDoS) attacks in IoT environments.

DOI: 10.1201/9781003434269-18

IoT represents the connectivity of physical items that are implanted with connectivity, software, and sensors that allow them to communicate and share data. For instance, many IoT gadgets, like health sensors, are linked to the body of a human in a health screening environment. These medical sensors transmit critical bodily signals to a hub, and then to a server either inside an organization or the cloud (Khang et al., 2023a). The internet of medical things (IoMT) is widely used today and is helping to intelligently revolutionize the healthcare sector. However, there are also significant security issues with IoT's widespread use (Vrushank and Khang, 2023).

One of the biggest security flaws in an IoT ecosystem is that it is susceptible to DDoS and botnet-based cyberattacks (Hayat et al., 2018). A well-known botnet assault in 2016 known as the "Mari botnet attack" caused global delays in digital communications and disruptions to internet-based businesses. The Mari botnet assault was followed by numerous subsequent attacks that were seen by the internet services, but in 2016 an internet hosting firm named OVH was subjected to a large-scale DDoS attack using CCTV cameras that the hackers had hijacked and used 145,607 cameras to conduct the DDoS attack. IoT devices are employed in important everyday applications such as home security, healthcare facilities, transport, industry automation, and industry monitoring and control (Khang et al., 2022a).

Hence, protecting IoT ecosystems against DDoS and botnet attacks is crucial. In addition to other IT amenities, IoT devices are deployed by organizations including hospitals, industries, and other important enterprises. They are produced offshore by third-party companies. IoT device tampering to build a bot for prospective DDoS assaults is a serious security risk. Much research has been presented to mitigate DDoS assaults launched by IoT-based bots, with the majority of those options relying on centralized defense systems to combat DDoS attacks (Dantas et al., 2020). Yet, it's crucial to offer a trustworthy and effective method of reducing DDoS assaults. Emerging innovations like distributed blockchain-based systems and smart contracts make it easier to create a reliable distributed framework, which can withstand DDoS assaults in this regard.

Significant research contributions have been made for the DDoS attacks' detection and mitigation employing blockchain technology. Mendez and Yang (2018) have created and tested a proof of concept built around the Ethereum blockchain to safeguard the edge of home networks. Its "gatekeeper" maintains a whitelist of permitted operations that was generated using data from an Ethereum smart contract as the basis for computation. To convey the local network data acquired by gatekeepers, the smart contract offers a tamper-proof ledger. This strategy has a number of drawbacks: Although having a blockchain foundation, it doesn't address the issue of policymaking that is raised above. Their strategy similarly relies on a smart contract that might be substituted by a database, as the point of interaction for gatekeepers.

Furthermore, this method is not intended for usage in the real world because its installation on the Ethereum public blockchain will necessitate payment in order to send policy modifications to the smart contract. Javaid et al. (2018) developed a blockchain-based approach to safeguard computer servers against DDoS assaults. The system's connectivity to several IoT devices, sink gadgets, and data transfers to the primary server via a gateway is the foundation for the environment assumptions.

The primary server receives data from the IoT devices. The devices lack security in this fictitious environment, are susceptible to turning into bots, and may thus launch DDoS attacks, according to the authors' argument. The Ethereum blockchain, which uses gas limits for transmitting transactions, is used to implement the authors' proposed gas-limit-based limited communication system. The gateway, which is considered to transfer information to a server in a safe method, is used by many IoT devices to build a cluster and transmit data to the primary server. The authors use smart contracts, which are customizable rules and conditions, to manage communication in the suggested framework.

If the smart contract regulations are not adhered to, a device does not participate in communication. Data is received by servers or mining equipment, which then verifies it utilizing smart contract agreements. Ahmed et al. (2019) proposed methods to protect IoT gadgets against Mirai botnet attacks using blockchain. The host is set up in a single autonomous system (AS) of the planned network. In this scenario, any host connected to AS is given a collection IP address, which is then utilized to save and convey data with other nodes so as to recognize hosts that are running dangerous software. By comparing the amount of packets a host or gadget transmits with the threshold value, each AS analyses the communication flow over the network to determine whether it contains dangerous software.

Although the authors argued that employing their proposed simulating techniques to identify the specific value for the hazardous threshold does not improve the time required for acknowledgment on the victimized target, the recommended strategy does successfully block the harmful packet from the damaged host. Java language was used to create this simulation. Their proposed method effectively stopped the Mirai botnet attack by detecting it earlier, with 95% accuracy for detection while the dangerous detection level was only 8. In this study, a privatized blockchain is shown along with a technically centralized verification system, which is managed by a single node. Essaid et al.'s (2019) analysis of the current DDoS attack cooperative defense mechanisms showed their high complexity and expense. They suggested developing an automated defense system using a blockchain infrastructure.

The system's first method for facilitating information interaction was a smart contract mechanism. They then used the long short-term memory (LSTM) paradigm as the DDoS attack detection algorithm. Finally, they were able to accurately target DDoS attack data. Rodrigues et al. (2017) suggested a model that combines blockchain and smart contracts technologies, offering new possibilities for adaptable and effective DDoS mitigation strategies among different areas. This approach's primary benefits are the use of a preexisting public as well as distributed framework to publish white-listed and blacklisted IP addresses; it also makes use of such framework as a supplementary security system to preexisting DDoS detection mechanisms, with no requirement to create specialized checklists or other distribution techniques.

This chapter demonstrates a multilevel blockchain-based DDoS mitigation technique which presents a technique that can be employed to deal with problems posed by compromised devices (Khang et al., 2022c). The following are the chapter's significant contents: the blockchain and how it works, blockchain technologies, the application of blockchain technologies in a multilayered DDoS mitigation strategy based on blockchain and an authentication system to prevent gadgets from turning into bots; the transaction in the intended system provides a control technique utilizing the smart contracts and gas limit characteristic of blockchain and blacklisting problematic gadgets to disconnect them from the main IoT ecosystem.

BLOCKCHAIN TECHNOLOGY

BLOCKCHAIN ARCHITECTURE

A blockchain is a peer-to-peer network made up of comparable nodes that offer anonymity, decentralization, permanence, and auditability. Blockchain technology offers a fully decentralized architecture without the need for any third parties, and it may be used to establish safe and transparent communication between many parties. To create a secure and reliable ecosystem across many nodes, blockchain employs a variety of consensus methods, including Proof-of-Stake and Proof-of-Work. The network's node entry and block addition procedures are governed by the consensus algorithms. The two primary groups of consensus algorithms include vote-based and proof-based methods.

A blockchain is often controlled via a peer-to-peer network. It makes use of a peer-to-peer protocol for internode interaction and transaction validation called Distributed Hash Table (DHT). The standard block structure, shown in Figure 18.1, is a linked list of blocks including a header block (Chaganti et al., 2023).

The integrity of the transactions in a particular block is upheld by the root hash. As a result, the transactions are protected by a blockchain and are unchangeable.

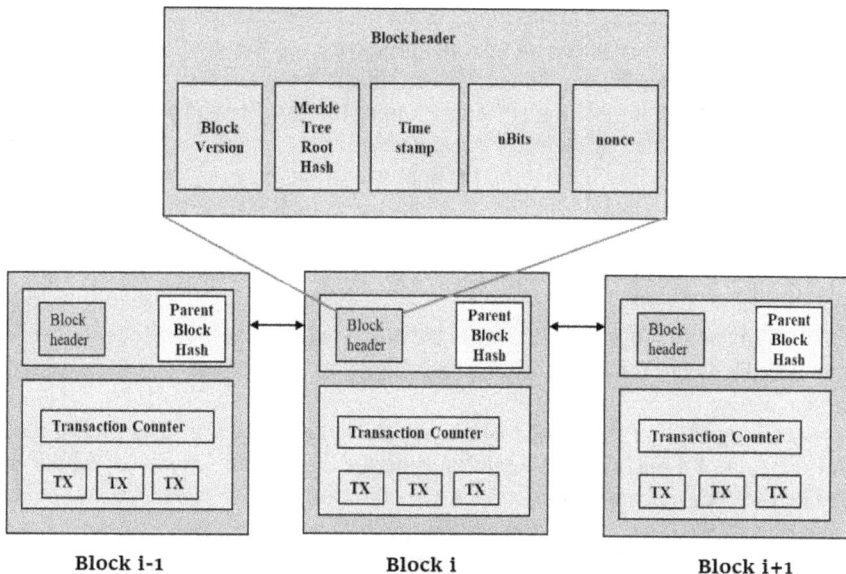

FIGURE 18.1   The linked list of blocks of the blockchain (Chaganti et al., 2023)

In order to expand a blockchain to keep track of new transactions, the block header additionally includes a Timestamp, which is the moment when the block was produced. The nBits parameter indicates the degree of difficulty being utilized for miner computations to append the transactions to the block. The nonce field, which may only be utilized once, contains a randomly generated number that was produced by the block's creator. To maintain the coherence of the chain of unmodified blocks between any two subsequent blocks, the parent block hash which is a cryptographic hash value of the parent block is employed. The latest block in a chain is pointed to via a specific data structure.

Blockchain demonstrates qualities like auditability, anonymity, persistency, and decentralization. Through the use of asymmetric cryptography, such as the RSA algorithm and digital signature, the crucial anonymity attribute is accomplished. Asymmetric cryptography employs a public and private key pair that each user has access to. The digital signature and user's validity will be verified using the hash values acquired from previous transactions. The user validation method consists of two steps: signing and verifying. There is no centralized node in the peer-to-peer blockchain technology. It utilizes consensus methods, which often call for winning a computation contest before allowing someone to append a block of validated transactions to a preexisting blockchain. The owner transmits a transaction to the following owner as a hash value. The blockchain's prior transaction value and the following owner's public key value are utilized to calculate the owner's hash value. The generated hash value is signed by the owner's private key to retain ownership.

In peer-to-peer blockchains, a consensus mechanism is employed to choose nodes to add a block of recent transactions to the current blockchain. Proof-of-Stake (POS), Proof-of-Work (POW), Practical Byzantine Fault Tolerance (PBFT), Delegated Proof-of-Stake (DPOS), and the ripple consensus method are some of the often utilized algorithms. Every node in POW, the protocol employed by Bitcoin, calculates the block header's hash value, and the algorithm states that the computed value must be lower than the required number. The accuracy of the other nodes' hash computations will be confirmed by the peer nodes. An authorized node to include the transaction in the block is chosen based on the approval of the majority of peer nodes. All other nodes on the chain are informed of this upgrade. It takes a lot of computation to calculate the value of the hash within the limitations; this process is known as mining. Users with larger holdings in POS are given the ability to upload transactions to the blockchain. Richer organizations will therefore continue to get richer, and a small number of users will control the blockchain management and transaction approval. However, this approach is less computationally intensive and is probably more effective (Geeksforgeeks, 2023).

According to the PBFT-based consensus mechanism, the transaction must have the approval of a majority of the nodes using the blockchain. The network will be updated with the approved transactions. One-third of the node malfunctions are tolerated by PBFT. The selection of a main node to carry out each transaction in a block is the first step in the consensus process. Pre-prepare, prepare, and commit are the three steps; if two-thirds of the nodes agree to the request, the transaction is included to the block. One application that uses PBFT as a consensus technique to finish network transactions is Hyperledger Fabric. For adding the transactions in DPOS, the delegated maximum currency stakeholder is selected. Some systems, like Tendermint, use the DPoS+PBFT algorithm combination to operate.

With the use of decentralized consensus techniques like POW and branching, competing parties may put forth a variety of transactions to add a new block to an existing blockchain. Due to the decentralized nature of mining, it is possible for a delay to occur when validating the transaction with 51% of the blockchain participants before adding it to the blockchain. Blockchain platforms may generally be divided into three categories. A public blockchain allows anybody to read the current transactions. Despite their lengthy computation times, the transactions offer a high degree of security and cannot be altered. A common illustration of a public blockchain is Bitcoin. Given that the address of the user's Bitcoin wallet is available, anybody may view the user's account balance and the transactions which the user's account involves.

Multiple organizations in one industry wish to employ the blockchain for commercial applications, which is an excellent example of consortium blockchain, where only chosen nodes have engaged in transactional processes. Each node stands for a different organization member. Only authorized users may read data from the blockchain, and the consensus process is quick. A private blockchain is often maintained within the organization and requires authorization to join the network. To communicate data within the organization, to keep records safely, and more, the nodes might be members of the same organization. If unreliable nodes engage in the mining process, the private blockchain often centralizes and the transactions can be altered. Because public blockchains take a long time to execute each transaction, they are less effective than private and consortium blockchains. The transaction processing time is longer on the public blockchain since there are more nodes that are openly linked.

Since the creation of Bitcoin, the community of blockchain developers has created a variety of other cryptocurrency currencies targeted at certain business uses. Ethereum, Litecoin, and Ripple are a few of the well-known currencies. Ethereum, which uses smart contract capabilities, is the second-most popular cryptocurrency and has the greatest market value. The use of Ethereum has been suggested as a solution for various Bitcoin scripting language drawbacks.

Since Ethereum allows the Turing-complete programming language, all operations, including loops, are possible. When specific criteria are satisfied, the smart contracts execute cryptographic rules. The nodes convert the smart contracts into EVM code, which is then executed by the nodes to finish the transaction (which results in the creation of a user account). In addition to having scalability problems, the Ethereum network has substantially higher transaction costs. Ethereum developers suggested Ethereum 2.0 as a solution to these issues. The POS consensus method is used by Ethereum 2.0 to process transactions and increase scalability. 2020 saw the debut of Ethereum 2.0's initial phase. The likelihood of widespread acceptance of the technology with lesser transaction costs may be increased by the complete deployment of Ethereum 2.0.

Hyperledger has received a lot of publicity recently due to the usefulness of its business standard abilities. It is well known that Hyperledger is systematically utilized in academic research to create blockchain applications and validate research findings. For the implementation of corporate blockchain applications, the Hyperledger suite of open-source tools, frameworks, and modules has been developed by the community. A significant tool for creating blockchain apps is Hyperledger Fabric, a distributed ledger platform that can include a private blockchain for providing the frameworks for certain services. The Fabric is a collection of access, script, model, and query files that have been compressed to create a business network bundle.

The Fabric employs "Chain code," a concept equivalent to Ethereum smart contracts, to conduct secure blockchain transactions. The Interplanetary File System (IPFS), a distributed file storage system, may also be connected to the Hyperledger Fabric. The blockchain's nodes may exchange data stored in the IPFS among themselves. For providing online content to consumers, a decentralized web application, for instance, can be hosted using material saved in IPFS. In general, Hyperledger is a highly helpful blockchain technology development platform that has been extensively utilized to create applications, including DDoS mitigation.

SMART CONTRACTS IN BLOCKCHAIN

A smart contract is a programmed transaction mechanism that carries out a contract's terms, according to Nick Szabo's definition from 1993. Blockchain-based applications called "smart contracts" run when certain criteria are satisfied. They are typically employed to streamline the execution of a deal so that both parties can become certain of the outcomes without the need for a mediator or further delay. It may also regulate a process such that it only takes place if a set of conditions are satisfied. Smart contracts are based on if–then rules that are written in blockchain code. Following the verification of predefined conditions, a network of computers will execute the actions. This might entail making the necessary payments, registering a car, sending out notifications, or drafting a fine. The blockchain is modified following the conclusion of a transaction. As a result, the transaction cannot be changed, and the only persons who are allowed to see the results are those who have been given permission.

One of a smart contract's key features is its capacity to carry out or self-perform contractual clauses. This was technically impractical before the development of blockchain technology (Figure 18.2). The right technology for facilitating smart contracts has emerged as blockchain. Additionally, smart contracts have considerably increased the effectiveness of blockchain. This combination has also resulted in a second era of blockchains, dubbed as Blockchain 2.0.

In essence, the blockchain serves as a repository for the smart contract's code. Clients simply need to submit a transaction to each contract's specific address so as to communicate with it. The accurate implementation of the contract is ensured by the blockchain consensus process. Smart contracts provide a number of benefits, including cost savings, accuracy, speed, transparency, and efficiency, which have sparked the creation of a large number of innovative applications in a range of different fields. Although Bitcoin offers a straightforward programming language, it

How does a smart contract work?

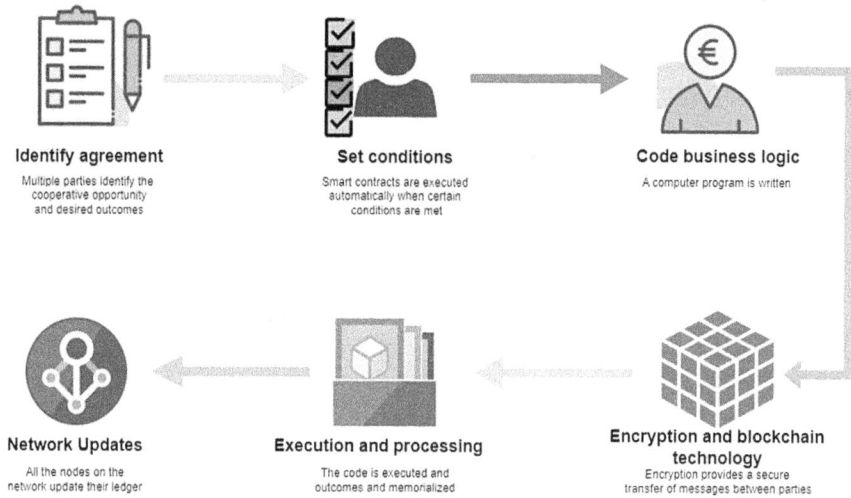

Identify agreement

Multiple parties identify the
cooperative opportunity
and desired outcomes

Set conditions

Smart contracts are executed
automatically when certain
conditions are met

Code business logic

A computer program is written

Network Updates

All the nodes on the
network update their ledger

Execution and processing

The code is executed and
outcomes and memorialized

**Encryption and blockchain
technology**

Encryption provides a secure
transfer of messages between parties

FIGURE 18.2   Workflow of smart contracts (Geeksforgeeks, 2023)

has been found to be inadequate, which has caused the rise of multiple blockchain systems with smart contract capability.

The most renowned blockchain for smart contracts is Ethereum. A Turing-complete programming language is built into the Ethereum blockchain, allowing for the creation of smart contracts and decentralized systems. A stack-based, low-level bytecode language called "Ethereum virtual machine code" is employed to create Ethereum contracts. Smart contracts for finance typically need to have access to data regarding actual circumstances and events. This data was gathered from claimed oracles. These entities are necessary for the effective implementation of smart contracts in the real world, but the complexity is increased by their authentication, security, and Oracle trust needs (Zhang et al., 2016).

The benefits of smart contracts cannot be obtained without a price, however, since they are susceptible to several assaults (Delmolino et al., 2016; Atzei et al., 2017; Christidis and Devetsikiotis, 2016) that present fresh, challenging problems. The delegation of contract execution to machines poses certain challenges since it leaves them open to technological difficulties like hackers, flaws, viruses, or communication breakdowns. Contract code errors are particularly dangerous due to the system's irreversible and unchanging nature. Smart contracts must have mechanisms in place to ensure their proper execution if they are to be extensively and securely adopted by users and service providers. In the next years, advancements are anticipated in the fields of formal verification of contract logic and its validity.

Furthermore, non-quantifiable phrases and conditions are frequently included in contracts in real life. More work has to be invested in modeling the conditions of contracts in smart contracts to ensure they can be defined and portrayed for computers to perform them. It is also necessary to make an effort to give consumers the tools they need to define and comprehend smart contracts. As illustrated in Figure 2.2, this is how a smart contract works (Geeksforgeeks, 2023): Agreement Identification: Several parties decide on the joint potential and intended results. Agreements can involve business procedures, asset exchanges, and so on.

- **Establish Conditions**: Smart contracts may be started by the parties involved or in response to certain events, such as GPS coordinates or stock market indexes.
- **Business Logic Code**: When the conditional factors are satisfied, a computer code is built that will run automatically (Tailor et al., 2022).
- **Blockchain Technology with Encryption**: Encryption offers safe message transit and authentication among parties related to smart contracts.
- **Execution and Processing**: In a blockchain iteration, the code is performed and the results are memorialized for compliance and validation whenever both sides reach an agreement on authentication and validation.
- **Network Updates**: Each node on the network updates its ledger to indicate the new status following the execution of a smart contract. The file is only in append mode, so once it has been posted and confirmed on the blockchain network; it cannot be changed at that point.

Gas Limit in Blockchain

The term "gas" refers to the metric used to express the amount of computing power necessary to carry out particular activities on the Ethereum network. Each Ethereum transaction has a cost since them all need computing resources to complete. Regardless of whether a transaction is successful or unsuccessful, Ethereum gas is the charge needed to carry out a transaction on the Ethereum blockchain (Ethereum gas is what users pay to process transactions or use smart contracts on the Ethereum network. Ethereum gas is denominated in gwei, short for gigawei, with one gwei equal to one billionth of an ETH). The native currency of Ethereum, ether (ETH), is used to pay for gas. Each gwei, which is used to represent gas costs and is itself a denomination of ETH, is equivalent to 0.000000001 ETH. The gas limit is the most work you anticipate a validator will do on a given transaction. By assigning gas limit to all nodes in a network, you can control a maximum of how many transactions or requests can be made at a particular interval of time.

IOT APPLICABILITY

Network Scenario of IoT

The internet of things, or IoT, represents the connectivity of physical items which are implanted with connectivity, software, and sensors that allow them to communicate and share data. IoT technology has become a fascinating field that allows a worldwide network of gadgets to exchange information with each other. The IoT architecture in a cyber-physical environment is mostly made up of a collection of distributed sensors (Hahanov et al., 2022). By 2026, there will be over 90 billion IoT devices worldwide, according to a study. An illustration of an IoT architecture with its key elements is shown in Figure 18.3.

In order to transmit data to IoT vendors for analysis, checking status, presenting in the user panel, etc., the IoT gateways are linked to the open internet. They are not restricted to utilizing IoT network protocols like HTTP, CoAP, MQTT, and AMP.

DDoS Attack Using IoT

A DDoS attack is an intentional effort to impede a service, server, or network's usual flow by saturating the target or its nearby hardware with an excessive amount of internet messages. A DDoS attack is a deliberate attempt to disrupt a server, service, or network's normal operation by overloading the target or its surrounding hardware with excessive internet traffic. A DDoS attack resembles an unforeseen traffic jam that blocks the road and prevents regular traffic from moving forward as intended.

The adversary compromises the devices by infecting the network with malware and transmitting spoofed requests to the target system (Figure 18.4). The perpetrator then seizes control of a server's resources, preventing the server from connecting to and interacting with IoT devices. IoT devices operate crucially, particularly in Industry 4.0 environments such as hospitals; consequently, these attacks must be effectively controlled and mitigated to minimize possible monetary and human life losses (Lohachab and Karambir, 2018).

Integration of IoT and Blockchain

The IoT is automating and improving manual processes, bringing them fully into the digital era and generating previously incomprehensible extents of knowledge. This information promotes the creation of clever apps that, for example, enhance citizen governance and the standard of life by digitizing services provided by the city. The IoT now has the ability to evaluate, interpret, and turn data into immediate actions

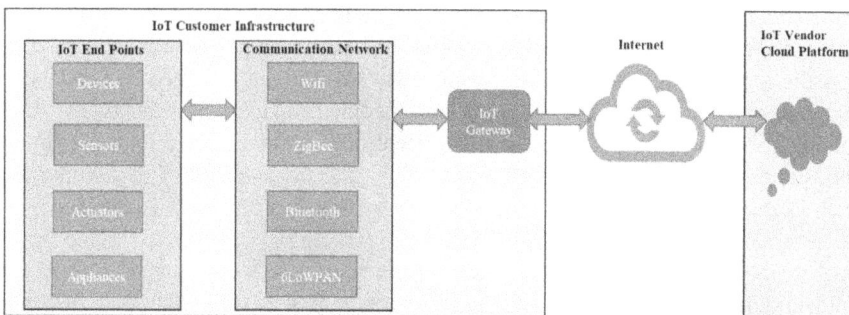

FIGURE 18.3   Working scenario of IoT

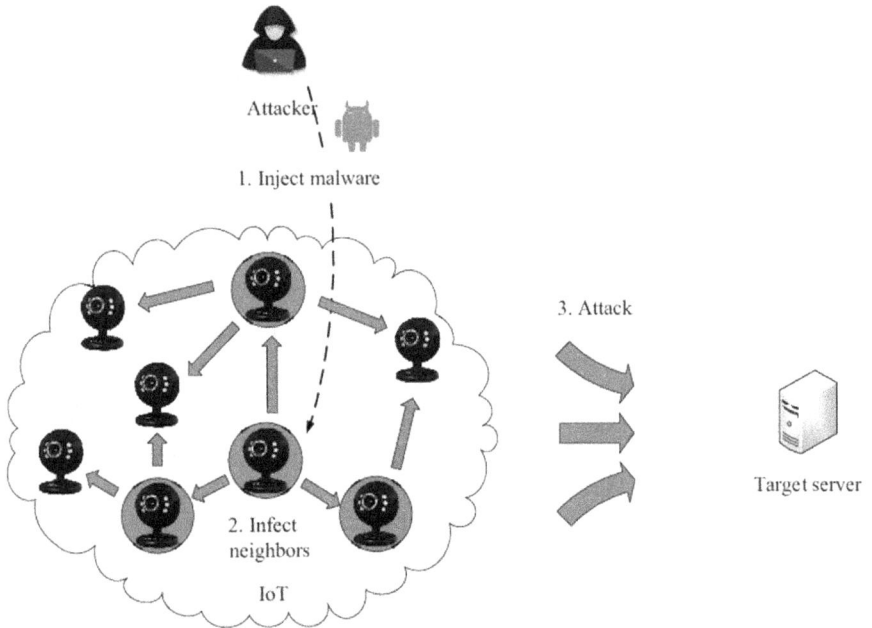

FIGURE 18.4 DDoS attack using IoT (Huang et al., 2020)

and information, thanks to advances in cloud computing technology (Bhambri et al., 2022). Unprecedented expansion in the IoTs has given the public new options, including new ways to gather and transmit information. These initiatives revolve around the idea of open data. An important flaw in these efforts is an abundance of confidence, though, as seems to be the case in many situations. The development of centralized systems like cloud computing has greatly helped the internet of things. They serve a secondary purpose with regard to data accessibility, and network participants are unclear about the manner in which the data they supply will be utilized (Rani et al., 2023).

Integration of cloud computing with IoT has shown to be quite beneficial. We understand that the internet of things might be completely transformed by blockchain. Blockchain can improve IoT by offering a reliable and trustworthy data-sharing solution. Data sources may be recognized at any moment, and because the data is static, its security is increased. When a lot of people need to safely transmit IoT data, this combination would be a big improvement. For example, the tracking of many foods is essential for guaranteeing food safety. The involvement of several parties, such as those involved in food production, handling, feeding, and distribution, may be necessary for food traceability.

A breach of information in any part of the supply network during a foodborne epidemic might result in fraud and hinder efforts to find the infection, both of which could have a serious negative effect on people's health and result in substantial monetary losses for companies, sectors, and countries. Greater control over these areas will raise food safety, improve consumer data sharing, and shorten the time needed to find a foodborne pandemic, perhaps saving lives. Furthermore, reliable data exchange might promote the acceptance of new players in ecosystems, strengthen their services, and accelerate their use in other industries like smart cities and smart cars. As a result, the implementation of blockchain can improve the data safety and dependability of the IoT. Blockchain technology is the answer to the scalability, privacy, and dependability issues of the IoT paradigm (Malviya, 2016).

The enhancements that this integration can provide are as follows:

- **Security**: The information and conversations can be safeguarded if they are kept as blockchain transactions. Blockchain technology has the potential to secure device communication by treating exchanges of messages as transactions that are verified by smart contracts. With the usage of blockchain, the IoT's present secure standard protocols can be improved (Khan and Salah, 2017).
- **Reliability**: Blockchain technology allows IoT data to be disseminated over time and stay unchangeable. Members of the system have the ability to confirm the veracity of the data and have confidence that it has not been altered with. Additionally, the system allows for the accountability and traceability of sensor data. The main feature of the blockchain that will enable IoT integration is reliability.
- **Decentralization and Scalability**: Centralized points of failures and bottlenecks will be eliminated when switching from a centralized structure to a P2P one (Veena et al., 2015). Additionally, it will assist to avoid situations in which a small number of strong corporations are in charge of handling and storing the data of a sizable number of individuals. Improvements in fault resilience and system scalability are further advantages of the decentralized design. In addition to reducing IoT silos, it would improve the scalability of the IoT.
- **Autonomy**: Blockchain technology enables next-generation application characteristics, opening the door to the creation of autonomous hardware and smart autonomous assets. With blockchain, devices may communicate with each other without the need for any servers. This capability might be used by IoT apps to offer decoupled and device-independent applications.
- **Identity**: Participants have the ability to identify each and every device by utilizing a shared blockchain system. Data supplied and put into the computer system is unchangeable and may be used to identify genuine data that was given by a device in a certain way. Additionally, for IoT applications, blockchain can offer reliable distributed device authentication and authorization. This would be a development for the IoT industry and its players.
- **Market for Services**: Blockchain might hasten the development of an IoT ecosystem with markets for services and data where peer-to-peer transactions are feasible without the involvement of authorities. Microservices can be quickly established, and micropayments can be securely processed in an

untrustworthy setting. It will enhance IoT connectivity and IoT data accessibility in blockchain.

- **Secure Code Deployment**: It allows for the safe and secure release of code into devices by utilizing blockchain-secure unchangeable storage. Manufacturers can monitor changes and statuses with the utmost assurance. This capability may be utilized by IoT middleware for updating IoT devices safely.

BLOCKCHAIN APPLICATIONS

In the financial world, there has been a considerable rise of alternative cryptocurrencies, which have created a new market and permitted new kinds of payment and investing. Ripple (Ripple, 2017), Litecoin (2011), Nxt (2014), Doge Markus (2013), Namecoin (2014), Dash (2017), and Monero (2017) are a few examples of them. Novel applications for trading, payment, and exchange infrastructures for these new currencies have also emerged as a result of this new market.

Beyond money, traceability systems have been suggested as a perfect application for the distributed ledger of blockchain. The network keeps the items and all of their modifications, enabling a thorough, transparent, and dependable tracking that is only made feasible by this technology. Numerous well-known corporations are working on continuing initiatives, such as Walmart, Nestle, IBM, and Unilever's collaboration on food traceability or Renault's tracking of vehicle maintenance records.

The application of blockchain technology for identity verification has also gained popularity. Blockchain offers an open, trustworthy distributed ledger that is capable of being used to store identities. This makes it feasible to handle identities globally. Many nations have advocated the adoption of blockchain technology in e-government, such as Dubai for passports, Estonia for e-identity, Illinois for digitizing birth certificates, and India for land registration. The utilization of smart contracts also offers a wide range of opportunities for several businesses, including those related to energy, insurance, music royalties, real estate, gaming, betting, etc. There have also been suggestions for blockchain applications in other fields such as e-health, education, and cloud storage (Khang et al., 2022b).

IoT-BLOCKCHAIN APPLICATIONS

Despite the fact that the IoT is just recently beginning to utilize blockchain, there are already several proposals for how to use this technology to improve the current state of IoT. In many situations, IoT and blockchain can work together. IoT gadgets may be included to improve almost all traceability operations. For example, sensors linked to the item being tracked can offer information about its distribution. Similar to this, blockchain technology may be used to improve several apps that digitalize the world by supplying sensed information. Additionally, it offers distributed identification, authentication, and authorization processes with no need for a centralized entity, which might assist in boosting data validity or trustworthiness (Khanh and Khang, 2021).

Blockchain has a strong chance of realizing the idea of the smart city. The idea of the "smart city" is built on intelligent, self-operating IoT gadgets. Since blockchain offers a distributed open database where gadgets may reliably query verified information, it facilitates communication and coordination, which can boost the autonomy of gadgets. Furthermore, the development of new IoT marketplaces is favored by the independence that blockchain offers. The most commonly utilized IoT-blockchain application platform is Ethereum. Compared to Bitcoin, Ethereum offers more functionality, and the addition of smart contracts considerably increases the range of potential uses.

DDOS ATTACK MITIGATION USING BLOCKCHAIN FOR IOT ENVIRONMENT

SYSTEM ARCHITECTURE

The approach's architecture proposed by Hayat et al. (2018) is shown in Figure 18.5. The method is built on the three-layer design described below. The layers are as follows: (1) perception; (2) edge computing; and (3) blockchain and storage layer. To achieve small-scale computing storage facilities, a variety of IoT devices are used in the perception layer that is linked to the higher-layer computing gadgets. A number of IoT ports and edge computing devices make up the second tier of the edge computing architecture. These devices' primary duties include supplying the interconnected IoT devices with program implementation logic, as well as associated computing and storage services. The third tier, which includes storage and blockchain, steps in for online storage and blockchain services.

Four lists make up our created smart contracts, one of which is for trustworthy devices and the others are for different uses. A gadget must send a signature message to communicate with the computer. The message signature and the presence of the device on the roster of trusted devices are both verified at the moment of interaction. Conversation can resume if the gadget successfully completes the validation process; if not, all operations are halted (and the gadget is appended to the roster of untrusted gadgets) and the conversation is completely cut off.

Devices without legitimate account identifiers are not allowed to interact as signature messages are employed as account identifiers. In addition to cutting off connection with potential invader devices, the signature message-based communication rejection shields IoT devices from being hacked. The blockchain-based solution prevents external devices from legitimately connecting to the computer servers, while the cloud storage service preserves data coming from the perception layer (Ahmad et al., 2023). Therefore, the intruder must first compromise some legitimate IoT device portions of the network in order to launch a DDoS attack in the IoT ecosystem. If the hacker is effective in taking control of an IoT device, a defense strategy is used. A gas limit is given to each IoT device to prevent the compromised device from flooding the computers with a large amount of messages thereby stopping possible DDoS attacks.

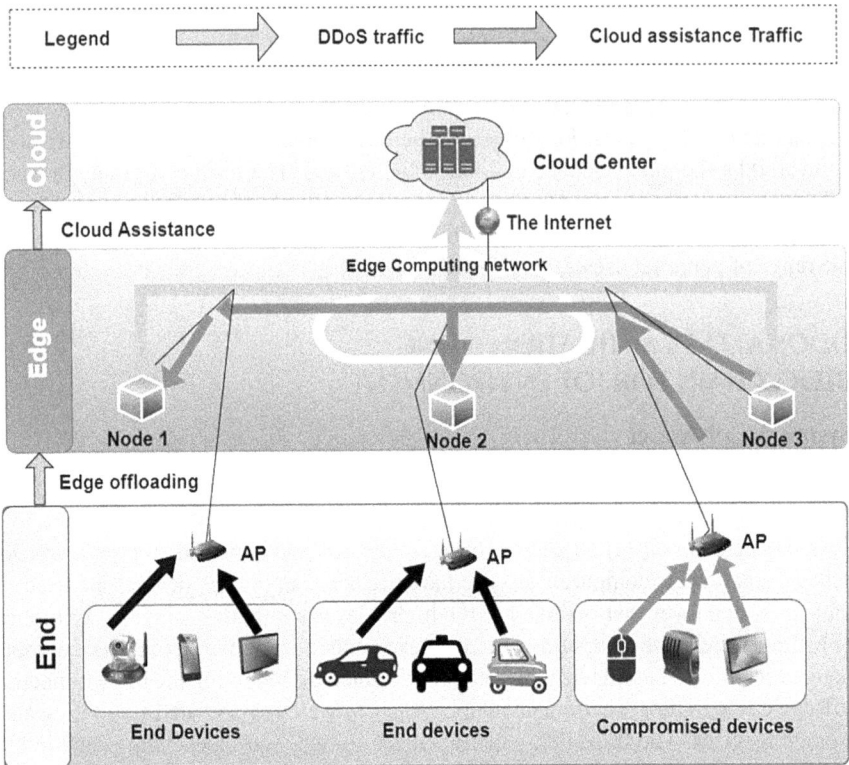

FIGURE 18.5   System architecture

EXECUTION FLOW AND ANALYSIS

There are four lists made: one for trustworthy devices, one for untrusted devices, one for signature messages, and one for banned devices. In the IoT network and server, these lists are exchanged via the Ethereum blockchain. The utilization of blockchain assures the highest levels of credibility, preventing any unauthorized entity from changing a particular hash. If the data are unchanged, the blockchain's record hash should match; any breach of the data's security can therefore be quickly identified.

A gadget first makes a query to the server, and then the server verifies if the inquiring device's ID is genuine (i.e., on the roster of trusted gadgets). Devices are appended to the roster of untrusted gadgets if their IDs are not identical to one of the permitted device IDs. After registering, a gadget must proceed to the next level of verification employing the signature message, a special and randomly generated data message provided to all authorized gadgets. The server checks the conveyed signature message from the device to validate its authenticity. A device is appended to the roster of untrusted gadgets and is barred from further contact if it is unable to authenticate via the message string. If a device is appended to the untrusted roster, it cannot be removed without specific executive powers and then added back into the IoT ecosystem. The untrusted roster is then distributed across the Ethereum network.

The strategy used employs a method to prevent the two potential bot attack trajectories. The approach addresses the problems posed by any rogue or bot device joining the IoT network as well as the potential corruption of an already existing legitimate device. The suggested technique for the second kind of bot attack reveals that every authorized IoT device employs a gas-limit quota that it uses to send messages over the network. The gas-limit approach aids in preventing the DDoS launched by a hacked gadget because of the possible transaction requirements (or responsibilities of IoT gadgets for their future communications). When an IoT device's gas supply falls low, an administrator can add more gas points to the device so it can keep running normally.

Additionally, a particular time-span number is also specified (in accordance with the device's particular transaction requirements). Take a temperature-tracking IoT gadget for an industrial setting, as an illustration. This device's time period (for transmitting the message with the temperature measurement) could be adjusted to 5 seconds. However, this is considered abnormal behavior and would be addressed if the gadgets try to flood the server or other gadgets with a lot of messages. As a result, abruptly behaving hacked devices are discovered and placed on the blacklist of gadgets. These lists are subsequently distributed around the IoT ecosystem utilizing the blockchain.

Now, breaking down the trusted gadget-related query verification procedure step-by-step. The server checks the gadget's ID when it receives a query from a gadget for the first time to make sure it is legitimate and on the trusted database. If a requesting gadget's ID does not correspond to one of the legitimate gadget IDs, it is placed to the untrusted table and has no ability to authenticate via the message contents. The server verifies the validity of the communicated signature message from the gadget. The gadget gets appended to the untrusted database if the verification procedure is unsuccessful.

The following shows the sequence of events in relation to a verification request made by a possible violating device. The server receives the request from the targeted device, which then determines whether or not it has transmitted a legitimate verification message string. Furthermore, whether a device is part of the trusted devices or not, it is verified for a legitimate ID. If the verification message does not match and the device is not presently on the trustworthy list, it is put to the blacklisted devices and all further contact with or to that device is forbidden. Once a device is added to the roster of blacklisted devices, it can only be removed with the admin's permission.

EVALUATION USING BOTNET-BASED SCENARIO

The Hyperledger caliper was chosen as the simulation framework for the blockchain ecosystem. The platform built on the Ethereum blockchain is supported in full by Hyperledger. The Hyperledger caliper is connected to the Ethereum blockchain's smart contract, and transactions are executed to deliver the desired results.

The experimental evaluation was carried out using the performance parameters listed below.

- **Throughput**: How much data can be delivered per unit of time using the given method? With Hyperledger Caliper2, a use case is established, several transactions are passed through it, and the achieved throughput is then measured.
- **Latency**: How much data can be sent using the suggested method in a given amount of time? Latency is the overall length of time needed for data to be collected, processed, transferred, or received at the target IoT gadget or server. To carry out the related processes and calculate the resulting latencies, multiple transactions are needed.
- **CPU Utilization**: The amount of work done by the CPU in accordance with its capability is referred to as CPU utilization. The amount of CPU time used depends on how computationally complex the task is. The compute intensity of the associated tasks has also been examined for this aspect. In this chapter, a hypothetical case is shown when a potential botnet is found and the DDoS attack is appropriately neutralized. These detection and mitigation processes' corresponding performance metrics have been assessed.

The following are the two particular scenarios that were tested: (1) when an authorized IoT gadget completes the necessary transaction; and (2) when a hacked IoT gadget initiates and completes the associated transactions (Hayat et al., 2018). During the throughput investigation, it was discovered that, in the bot-based scenario, the throughput related to transaction count significantly increases once the attack is conducted. However, because this technology proactively identifies botnets (or compromised devices), a DDoS scenario is prevented, leading to lower CPU usage, normal throughput, and decreased latency (due to the gadgets being rapidly discovered and prohibited) (Hayat et al., 2018).

The latency data computed before and after the device was hacked (i.e., in the instance when the gadget became a bot and was then controlled) (Figure 18.6). In the given scenario, an example interval of 5 seconds is established for each device. For instance, if an intruder builds a bot on the system and sends extra messages, a DDoS attack could be initiated (DDoS is a distributed denial-of-service). By limiting its transactions, identifying it as a bot, and then disconnecting it from the IoT system, it may be stopped (Hayat et al., 2018).

Alt-text: Figure 18.6 illustrates the average computed latency of the system. The graph illustrates that average observed latencies rise in routine IoT activities, for example, when compared to other circumstances (i.e., the existence of bots), which are then addressed by the indicated technique.

The delay is larger when there are more devices (Figure 18.7); when there are fewer devices, the delay is reduced (blacklisted and taken off from the system) (Hayat et al., 2018).

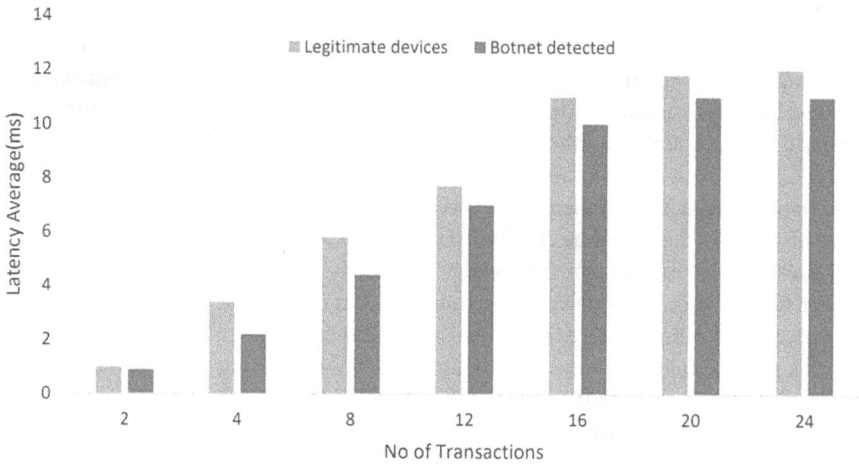

FIGURE 18.6   Average latency for authorized devices and when a botnet is found

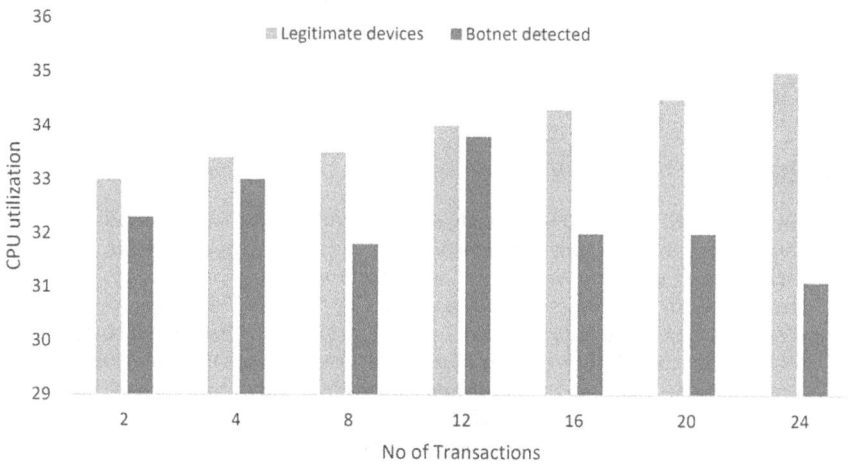

FIGURE 18.7   Average CPU utilization for authorized gadgets and when a botnet is found

CONCLUSION

The blockchain is recognized and approved for its peer-to-peer architecture and decentralized design. Bitcoin shields many blockchain investigations. But blockchain has a lot more applications than just Bitcoin. Blockchain technology's core characteristics, such as anonymity, persistence, auditability, and decentralization, have shown that it has the potential to disrupt a number of well-established sectors. This chapter demonstrates blockchain technologies and a method to safeguard IoT gadgets and other computer tools or machines by employing a blockchain-based architecture to address these issues.

The main idea behind the plan was to keep compromised devices (i.e., bots) out of IoT settings by adopting a device-based verification method using blockchain. Utilizing the blockchain benchmarking tool Hyperledger caliper, the efficiency of the system was evaluated utilizing three benchmark applications. The method uses a smart contract and the Ethereum blockchain to replace centralized architecture with decentralized architecture. The administrator can check the legitimacy of the communicating gadgets if the gadget was registered in the network since each device has a distinct ID that could be checked. The framework either excludes any tempered and hacked IoT devices from the registry or actively detects them.

For future work, both power usage and cyberattacks on IoT gadgets may be further decreased. Furthermore, additional attack types, such as internal and external attacks, may be taken into account by dynamically modifying the nodes to address security concerns and enhance the flexibility and security of the approach. This discussion shows that the area of study is in its infancy, and blockchain's full potential for addressing problems related to DDoS mitigation has not yet been exploited. While other works have categorized solutions to IoT security challenges, this one focuses solely on DDoS attack detection and mitigation. In addition, the scope of this article is limited to blockchain-based solutions, but in future research, it can also be used to investigate whether or not the challenges of DDoS security and privacy have been addressed through the implementation of machine learning or other technologies (Khang et al., 2023c).

REFERENCES

Al Yakin, Ahmad, Khang A. Muthmainnah, Abdul Mukit, Mohd Zuber, Personalized social-collaborative IoT-symbiotic platforms in smart education ecosystem. *Smart Cities: IoT Technologies, Big Data Solutions, Cloud Platforms, and Cybersecurity Techniques* (1st Ed.) (2023). CRC Press. https://doi.org/10.1201/9781003376064-15.

Ahmed Z., S. M. Danish, H. K. Qureshi, M. Lestas, Protecting IoTs from Mirai botnet attacks using blockchains. In *Proceedings of the IEEE International Workshop on Computer Aided Modeling and Design of Communication Links and Networks*, 2019, pp. 1–6.

Atlam H. F., G. B. Wills, Technical aspects of blockchain and IoT. In *Advances in Computers*, Amsterdam, The Netherlands: Academic Press, 2019, vol. 115, pp. 1–39.

Atzei N., M. Bartoletti, T. Cimoli, A survey of attacks on Ethereum smart contracts (sok). In *International Conference on Principles of Security and Trust*, Uppsala: Springer, 2017, pp. 164–186.

Bhambri P., S. Rani, G. Gupta, A. Khang, *Cloud and Fog Computing Platforms for Internet of Things* (2022). CRC Press. https://doi.org/ 10.1201/9781003213888.

Chaganti R., B. Bhushan, V. Ravi, A survey on Blockchain solutions in DDoS attacks mitigation: Techniques, open challenges and future directions. *Computer Communications* 197 (2023) 96–112. https://doi.org/10.1016/j.comcom.2022.10.026.

Christidis K., M. Devetsikiotis, Blockchains and smart contracts for the internet of things. *IEEE Access* 4 (2016), 2292–2303.

Dantas Silva F. S., E. Silva, E. P. Neto, M. Lemos, A. J. Venancio Neto, F. Esposito, A taxonomy of DDoS attack mitigation approaches featured by SDN technologies in IoT scenarios. *Sensors* (2020), 20(11), 3078.

Dash (2017). https://www.dash.org/es/.

Delmolino K., M. Arnett, A. Kosba, A. Miller, E. Shi, Step by step towards creating a safe smart contract: Lessons and insights from a cryptocurrency lab. In *International Conference on Financial Cryptography and Data Security*, Christ Church, Barbados: Springer, 2016, pp. 79–94.

Essaid M., D. Kim, S. H. Maeng, S. Park, H. T. Ju, A collaborative DDoS mitigation solution based on Ethereum smart contract and RNN- LSTM. In *Proceedings of the Asia-Pacific Network Operations and Management Symposium*, 2019, pp. 1–6.

Geeksforgeeks (2023). https://media.geeksforgeeks.org/wp-content/uploads/20220527093234 /Howdoesasmartcontractwork1.jpg.

Hahanov V., A. Khang, E. Litvinova, S. Chumachenko, V. A. Hajimahmud, V. A. Alyar, The key assistant of smart city – sensors and tools. In *AI-Centric Smart City Ecosystems: Technologies, Design and Implementation* (1st Ed.), 17(10), (2022). CRC Press. https:// doi.org/10.1201/9781003252542-17.

Hayat R. F., S. Aurangzeb, M. Aleem, G. Srivastava, J. C.-W. Lin, ML-DDoS: A blockchain based multilevel DDoS mitigation mechanism for IoT environments. In *IEEE Transactions on Engineering Management* (2018). https://doi.org/10.1109/TEM.2022 .3170519.

Huang, K. Y., L.-X. Yang, X. Xiang, Y. Yong Tang, A LowCost distributed denial-of-service attack architecture. *IEEE Access* (2020), 1–1. https://doi.org/10.1109/ACCESS.2020 .2977112.

Javaid U., A. K. Siang, M. N. Aman, B. Sikdar, Mitigating IoT device based DDoS attacks using blockchain. In *Proceedings of the Workshop on Cryptocurrencies and Blockchains for Distributed Systems*, 2018, pp. 71–76. https://dl.acm.org/doi/abs/10 .1145/3211933.3211946

King S., S. Nadal, Peercoin–secure sustainable cryptocoin (2012). https://peercoin.net/ whitepaper.

Khan M. A., K. Salah, Iot security: Review, blockchain solutions, and open challenges. *Future Generation Computer Systems* (2017). https://www.sciencedirect.com/science /article/pii/S0167739X17315765

Khang A., S. Chowdhury, S. Sharma, *The Data-Driven Blockchain Ecosystem: Fundamentals, Applications, and Emerging Technologies* (1st Ed.) (2022a). CRC Press. https://doi.org /10.1201/9781003269281.

Khang A., N. A. Ragimova, V. A. Hajimahmud, V. A. Alyar, Advanced technologies and data management in the smart healthcare system. In *AI-Centric Smart City Ecosystems: Technologies, Design and Implementation* (1st Ed.), 16(10) (2022b). CRC Press. https:// doi.org/10.1201/9781003252542-16.

Khang A., V. Hahanov, G. L. Abbas, V. A. Hajimahmud, Cyber-physical-social system and incident management. In *AI-Centric Smart City Ecosystems: Technologies, Design and Implementation* (1st Ed.), 2(15) (2022c). CRC Press. https://doi.org/10.1201 /9781003252542-2.

Khang A., G. Rana, R. K. Tailor, V. A. Hajimahmud (Eds.), *Data-Centric AI Solutions and Emerging Technologies in the Healthcare Ecosystem* (2023a). CRC Press. https://doi .org/10.1201/9781003356189.

Khang A., S. K. Gupta, V. Shah, A. Misra (Eds.), *AI-Aided IoT Technologies and Applications in the Smart Business and Production* (2023b). CRC Press. https://doi.org/10.1201 /9781003392224.

Khang A., *Advanced Technologies and AI-Equipped IoT Applications in High-Tech Agriculture* (1st Ed.) (2023c). IGI Global Press. https://doi.org/10.4018/978-1-6684 -9231-4.

Khanh H. H., A. Khang, The role of artificial intelligence in blockchain applications. *Reinventing Manufacturing and Business Processes Through Artificial Intelligence* (2021), 2, 20–40. CRC Press. https://doi.org/10.1201/9781003145011-2.

Litecoin (2011). https://litecoin.org/.

Lohachab A., B. Karambir, Critical analysis of DDoS—An emerging security threat over IoT networks. *Journal of Communications and Information Networks* (2018), 3(3), 57–78. https://link.springer.com/article/10.1007/s41650-018-0022-5

Malviya H., How blockchain will defend IOT (2016). https://ssrn.com/abstract=2883711.

Markus B., Dogecoin (2013). http://dogecoin.com/.

Mendez Mena D. M., B. Yang, Blockchain-based whitelisting for consumer IoT devices and home networks. In *Proceedings of the Annual SIG Conference on Information Technology Education*, 2018, pp. 7–12. https://ieeexplore.ieee.org/abstract/document/8528677/

Monero (2017). https://getmonero.org/.

Nakamoto S., Bitcoin: A peer-to-peer electronic cash system. *Decentralized Business Review* (2008), 21260. https://assets.pubpub.org/d8wct41f/31611263538139.pdf

Namecoin (2014). https://namecoin.org/.

Nxt White Paper (2014). https://bravenewcoin.com/assets/ Whitepapers/NxtWhitepaper-v122 -rev4.pdf.

Rani S., P. Bhambri, A. Kataria, A. Khang, A. K. Sivaraman, *Big Data, Cloud Computing and IoT: Tools and Applications* (1st Ed.) (2023). Chapman and Hall/CRC. https://doi .org/10.1201/9781003298335.

Rani S., M. Chauhan, A. Kataria, A. Khang (Eds.), IoT equipped intelligent distributed framework for smart healthcare systems. *Networking and Internet Architecture* (2021), vol. 2, 30. https://doi.org/10.48550/arXiv.2110.04997.

Ripple (2017). https://ripple.com/.

Rodrigues B., T. Bocek, A. Lareida, D. Hausheer, S. Rafati, B. Stiller, A blockchain-based architecture for collaborative DDoS mitigation with smart contracts. In *Proceedings of the IFIP International Conference on Autonomous Infrastructure, Management, and Security*, 2017, pp. 16–29.

Tailor R. K., Khang A. Ranu Pareek, Robot process automation in blockchain. In Khang A., S. Chowdhury, S. Sharma (Eds.), *The Data-Driven Blockchain Ecosystem: Fundamentals, Applications, and Emerging Technologies* (1st Ed.), 8(13), 149–164, (2022). CRC Press. https://doi.org/10.1201/9781003269281-8.

Veena P., S. Panikkar, S. Nair, P. Brody, Empowering the edge-practical insights on a decentralized internet of things. In *Empowering the Edge Practical Insights on a Decentralized Internet of Things*, vol. 17. IBM Institute for Business Value, 2015.

Vrushank S., A. Khang, Internet of medical things (IoMT) driving the digital transformation of the healthcare sector. *Data-Centric AI Solutions and Emerging Technologies in the Healthcare Ecosystem*, P (1) (1st Ed.) (2023). CRC Press. https://doi.org/10.1201 /9781003356189-2.

Zhang F., E. Cecchetti, K. Croman, A. Juels, E. Shi, Town crier: An authenticated data feed for smart contracts. In *Proceedings of the 2016 ACM SIGSAC Conference on Computer and Communications Security*, Vienna: ACM, 2016, pp. 270–282.

19 Agricultural Internet of Things (AIoT)
Architecture, Applications, and Challenges

Kavitha Rajamohan, Sangeetha Rangasamy, Libin Baby, Niveditha G., and Muktha D. S.

INTRODUCTION

INTERNET OF THINGS

The internet of things (IoT) is a system that connects physical devices, household appliances, vehicles, and other objects through the use of sensors, software, and diverse technologies. This connectivity enables these devices to gather and share data with each other via the internet (Sekaran et al., 2020). This connectivity facilitates remote device monitoring and control, automation of routine tasks, and data collection for performance analysis. IoT technology has the potential to revolutionize various industries, including healthcare, agriculture, transportation, and manufacturing. The use of IoT devices in manufacturing can improve efficiency, reduce costs, and minimize downtime by monitoring equipment performance and predicting maintenance needs (Alharbi & Aldossary, 2021).

AGRICULTURE

Agriculture is the backbone of our civilization, providing food, fiber, and fuel for sustenance and economic growth. It encompasses a wide range of practices, from crop cultivation and animal husbandry to technological advancements that optimize productivity and promote sustainable farming methods. The agriculture industry is facing new and evolving requirements due to changing consumer preferences, increasing demand for food, and sustainability concerns. These include the need for sustainable practices, precision agriculture utilizing data analysis, and navigating labor shortages and immigration regulations (Chen et al., 2020). Farmers and agribusinesses must remain adaptable and invest in new technologies and practices to meet these demands and stay competitive.

DOI: 10.1201/9781003434269-19

AIoT: AGRICULTURAL INTERNET OF THINGS

Agricultural IoT utilizes internet-connected sensors and devices in farming to improve productivity, minimize wastage, and optimize resource utilization. Using IoT, farmers have the ability to monitor and control various aspects of their operations in real-time, including soil moisture, temperature, and livestock health (Quy et al., 2022). By collecting and analyzing data, farmers can make informed decisions about when and how to water crops, apply fertilizers and pesticides, and harvest their yields.

The benefits of agricultural IoT include reducing water waste and the use of harmful chemicals, increasing yields and profits, and improving the efficiency of farming operations. IoT devices can also help farmers detect problems with their crops or livestock early, enabling more effective and timely interventions that can lead to higher yields and greater food security (Chen et al., 2020). By utilizing IoT, farmers can also minimize their environmental impact, make their operations more sustainable, and keep up with the demands of an ever-changing and evolving agricultural industry. However, the implementation of IoT in agriculture also poses challenges, such as the need for strong connectivity in rural areas and concerns about data security and privacy.

AIoT IN SMART BUSINESS AND PRODUCTION

AIoT in smart business and production refers to the integration of artificial intelligence (AI) and IoT technologies to enhance efficiency, productivity, and decision-making in business operations and manufacturing processes. By connecting IoT devices and applying AI algorithms, businesses can gather real-time data and gain valuable insights for optimization (Rana et al., 2021).

AIoT improves operational efficiency by collecting data from sensors and smart machines, enabling real-time analysis to identify patterns and inefficiencies. Predictive maintenance using AIoT detects equipment failures in advance, reducing downtime and maintenance costs. Intelligent decision-making is facilitated through machine learning (ML) algorithms that derive insights from data, aiding in demand forecasting, inventory optimization, and quality control. Process automation is another key application of AIoT. By automating repetitive tasks using AI algorithms, businesses can save time and resources (Khang et al., 2023a).

For example, AIoT systems automate quality control processes, improving product quality and production cycles. AIoT enables the creation of smart and interconnected ecosystems. By integrating AI and IoT, businesses establish networks of devices, systems, and stakeholders, fostering seamless communication and collaboration. Real-time tracking and monitoring in supply chain management optimize operations and responsiveness.

In summary, AIoT revolutionizes smart business and production by optimizing efficiency, enabling informed decision-making, automating processes, and creating interconnected ecosystems. Embracing AIoT empowers businesses to stay competitive in a digital and data-driven world, unlocking productivity and innovation.

AIOT ARCHITECTURE FOR SMART AGRICULTURE

IoT Devices

Specialized sensors and devices are used in agricultural IoT to collect data in farming and agricultural practices. Farmers gather information on parameters like soil moisture, temperature, humidity, crop health, and livestock monitoring. This data enables farmers to make informed decisions, resulting in improved resource utilization and increased productivity in agriculture. There are several IoT devices used in agricultural IoT architectures, including the following:

- **Sensors**: Sensors are the backbone of an IoT architecture. They can detect various environmental factors like temperature, humidity, soil moisture, rainfall, and light levels.
- **Drones**: Drones can be used for crop monitoring, for spraying pesticides, and to capture aerial images of crop fields.
- **Smart Irrigation Systems**: These systems are equipped with sensors that can detect soil moisture levels and can automate the watering process.
- **Livestock Monitoring**: IoT devices like GPS trackers and activity monitors can be attached to livestock to track their movements, health, and behavior.
- **Automated Greenhouses**: Automated greenhouses use IoT technology to monitor and control temperature, humidity, and lighting levels to create the ideal growing environment for crops.
- **Smart Tractors**: Smart tractors are equipped with sensors and GPS technology to automate tasks like plowing, planting, and harvesting.

Network Communication

Network communications in IoT enable the connection and data exchange between a multitude of devices and systems. It forms the backbone of IoT, allowing seamless communication, data transfer, and coordination among various interconnected devices and platforms. There are several network communications used in agricultural IoT architectures, including the following:

- **Wi-Fi**: It is a common wireless network communication used in agricultural IoT systems. It enables devices to connect to the internet and transmit data to a central database.
- **Cellular Networks**: Cellular networks are used to connect IoT devices in remote areas where Wi-Fi signals may not be available. They allow farmers to collect and transmit data from their fields to a central system, such as weather stations or soil moisture sensors.
- **Bluetooth**: Bluetooth is used for short-range communication between IoT devices. It is often used in conjunction with other wireless networks to collect data from sensors and send it to a central system.

- **Zigbee**: Zigbee is a wireless mesh network protocol that is commonly used in agricultural IoT systems. It is a low-power, low-bandwidth network that is designed for devices that don't need to transmit large amount of data.
- **LoRa**: Long Range (LoRa) is a wireless technology that enables long-range communication between IoT devices. It is commonly used in rural areas to connect devices over long distances.
- **MQTT**: Message Queuing Telemetry Transport (MQTT) is a messaging protocol that is used to transmit data between IoT devices and central servers. It is designed for low-power devices and unreliable networks, making it a good choice for agricultural IoT systems.

ANALYTICS AND STORAGE SOLUTIONS

Analytics and storage solutions are essential components in agricultural IoT architectures as they facilitate the gathering, processing, and analysis of the large volume of data generated by IoT devices in agricultural settings (Table 19.1).

Common IoT analytics and storage solutions used in agriculture are summarized in Table 19.1. Some of these solutions are discussed below (Kristen et al., 2021):

- **Cloud Computing**: The utilization of cloud-based solutions such as Amazon Web Services (AWS), Microsoft Azure, and Google Cloud Platform is prevalent in agricultural IoT for data storage and processing purposes (Rani et al., 2023). These solutions offer scalable and cost-effective features, enabling data analysis and informed decision-making for farmers.
- **Edge Computing**: Edge computing refers to a decentralized infrastructure that brings computation and data storage closer to the location where it is needed, such as on a farm or in a field. It provides rapid analytics and decision-making capabilities, particularly in regions with restricted internet connectivity.
- **Data Lakes**: Data lakes serve as a central repository for storing and analyzing large volumes of raw data. They accommodate various data types from different sources, enabling comprehensive data storage and analysis.

TABLE 19.1
AIoT Architecture for Smart Agriculture

IoT devices	Network communications	Analytics and storage solutions
Sensors	Wi-Fi	Cloud computing
Drones	Cellular networks	Edge computing
Smart irrigation systems	Bluetooth	Data lakes
Livestock monitoring	Zigbee	Predictive analytics
Automated greenhouses	LoRa	Data visualization
Smart tractors	MQTT	

- **Predictive Analytics**: Predictive analytics utilizes statistical algorithms and machine learning to analyze historical data and forecast future events. In agricultural IoT, it can predict weather patterns, crop yields, and pest infestations, assisting farmers in making informed decisions regarding crop management.
- **Data Visualization**: Data visualization tools like Tableau and Power BI allow farmers to create visual representations of their data, enhancing understanding and interpretation. By identifying trends and patterns quickly, data visualization empowers farmers to make better decisions.

APPLICATION DOMAINS OF AIOT

The emergence of the internet of things (IoT) has had a substantial impact on the tracking, monitoring, and control aspects of contemporary agriculture. Below is a brief description of these concepts and their implementations in the agricultural industry.

TRACKING

In agriculture, tracking refers to the process of monitoring the location and growth of crops, livestock, and farm equipment. This information is critical for optimizing operations and improving the efficiency of farming activities. IoT-enabled devices and sensors, such as GPS and RFID tags, are increasingly being used to track the location and movement of crops, livestock, and farm equipment in real-time. For example, IoT-enabled sensors and GPS devices can be used to track the location and usage of farm equipment. This information helps farmers optimize equipment utilization, reduce maintenance costs, and improve overall efficiency (Rekha et al., 2020).

MONITORING

Monitoring involves the collection and analysis of data pertaining to various aspects of agricultural operations, such as soil moisture, temperature, humidity, light levels, and weather conditions. IoT-enabled sensors and devices are used to gather this data, providing farmers with real-time insights into the factors influencing their crops and livestock. This valuable information helps farmers make informed decisions regarding crucial tasks like planting and harvesting, resulting in increased efficiency and productivity.

Examples of IoT applications in agriculture include equipment tracking, crop monitoring, and the use of IoT-based soil moisture sensors. By implementing moisture sensors in their fields, farmers can monitor soil conditions and receive updated information on moisture levels. This data enables them to make informed irrigation decisions, leading to enhanced crop yields and reduced water waste.

CONTROLLING

Controlling involves the adjustment and management of different elements in agricultural operations to optimize performance and enhance efficiency. IoT-enabled devices and systems can be utilized to control irrigation systems, greenhouse environments, and other vital aspects of farming. For instance, an IoT-enabled irrigation system (Vunnava et al., 2021) can be programmed to automatically regulate watering levels based on real-time soil moisture data, minimizing the chances of over- or under-watering.

APPLICATIONS OF AIOT

In agriculture, the utilization of the internet of things (IoT) can lead to enhanced crop yields, cost reduction, and improved resource management. Several examples of IoT applications in agriculture include the following:

- Smart irrigation systems that leverage weather data and soil moisture sensors to optimize water usage.
- Precision agriculture techniques that utilize sensors and data analysis to optimize planting, fertilization, and pest control practices.
- Livestock monitoring systems incorporate sensors to track and monitor the health and well-being of animals.
- Climate control systems for greenhouses and barns, employing IoT devices to monitor factors like temperature, humidity, and other environmental conditions to optimize the growth environment.

These IoT applications play a crucial role in optimizing agricultural processes and achieving higher productivity while ensuring efficient resource utilization.

SMART IRRIGATION SYSTEMS

Smart irrigation is an innovative technique that utilizes technology to effectively manage and optimize agricultural irrigation systems. This method incorporates sensors, weather data, and algorithms to determine the optimal timing and quantity of water needed for crops. As a result, water waste is reduced, and crop yields are improved. Smart irrigation systems can be integrated with other agricultural technologies, such as precision agriculture, to make informed decisions about water usage and crop management. The main goal of smart irrigation is to maximize crop yields while minimizing water consumption and expenses, promoting sustainable and efficient agricultural practices.

   The system of design and implementation provides an improved method for analyzing and monitoring water quality parameters; various parameters such as pH, total dissolved solids (TDS), turbidity, biochemical oxygen demand (BOD), and temperature are measured. The values are shown on an LCD screen, and the data is stored in the cloud and transmitted to users via a GSM module. The system sends

Greenhouse Monitoring

Livestock Monitoring

Applications of IoT in Agriculture

Drone Spray

Crop Monitoring

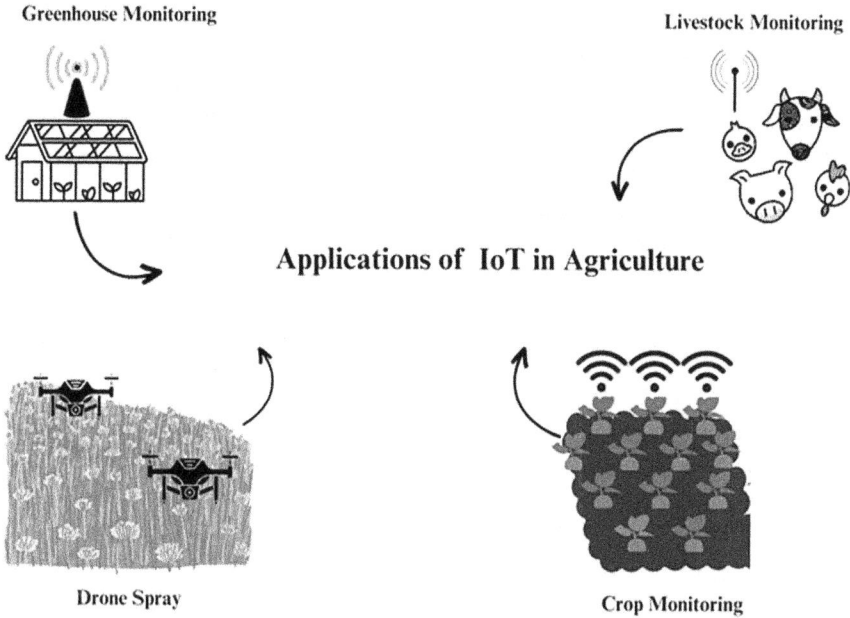

FIGURE 19.1 IoT applications in agriculture

alerts and offers solutions to improve water quality (Rekha et al., 2020). Existing methods and devices for irrigation scheduling are outdated or limited to research labs, and farmers, especially in developing nations, seldom use them.

A portable and affordable device known as the electronic wetting front detector (EWFD) has been created as a solution to tackle this problem. The EWFD interprets real-time telemetry sensor readings and notifies farmers about over or under-irrigation. This user-friendly device helps farmers comprehend soil water content and enables them to schedule irrigation according to the specific requirements of their crops. The EWFD underwent laboratory testing and its performance was validated under real field conditions (Hasnain & Singh, 2022).

PRECISION AGRICULTURE

IoT-enabled sensors and devices have the capability to monitor crucial factors such as soil moisture, temperature, humidity, and light levels, which have a significant impact on crop growth and productivity. Peddinti et al. (2020) highlights the accuracy and affordability of a wireless soil moisture sensor, particularly after calibration, making it suitable for precise irrigation and detailed soil moisture monitoring.

Calibration involves considering factors like soil solution electrical conductivity and temperature. Similarly, Tocco et al. (2023) emphasizes the cost-effectiveness and reliability of wearable sensors in monitoring plant growth and microclimate in agriculture. Challenges still need to be addressed, such as securely attaching sensors to plants, compatibility with different plant types, and scalability. Nonetheless, the authors express optimism about the future of wearable sensors in precision agriculture and smart farming. They aim to improve the device by integrating additional sensors and creating a versatile wearable system to optimize agronomic procedures and promote sustainability.

Within the AREThOU5A IoT platform, an experiment was conducted to assess the performance of the RF energy harvesting module. This experiment involved using a dual E-shaped patch antenna operating in the frequency bands of LoRaWAN and EGSM-900 mobile networks. The copper antenna was positioned on an FR-4 substrate, with a microstrip line connected to ensure efficient energy transfer to the rectifier. The length of the microstrip line was optimized to maximize power transfer. The antenna design was enhanced using a population-based algorithm (Boursianis et al., 2021).

Hundal et al. (2023) highlight the significance of low data latency in autonomous applications such as smart irrigation and farm machinery navigation, while accepting higher latency for monitoring purposes. The scalability of a system is influenced by factors like data interoperability, scalability, and the cost of data collection. The study offers valuable insights for the development of cost-effective and energy-efficient IoT systems in precision agriculture, although there may be limitations in interpreting the results.

LIVESTOCK MONITORING SYSTEMS

IoT-enabled wearables and sensors offer a new livestock monitoring solution, enabling farmers to track animal health and location. Real-time monitoring, improved feed efficiency, and disease outbreak prevention are achieved, addressing challenges in remote areas with limited access to veterinary services. Non-contact temperature measurement and behavior monitoring systems enable early disease detection and enhanced animal well-being. Leveraging IoT and data analytics empowers farmers to make informed decisions, enhancing productivity and cost reduction. The authors propose using IC tags for pig behavior monitoring, an effective solution for large-scale pig farms, improving livestock management and ensuring animal welfare (Quy et al., 2022).

CLIMATE CONTROL SYSTEMS

By utilizing IoT weather stations and sensors, farmers can access real-time weather data, including temperature, humidity, and precipitation. This information allows them to make informed decisions about farming activities such as planting and harvesting. Multiple data points collected by the sensors are transmitted to a central unit for analysis, control of actuators, problem diagnosis, and alerting farmers in case of deviations. Timely detection and resolution of issues help reduce the risk of crop failure, while remote management saves time and minimizes manual labor (Tripathy et al., 2021).

Climate smart agriculture (CSA) aims to achieve these objectives by implementing three interconnected and mutually reinforcing pillars: enhancing agricultural productivity and incomes sustainably, strengthening the resilience of agricultural systems to climate change, and reducing greenhouse gas emissions from agriculture. Various practices and technologies like conservation agriculture, agroforestry, integrated water management, precision agriculture, and the use of drought-resistant crops contribute to the realization of these goals (Yang et al., 2021).

IOT-ENABLED TECHNOLOGIES FOR AGRICULTURE

Agriculture has witnessed significant advancements and holds promise for the future with innovative technologies (Ait Issad et al., 2019). IoT, originally an emerging technology in agriculture, has now become mainstream. Agricultural IoT enables autonomous processing, analysis, and access to data on environmental conditions, crop development, and agricultural processes. The IoT architecture comprises three layers: perception, transmission, and application, as described by Yang et al. (2021). In the perception layer, sensors collect data on the field environment, crop growth, and environmental state. Robotic and drone-based technologies are also employed. The transmission layer transfers data from perception to application using various communication protocols. Agriculture heavily relies on the application layer, which can be locally based (edge computing) or cloud-based. Different technologies are used in each layer of the IoT architecture, catering to agricultural needs (El-Basioni & El-Kader, 2020).

DATA ACQUISITION TECHNOLOGY

Data acquisition involves gathering external data and integrating it into an internal system using various devices and methods such as serial ports, USB interfaces, wireless communication modules, sensors, and actuators (Yang et al., 2021). In the perception layer of the internet of things (IoT), sensors have a vital role in gathering data concerning the field environment, crop growth, and environmental conditions. Recent advancements in sensor technology have simplified crop monitoring, allowing for real-time monitoring and control of agricultural tasks such as plowing, sowing, irrigation, fertilization, pesticide usage, harvesting, and livestock management (Shaikh et al., 2022). Wireless sensor networks (WSNs) are commonly employed, and drones equipped with advanced cameras and sensors are utilized for monitoring large agricultural areas. Agricultural robots, also known as agbots, are also being deployed to streamline farming processes, including the identification and removal of weeds (Shaikh et al., 2022).

COMMUNICATION TECHNOLOGY

Effective implementation of IoT systems heavily relies on communication technology, which can be categorized into standards, spectrum, and application scenarios according to Elijah et al. (2018). IoT systems utilize both licensed and unlicensed spectrums for communication. The unlicensed spectrum, referred to as the ISM band, is commonly used but has drawbacks such as security concerns, potential interference, and costly infrastructure.

On the other hand, the licensed spectrum allocated to cellular networks offers advantages such as improved traffic management, reduced interference, and reliable service, enhanced quality of service (QoS), high security, extensive coverage, and lower infrastructure costs. However, it requires a data transmission subscription and leads to higher power consumption for IoT devices (Elijah et al., 2018). Wireless communication standards for IoT can be classified into short-range and long-range categories. Short-range standards like NFC, Bluetooth, Zigbee, Z-Wave, and passive/active RFID have a maximum range of approximately 100 meters.

Long-range standards, known as low power wide area (LPWA) technologies, including LoRa, Sigfox, and NB-IoT, can cover distances up to tens of kilometers while operating with low power consumption (Yang et al., 2021). The choice of communication technology for IoT devices depends on specific application scenarios. Some devices function as nodes, transmitting small data volumes over short distances with low power consumption. Others serve as backhaul networks, handling higher data rates and covering longer distances. Certain communication technologies support bidirectional links, facilitating features such as error correction, data reliability, encryption, firmware updates, and device communication (Wójcicki et al., 2022).

CLOUD COMPUTING TECHNOLOGY

The application layer is a critical element in agriculture and can be implemented through cloud-based systems, which utilize multiple servers, or through local systems that leverage edge computing facilitated by a gateway (Bhambri et al., 2022). This layer encompasses various functions, including data storage (e.g., cloud platforms or Hadoop file systems), data management, data analytics (such as decision-making systems, yield models, and instructions for automatic agricultural control), and data marketing (including data visualization and traceability systems for ownership and privacy, as well as the creation of new business models) (Yang et al., 2021). Cloud-based solutions provide cost-effective and scalable storage and processing capabilities.

The next tier is Fog computing, which extends cloud capabilities by incorporating nodes like base stations, access points, switches, routers, and gateways between the cloud and edge tiers. The foundational layer is edge computing, which operates with a distributed architecture, offering low latency and bandwidth requirements while maintaining high scalability. Edge computing processes information at the edge gateway, closer to the device level, reducing the volume of data that needs to be transmitted to the cloud (Wójcicki et al., 2022).

EMERGING TECHNOLOGIES IN AIOT

Smart agriculture integrates sensing, actuating, and computing technologies into traditional farming practices. It employs IoT technologies like agbots, drones, remote sensing, and artificial intelligence to tackle challenges in plowing, planting, irrigation, fertilizing, harvesting, and livestock management. By leveraging these technologies, smart agriculture enhances drought response, crop optimization, environmental sustainability, pest control, and irrigation efficiency (Shaikh et al., 2022). Figure 19.2 showcases prominent equipment and emerging IoT technologies (Nassereddine and Khang, 2024).

IoT Devices

IoT devices in agriculture possess traits like low power consumption, ample memory, efficient processing, mobility, robustness, broad coverage, dependability, and affordability (Elijah et al., 2018).

Sensors

Agricultural sensors are essential components in smart agriculture, as they collect and transmit data on environmental conditions and crop growth. These sensors can be categorized into different types, including location sensors, photoelectric sensors, mechanical sensors, electrochemical sensors, airflow sensors, and optical sensors. They gather crucial information such as meteorological data (temperature, humidity, CO_2 concentration), crop-related information (growth conditions, diseases), soil information (type, moisture level), and location information for precise crop placement (Yang et al., 2021).

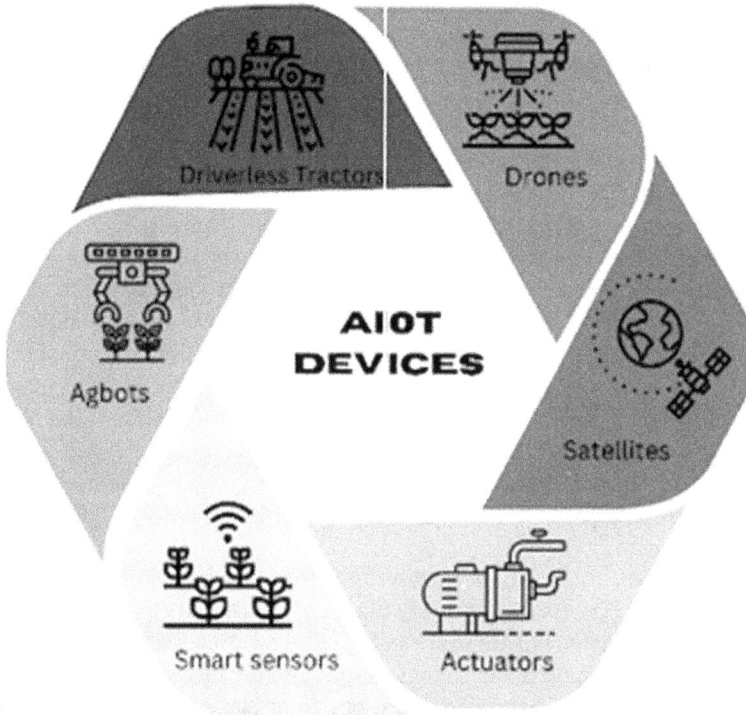

FIGURE 19.2 IoT devices with agbots, drones, remote sensing, and artificial intelligence

The integration of artificial intelligence (AI) has significantly enhanced the capabilities of smart sensors in remote sensing and agriculture. AI-enabled sensors find extensive applications in robotics, navigation, automation, remote sensing, and underwater imaging. They provide real-time data detection and analysis, resulting in improved remote sensing applications. Active and passive sensors are utilized in remote sensing for various purposes, including navigation, surveillance, communication, forecasting, and earth resource monitoring (Ullo and Sinha, 2021). With the integration of the internet of things (IoT), agricultural sensors have become more intelligent and interconnected. This integration enables the development of sensor-based equipment and devices that enhance the efficiency and productivity of farming practices. IoT-enabled sensors, combined with AI techniques, play a significant role in revolutionizing the agriculture sector and enabling precision farming (Yang et al., 2021; Ullo and Sinha, 2021).

Actuators

Once the sensors have gathered the data, it is transmitted to the cloud for analysis and processing. Using predefined criteria, the collected data triggers the activation or deactivation of agricultural equipment through actuators. In facility agriculture, common types of actuators used include irrigation equipment, which adjusts moisture levels to promote optimal crop growth; lighting control systems, which ensure appropriate lighting conditions; air circulation mechanisms, which regulate carbon dioxide concentration in enclosed growing environments; and crop disease control mechanisms, which protect the health of the crops (Yang et al., 2021).

Agricultural Drones

Various types of agriculture drones are available for different purposes within the farming industry. Unmanned aerial vehicles (UAVs) have introduced notable advancements to conventional agricultural methods, providing advantages in terms of cost-effectiveness and time efficiency (Dileep et al., 2020). Unlike labor-intensive and time-consuming ground-based methods, drones have the ability to cover larger areas more efficiently. They come in different configurations, including computer-controlled, fully automated, semi-automated, and remote-accessible designs. In agriculture, drones are utilized for a range of tasks such as monitoring weather conditions, detecting crop infections, and assessing land fertility. Advanced drone technology incorporates features like automatic adjustment of solar panels to optimize solar energy harvesting based on sun angle and optimized flight paths to take advantage of favorable wind conditions for extended flight duration (Rehman et al., 2022).

Different types of drones are employed for specific applications in agriculture. Multirotor drones possess the capability to capture aerial photography from various angles, resulting in more precise outcomes. Fixed-wing drones have wings similar to airplanes and require external force for take-off, limiting their application to specific tasks. Single-rotor helicopters, with their long rotor blades, are efficient and well-suited for agricultural purposes. Fixed-wing hybrid vertical take-off and landing (VTOL) drones combine the characteristics of both fixed-wing aircraft and VTOL capabilities, allowing them to perform take-off and hover functions. As a result, they are well-suited for extended flight operations in a specific location (Yang et al., 2021).

Agricultural Robots

Agricultural robots, known as agbots, have significantly benefited crop productivity and farming costs. However, there are still challenges to overcome in order to fully utilize their potential. While robots have addressed labor shortages and declining profits in agriculture, their capabilities are currently limited to specific tasks and they are not fully adapted to outdoor conditions. Achieving standardization and interoperability among agbots and drones is necessary, and the cost and infrastructure requirements present challenges, particularly in developing countries (Zhang et al., 2021).

Researchers have made progress in developing a harvesting robot capable of autonomously collecting fruits under specific weather conditions. Although the vision-based mechanism of this robot is still in its early stages, it shares a similar structure with other agricultural robotic systems. The system consists of a self-governing mobile platform, a versatile mechanical arm with multiple degrees of motion, a sensory machine vision system, an intelligent decision-making and control system, and supplementary hardware and software elements (Rehman et al., 2022). Furthermore, a sophisticated neural network has been created for automating the process of apple harvesting. This system utilizes computer vision techniques, including Mobile-DasNet, an efficient instance segmentation network, and an improved PointNet model for fruit modeling and grip estimation using RGB-D camera data. These features have been integrated to create a precise autonomous fruit-picking system (Rehman et al., 2022).

Satellites and Aircraft

Agricultural satellite remote sensing uses technology to observe various aspects of agriculture, including different agricultural systems, production processes, and elements of agriculture like production, environment, and ecology. It is applied in fields such as area estimation, monitoring crop growth, detecting pests, predicting yields, monitoring grassland vegetation, and mapping agricultural resources (Yang et al., 2021). Advancements in satellite-based remote sensing for agriculture involve the collaborative observation and monitoring of agricultural activities using satellites that offer enhanced spatial and temporal resolution.

Driverless Tractor

Driverless tractors, a fusion of traditional tractors and advanced autonomous technology, have emerged as a significant advancement in farming equipment. They serve as the power source for various farming tools, including cultivators, pesticide sprayers, and harvesters. The incorporation of autonomous technology offers numerous advantages, including improved operational efficiency achieved through optimized route mapping and real-time obstacle avoidance using Lidar (Lidar is a method for determining ranges by targeting an object or a surface with a laser and measuring the time for the reflected light to return to the receiver). These advancements enhance the overall effectiveness and output of agricultural practices, while also maximizing the utilization of valuable resources like land, seeds, and fertilizers.

Cloud/Edge Computing

Cloud computing plays a vital role in the agricultural sector by providing farmers with storage space and access to their agricultural data. This enables effective data storage and analysis. With its flexible and virtualized resources, cloud computing has become widely adopted in various agricultural applications, including planting, aquaculture, animal husbandry, and indoor agriculture.

In contrast, edge computing is a computing model that processes data near its source, reducing latency and enabling real-time processing. It eases the computational load and data transmission to and from fog or cloud data centers. If edge resources are insufficient, fog or cloud processing can be requested hierarchically. Fog computing, introduced by Cisco, extends cloud computing resources to the edge of the network, improving latency and reducing costs. It provides computation, networking, and storage services in the fog layer, while complex data processing and scalable resources are available in the cloud layer.

The combination of fog, edge, and cloud computing enhances data processing performance, security, and reliability. Stream data processing is handled by distributed fog and edge nodes, while big data processing occurs in the cloud. However, the centralized nature of cloud computing faces challenges due to the increasing volume of agricultural data, resulting in cost, latency, and privacy concerns. To overcome these limitations, research is focusing on leveraging fog computing resources to provide cost-effective solutions in underdeveloped areas while meeting performance requirements (Fawcett et al., 2020).

Advancements in electronics, such as System on Chip (SoC), empower edge devices to make autonomous decisions without relying on central servers running AI algorithms (Khang et al., 2023b). Embedded systems with improved processing power and memory, coupled with AI-based computer vision algorithms, enable real-time pest detection and image compression techniques to overcome data transmission rate limitations in wireless sensor networks. This shift from cloud to edge computing enables real-time threat detection and timely prevention in AI-based smart agriculture systems (SAS).

Internet Protocols

In smart agriculture systems, sensor data is transmitted to a server or database using either wired or wireless networks. Wireless communication technologies are crucial for ensuring the reliable, accurate, and secure transmission of collected information to relevant stakeholders. Smart agriculture utilizes a variety of wireless communication technologies, including RFID, Bluetooth, Zigbee, Wi-Fi, LoRa, Sigfox, WiMAX, NB-IoT, and cellular networks. These communication protocols enable devices within the system to exchange information, communicate, and make informed decisions to monitor and control farming conditions, ultimately leading to improved yields and production efficiency. In terms of communication range, low-power communication protocols in smart agriculture can be classified into short-range and long-range categories.

- The preferred protocol for wired networks was Ethernet (Omar, 2021).
- **Short-Range Protocols**: NFMI (near-field magnetic induction), Bluetooth, Zigbee, terahertz (Z-Wave), RFID (radio frequency identification) (Rosa et al., 2022).
- **Long-Range Protocols**: LoRa, Sigfox, NB-IoT (Narrowband IoT) (Quy et al., 2022).

Smart agriculture faces challenges related to limited coverage of short-range wireless technologies and the need for improvements in cellular networks to handle IoT network traffic. However, Low-Power Wide Area Network (LPWAN) technology, including LoRa, offers wide spatial coverage and is suitable for small data exchange in IoT applications. Commonly used communication protocols in smart farming systems include Zigbee, Wi-Fi, and MQTT (Bayih et al., 2022; Omar, 2021).

The future of smart agriculture will be influenced by the emergence of 5G wireless technology, which provides high data rates, extensive coverage, and supports real-time crop monitoring and vehicle communications. The low latency and new frequency bands offered by 5G will be instrumental in connecting farming machinery such as smart tractors and drones, enabling vehicle-to-vehicle and vehicle-to-infrastructure communications (Qazi et al., 2022).

ARTIFICIAL INTELLIGENCE

As technology advances, AI's role in smart agriculture is becoming increasingly important. With the development of energy-efficient hardware, researchers are exploring distributed AI approaches instead of relying solely on centralized AI systems that rely on the cloud (Khang et al., 2024a). AI can be harnessed by edge networks and IoT devices to provide accurate forecasts on weather patterns, plant growth, water requirements, and more. Researchers are actively developing AI techniques that require reduced data and computational resources to improve efficiency in smart agriculture. Swarm intelligence-based machine learning and deep learning (ML) algorithms are being explored to make accurate predictions in agricultural applications. Furthermore, the advancement of natural language processing systems and chatbots can assist in training farmers for increased productivity and secure data analysis (Shaikh et al., 2022).

Artificial intelligence finds application in various aspects of agriculture (Elbasi et al., 2023; Yang et al., 2021). Some examples include the following:

- **Crop and Soil Monitoring**: Real-time monitoring of crops and soil health using AI and machine learning allows companies to estimate crop yields and determine the optimal harvesting time (Khang et al., 2023c).

- **Disease Diagnosis**: AI enables early detection of plant diseases, empowering farmers to implement effective strategies for disease control and prevention.
- **Agricultural Robots**: AI-powered robots equipped with sensors, machine vision, and AI models can perform harvesting processes with improved accuracy and speed.
- **Predictive Insights**: AI provides valuable insights that inform decision-making in agriculture.
- **Crop Yield Prediction**: AI techniques help predict the optimal timing for fertilizing fields and sowing seeds, maximizing yield and improving profitability.
- **Intelligent Spraying**: AI-powered intelligent spraying optimizes the application of chemicals, reducing waste and enhancing pest and disease control.
- **Data Collection and Analysis**: Farmers can track and analyze various data points using AI-based analytics, leading to better-informed decisions and the discovery of innovative solutions.
- **Predictive Analytics**: AI tools enable predictions of changes in weather, soil conditions, pest infestations, and facilitate soil quality improvement planning and overall farm management.

Machine Learning

Machine learning (ML) is of utmost importance in the field of agriculture as it enhances productivity and crop quality. ML employs algorithms to monitor and manage various agricultural activities, aiming to develop computer programs capable of making predictions based on input data. These algorithms are utilized for tasks like categorizing agricultural datasets, predicting suitable crops for specific regions, determining optimal timing for sowing and harvesting, and selecting appropriate irrigation methods and soil types. The choice of algorithms depends on factors such as data availability, output accuracy, interpretability, training time, and specific features. In the domain of smart farming, ML systems utilize computer vision techniques to identify and analyze objects in agricultural fields, collecting data through a range of sensors (Elbasi et al., 2023).

Deep Learning

Deep learning (DL) networks differ from neural networks due to their deeper structure, enabling them to uncover hidden structures in unlabeled and unstructured data. Training DL algorithms with large datasets demands significant computational resources, often utilizing powerful GPUs. In agriculture, DL systems are employed for tasks such as detecting plant diseases, identifying pests in crops, providing agricultural advisory services, and facilitating real-time pest threat detection. DL is also used for plant phenotyping and weed detection, addressing factors that decrease farmland productivity. Image-processing-based DL systems, particularly convolutional neural networks (CNNs), find practical application in agriculture (Qazi et al., 2022). However, challenges remain concerning investment costs, compatibility with existing technology, skills, privacy, security, and regulations. Farmers must adapt to new approaches and collaborate with stakeholders to reduce technology deployment costs. Implementing AI and robotics can offer economies of scale, but proper training and education are necessary to stay updated with the latest advancements in AI (Khang et al., 2024b).

BLOCKCHAIN

Blockchain technology, which was introduced in 2008 by individuals using the pseudonym 'Satoshi Nakamoto', is a decentralized system for storing data. It relies on consensus mechanisms and cryptography to ensure secure data transmission. By enabling the deployment of smart contracts across multiple nodes, blockchain facilitates intelligent data management. In the agricultural sector, blockchain is widely used in traceability systems to establish mechanisms for tracking agricultural products, enhance transparency of information, and create digital marketplaces. The security issues related to the adoption of blockchain in agriculture are currently under investigation, as the technology is still in its early developmental phase (Yang et al., 2021; Wang et al., 2021).

Application of blockchain technology in food supply chain is the utilization of blockchain technology in the agricultural industry has the potential to transform the food supply chain. It enables the creation of a digital ledger that tracks and manages the movement of resources throughout the supply chain, ensuring transparency, traceability, and real-time monitoring of transactions. All parties involved can access specific information about products and transactions, gaining a comprehensive understanding of each product's journey (Khanh and Khang, 2021).

The implementation of blockchain in the food supply chain enhances efficiency and product quality, and empowers consumers to verify the authenticity of products. By maintaining a continuous flow of data, blockchain facilitates regulatory compliance and holds accountable those responsible for any breaches in product authenticity. However, the absence of a robust food traceability system poses challenges to food safety (Khang et al., 2022). To address this, integrating blockchain with the internet of things (IoT) can create a decentralized and transparent system. This integration allows for the creation of digital identities for food products, providing detailed information from origin to retailer and fostering consumer trust. Ultimately, the integration of blockchain and IoT technologies offers significant potential for establishing a reliable, transparent, and environmentally conscious smart agriculture system in the future (Awan et al., 2021).

SECURITY AND PRIVACY

The agriculture industry encounters a range of security challenges that require careful attention and mitigation. In smart farming, IoT devices are vulnerable to physical tampering, such as theft or damage by animals, which can compromise their functionality and integrity. The limited capabilities of IoT devices, including memory, communication, and energy consumption, present difficulties in implementing complex security algorithms. Gateways, which facilitate communication between devices, are susceptible to congestion, denial of service (DoS) attacks, and forwarding attacks. The accuracy of location information crucial for precision farming can be compromised by attacks such as device capture, wireless signal jamming, and man-in-the-middle attacks (Kumar et al., 2021).

Cloud servers used in automated farming processes are at risk of data tampering, unauthorized access, and session hijacking. The integration of sensors and IoT devices in smart farming raises concerns about security and privacy, as they can be exploited for malicious purposes. Protecting vast rural areas from security threats, agro terrorism, and environmental risks is a challenging and costly endeavor. To establish a secure system, blockchain technology can be utilized to facilitate transparent communication and agreements, while robust measures are necessary to combat DoS attacks and ensure prompt responses to natural disasters. Overall, it is essential to address these security challenges effectively to safeguard the agricultural sector (Elijah et al., 2018; Elbasi et al., 2023; Qazi et al., 2022).

AIOT-ENABLED AGRICULTURE IN VARIOUS COUNTRIES

Artificial intelligence of things (AIoT) is gaining momentum in the agricultural sector worldwide, with different countries adopting it to enhance their farming practices and productivity. Various examples of AIoT implementation in agriculture can be observed globally. In the USA, AIoT is utilized to optimize resource utilization and improve crop yields. Taranis, a company utilizing artificial intelligence (AI) and computer vision technology, provides a prime example. Their innovative approach enables the early detection of plant diseases and pests, empowering farmers to take prompt measures to mitigate potential crop damage. The United States Department of Agriculture (USDA) is also funding research to explore the integration of AI and IoT technologies, aiming to enhance farm operations' efficiency and boost yields. Additionally, autonomous farming systems have been developed in the country, leveraging AIoT to automate tasks such as planting, harvesting, and irrigation.

Japan is one of the world leaders in agricultural technology, and AIoT is no exception. Japanese companies are developing a range of solutions, such as AI-powered tractors and drones that can be used for precision agriculture. Japan is also leveraging AIoT technologies to address the challenges of an aging farmer population and labor shortages. The country has implemented autonomous tractors, robots for crop monitoring, and AI-powered systems for seed selection and fertilization.

China is one of the largest adopters of AIoT in agriculture. The country has implemented a range of technologies such as precision agriculture, automated irrigation, and livestock monitoring, to improve crop yields, reduce waste, and enhance farm efficiency. Chinese companies have also developed AIoT-powered solutions for crop protection, pest and disease detection, and forecasting. For example, the company XAG uses drones and AI to monitor crop health and apply pesticides and fertilizers more precisely, reducing waste and increasing yields (Rani et al., 2021).

Israeli efforts in AIoT agriculture have been focused on developing technologies and practices that enable farmers to optimize resource usage and increase yields while minimizing environmental impact. Some of the key areas of focus include precision irrigation, precision fertilization, and precision pest control. Israel's harsh desert climate and limited water resources have driven the need for innovative solutions that are efficient, effective, and sustainable.

One example of Israeli innovation in AIoT agriculture is Netafim, a company that has developed precision irrigation systems that use sensors and AI algorithms to determine the precise water needs of crops. By providing crops with the exact amount of water they need, Netafim's systems can reduce water waste and increase crop yields. Another company, Taranis, uses AI and image recognition technology to detect crop diseases and pest infestations early on, allowing farmers to take action before the problem spreads. Israel's government has also been supportive of the development of AIoT agriculture. In 2019, the government launched a national initiative to promote the development of innovative technologies and practices in agriculture, with a particular focus on precision agriculture and smart farming.

India is actively embracing artificial intelligence of things (AIoT) in its agricultural sector, driven by various government initiatives aimed at enhancing technology adoption and improving farmers' income. Noteworthy examples include the following:

National Agriculture Market (eNAM) Platform: The eNAM platform serves as a digital marketplace connecting farmers, traders, and buyers nationwide. By leveraging AI and machine learning, it analyzes market trends, predicts demand, and optimizes the supply chain (Narwane et al., 2022).

Precision Agriculture: Startups in India are dedicating their efforts to the development of precision agriculture solutions that combine artificial intelligence of things (AIoT) technologies. These solutions gather and evaluate data concerning crop health, soil moisture levels, and weather conditions. By doing so, they aim to enhance the efficient utilization of resources like water, fertilizers, and pesticides, ultimately leading to increased crop yields.

Smart Irrigation Systems: Indian companies are actively working on smart irrigation systems that utilize AIoT to monitor soil moisture levels and automate irrigation processes. These systems help minimize water wastage and enhance irrigation efficiency.

Crop Monitoring and Disease Detection: Indian startups are leveraging AI-powered solutions for crop monitoring and disease detection. By utilizing image recognition and machine learning algorithms, these solutions can identify crop diseases and pests, providing farmers with treatment recommendations.

India's efforts in implementing AIoT in agriculture showcase a commitment to leveraging advanced technologies for the benefit of farmers and the agricultural industry as a whole.

CONCLUSION

Agriculture IoT enables real-time data collection, analysis, and automation, leading to smarter decision-making, improved production outcomes, and transforming smart business. Livestock monitoring is an area transformed by IoT solutions. Through the use of smart collars or tags equipped with sensors, farmers can track and monitor the health and behavior of their animals in real-time (Rejeb et al., 2022). This enables early detection of illnesses, effective management of feeding schedules, and improved breeding practices, resulting in better animal welfare and increased productivity. IoT technology also facilitates supply chain optimization in agriculture. By utilizing sensors and RFID tags, agricultural products can be tracked throughout their journey from farm to market. This enables stakeholders to ensure proper storage conditions, prevent spoilage, and reduce food waste.

The transparency provided by IoT in the supply chain allows for informed decision-making, improved logistics, and enhanced overall efficiency. By employing connected devices like automated irrigation systems, robotic harvesters, and drones, farmers can reduce manual labor and boost productivity. The integration of IoT technology in agriculture brings numerous benefits such as time savings, lower labor costs, and improved resource management. The substantial volume of data produced by IoT devices in the agricultural sector facilitates decision-making based on data analysis. By employing sophisticated analytics and machine learning algorithms, this data can be examined to generate predictive models, forecasts related to crop yields, timely warnings about disease outbreaks, and other valuable insights. These insights empower farmers to allocate resources effectively, mitigate risks, and make informed decisions, ultimately leading to enhanced agricultural outcomes. By embracing IoT technologies, farmers and stakeholders in the agricultural sector can achieve higher efficiency, cost reductions, increased productivity, and the promotion of sustainable practices. This utilization of IoT technology enables resource optimization, adaptation to market demands, and the attainment of more efficient and profitable results in agriculture.

CHALLENGES IN AIOT AGRICULTURE SERVICES

The implementation of artificial intelligence of things (AIoT) in agriculture services presents various challenges that must be overcome for successful adoption and implementation. These challenges include the following:

- **Connectivity**: Ensuring reliable connectivity is crucial for the effective utilization of AIoT in agriculture, although it can be challenging to establish in rural and remote areas. Limited or unreliable connectivity can result in data loss or delays, negatively impacting data analysis and decision-making processes.
- **Data Management and Security**: Managing and securing the substantial volume of data generated by AIoT devices require robust measures. It is essential to protect the data from loss, theft, or misuse by implementing

strong security protocols and establishing secure infrastructure (Saad et al., 2020).

- **Cost**: The implementation of AIoT can be financially demanding, particularly for small-scale farmers with limited resources. The costs associated with acquiring the necessary technology, sensors, and data analytics tools can serve as barriers to entry for many farmers.
- **Expertise**: Effectively utilizing AIoT in agriculture services requires technical expertise that may not be readily available in all areas. Farmers may need access to specialized training and support to fully leverage AIoT technology.
- **Adoption and Acceptance**: Some farmers may be resistant to adopting new technology or fail to recognize its value. Overcoming this challenge requires effective communication and education to promote the benefits of AIoT and encourage widespread adoption (Rettore et al., 2020).

Addressing these challenges is crucial to realizing the potential benefits of AIoT in agriculture services and ensuring its successful integration into the agricultural industry.

RESEARCH ISSUES, SOLUTIONS, AND OPPORTUNITIES OF AIOT AGRICULTURE

Some research issues and their solutions are as follows:

- Handling the substantial amount of data produced by sensors and devices poses a significant challenge in AIoT agriculture. Efficiently processing and leveraging this data necessitate the use of techniques such as machine learning to extract valuable insights and facilitate informed decision-making.
- Ensuring interoperability between different devices and systems is another challenge. Standardizing communication protocols and data formats is necessary to enable seamless data sharing and collaboration.
- The security and privacy of sensitive data, including crop yields and environmental information, are major concerns in AIoT agriculture. Protecting against cyber threats and safeguarding the confidentiality of data are ongoing areas of research and development in this field.

Opportunities of AIoT include,:

- AIoT integration in agriculture enables precision farming practices that enhance crop growth, optimize resource utilization, and maximize yield through advanced technologies.
- Furthermore, AIoT facilitates early detection and monitoring of pests and diseases, empowering farmers to take timely and efficient actions to mitigate their impact.

- Sustainable Agriculture: By collecting and analyzing data on resource usage, AIoT can help farmers adopt sustainable practices such as reducing water and pesticide usage.
- Supply Chain Optimization: AIoT can be used to optimize supply chains by tracking produce from farm to consumer, and supporting more efficient and cost-effective distribution.
- Farm Automation: AIoT can be used to automate routine tasks such as irrigation and fertilization, reducing labor costs and improving efficiency.

The integration of AI and IoT in agriculture holds great potential for improving productivity, reducing waste, and increasing crop yields. However, to fully realize the benefits of AIoT in agriculture, it is essential to conduct further research and exploration in areas such as data management, interoperability, security, and privacy (Kali et al., 2024).

REFERENCES

Ait Issad, H., Aoudjit, R., & Rodrigues, J. J. P. C. (2019). *A Comprehensive Review of Data Mining Techniques in Smart Agriculture*. Asian Agricultural and Biological Engineering Association. https://doi.org/10.1016/j.eaef.2019.11.003

Alharbi, H. A., & Aldossary, M. (2021). *Energy-Efficient Edge-Fog-Cloud Architecture for IoT-Based Smart Agriculture Environment*. Institute of Electrical and Electronics Engineers (IEEE). https://doi.org/10.1109/access.2021.3101397

Awan, S., Ahmed, S., Ullah, F., Nawaz, A., Khan, A., Uddin, M. I., … Alyami, H. (2021). *IoT with BlockChain: A Futuristic Approach in Agriculture and Food Supply Chain* (S. Khan, Ed.). Hindawi Limited. https://doi.org/10.1155/2021/5580179

Bayih, A. Z., Morales, J., Assabie, Y., & de By, R. A. (2022). *Utilization of Internet of Things and Wireless Sensor Networks for Sustainable Smallholder Agriculture*. MDPI AG. https://doi.org/10.3390/s22093273

Bhambri, P., Rani, S., Gupta, G., & Khang, A. (2022). *Cloud and Fog Computing Platforms for Internet of Things*. CRC Press. https://doi.org/ 10.1201/9781003213888

Boursianis, A. D., Papadopoulou, M. S., Gotsis, A., Wan, S., Sarigiannidis, P., Nikolaidis, S., & Goudos, S. K. (2021). *Smart Irrigation System for Precision Agriculture—The AREThOU5A IoT Platform*. Institute of Electrical and Electronics Engineers (IEEE). https://doi.org/10.1109/jsen.2020.3033526

Chen, C.-J., Huang, Y.-Y., Li, Y.-S., Chang, C.-Y., & Huang, Y.-M. (2020). An AIoT Based Smart Agricultural System for Pests Detection. In *IEEE Access* (Vol. 8, pp. 180750–180761). Institute of Electrical and Electronics Engineers (IEEE). https://doi.org/10 .1109/access.2020.3024891

Climate-Smart Agriculture (worldbank.org).

Dileep, M. R., Navaneeth, A. V., Ullagaddi, S., & Danti, A. (2020). A Study and Analysis on Various Types of Agricultural Drones and Its Applications. Presented at the 2020 Fifth International Conference on Research in Computational Intelligence and Communication Networks (ICRCICN). https://doi.org/10.1109/icrcicn50933.2020.9296195

Elbasi, E., Mostafa, N., AlArnaout, Z., Zreikat, A. I., Cina, E., Varghese, G., … Zaki, C. (2023). *Artificial Intelligence Technology in the Agricultural Sector: A Systematic Literature Review*. Institute of Electrical and Electronics Engineers (IEEE). https://doi .org/10.1109/access.2022.3232485

El-Basioni, B. M. M., & El-Kader, S. M. A. (2020). Laying the Foundations for an IoT Reference Architecture for Agricultural Application Domain. In *IEEE Access* (Vol. 8, pp. 190194–190230). Institute of Electrical and Electronics Engineers (IEEE). https:// doi.org/10.1109/access.2020.3031634

Elijah, O., Rahman, T. A., Orikumhi, I., Leow, C. Y., & Hindia, M. H. D. N. (2018). *An Overview of Internet of Things (IoT) and Data Analytics in Agriculture: Benefits and Challenges.* Institute of Electrical and Electronics Engineers (IEEE). https://doi.org/10 .1109/jiot.2018.2844296

Fawcett, D., Panigada, C., Tagliabue, G., Boschetti, M., Celesti, M., Evdokimov, A., … Anderson, K. (2020). *Multi-scale Evaluation of Drone-Based Multispectral Surface Reflectance and Vegetation Indices in Operational Conditions.* MDPI AG. https://doi .org/10.3390/rs12030514

Hasnain, S., & Singh, A. (2022). *Development of Electronic Wetting Front Detector for Irrigation Scheduling.* Elsevier BV. https://doi.org/10.1016/j.agwat.2022.107980

Hundal, G. S., Laux, C. M., Buckmaster, D., Sutton, M. J., & Langemeier, M. (2023). *Exploring Barriers to the Adoption of Internet of Things-Based Precision Agriculture Practices.* MDPI AG. https://doi.org/10.3390/agriculture13010163

Khang, A., Chowdhury, S., & Sharma, S. (2022). *The Data-Driven Blockchain Ecosystem: Fundamentals, Applications, and Emerging Technologies* (1st Ed.). CRC Press. https:// doi.org/10.1201/9781003269281

Khang, A., Gupta, S. K., Shah, V., & Misra, A. (Eds.). (2023a) *AI-Aided IoT Technologies and Applications in the Smart Business and Production.* CRC Press. https://doi.org/10 .1201/9781003392224

Khang, A., Gupta, S. K., Hajimahmud, V. A., Babasaheb, J., & Morris, G. (2023c). *AI-Centric Modelling and Analytics: Concepts, Designs, Technologies, and Applications* (1st Ed.). CRC Press. https://doi.org/10.1201/9781003400110

Khang, A., Rana, G., Tailor, R. K., & Hajimahmud, V. A. (Eds.). (2023b). *Data-Centric AI Solutions and Emerging Technologies in the Healthcare Ecosystem.* CRC Press. https://doi.org/10.1201/9781003356189

Khang, A., Gujrati, R., Uygun, H., Tailor, R. K., & Gaur, S. S. (2024b). *Data-Driven Modelling and Predictive Analytics in Business and Finance.* CRC Press. https://doi.org/10.1201 /9781032600628

Khang, A., Abdullayev, V., Hahanov, V., & Shah, V. (2024b). *Advanced IoT Technologies and Applications in the Industry 4.0 Digital Economy* (1st Ed.). CRC Press. https://doi.org /10.1201/9781003434269

Khanh, H. H., & Khang, A. (2021). The Role of Artificial Intelligence in Blockchain Applications. In: *Reinventing Manufacturing and Business Processes Through Artificial Intelligence* (Vol. 2, pp. 20–40). CRC Press. https://doi.org/10.1201/9781003145011-2

Kristen, E., Kloibhofer, R., Díaz, V. H., & Castillejo, P. (2021). Security Assessment of Agriculture IoT (AIoT) Applications. In: *Applied Sciences* (Vol. 11, Issue 13, p. 5841). MDPI AG. https://doi.org/10.3390/app11135841

Kumar, R., Mishra, R., Gupta, H. P., & Dutta, T. (2021). *Smart Sensing for Agriculture: Applications, Advancements, and Challenges.* Institute of Electrical and Electronics Engineers (IEEE). https://doi.org/10.1109/mce.2021.3049623

Nassereddine, M., & Khang, A. (2024). Applications of Internet of Things (IoT) in Smart Cities. *Advanced IoT Technologies and Applications in the Industry 4.0 Digital Economy* (1st Ed.). CRC Press. https://doi.org/10.1201/9781003434269-6

Narwane, V. S., Gunasekaran, A., & Gardas, B. B. (2022). Unlocking Adoption Challenges of IoT in Indian Agricultural and Food Supply Chain. In: *Smart Agricultural Technology* (Vol. 2, p. 100035). Elsevier BV. https://doi.org/10.1016/j.atech.2022.100035

Omar, S. (2021). Internet of Things (IoT) for Smart Farming: A Systematic Review. *Foundation of Computer Science.* https://doi.org/10.5120/ijca2021921182

Peddinti, S. R., Hopmans, J. W., Abou Najm, M., & Kisekka, I. (2020). *Assessing Effects of Salinity on the Performance of a Low-Cost Wireless Soil Water Sensor.* MDPI AG. https://doi.org/10.3390/s20247041

Qazi, S., Khawaja, B. A., & Farooq, Q. U. (2022). *IoT-Equipped and AI-Enabled Next Generation Smart Agriculture: A Critical Review, Current Challenges and Future Trends.* Institute of Electrical and Electronics Engineers (IEEE). https://doi.org/10.1109/access.2022.3152544

Quy, V. K., Hau, N. V., Anh, D. V., Quy, N. M., Ban, N. T., Lanza, S., ... Muzirafuti, A. (2022). *IoT-Enabled Smart Agriculture: Architecture, Applications, and Challenges.* MDPI AG. https://doi.org/10.3390/app12073396

Rana, G., Khang, A., Sharma, R., Goel, A. K., & Dubey, A. K. (Eds.). (2021)*Reinventing Manufacturing and Business Processes through Artificial Intelligence.* CRC Press. https://doi.org/10.1201/9781003145011

Rani, S., Chauhan, M., Kataria, A., & Khang, A. (Eds.). (2021). IoT Equipped Intelligent Distributed Framework for Smart Healthcare Systems. *Networking and Internet Architecture,* Vol. 2, p. 30. https://doi.org/10.48550/arXiv.2110.04997

Rani, S., Bhambri, P., Kataria, A., Khang, A., & Sivaraman, A. K. (2023). *Big Data, Cloud Computing and IoT: Tools and Applications* (1st Ed.). Chapman and Hall/CRC. https://doi.org/10.1201/9781003298335

Rath, K. C., Khang, A., & Roy, D. (2024). The Role of Internet of Things (IoT) Technology in Industry 4.0. *Advanced IoT Technologies and Applications in the Industry 4.0 Digital Economy* (1st Ed.). CRC Press. https://doi.org/10.1201/9781003434269-1

Rehman, A., Saba, T., Kashif, M., Fati, S. M., Bahaj, S. A., & Chaudhry, H. (2022). *A Revisit of Internet of Things Technologies for Monitoring and Control Strategies in Smart Agriculture.* MDPI AG. https://doi.org/10.3390/agronomy12010127

Rejeb, A., Rejeb, K., Abdollahi, A., Al-Turjman, F., & Treiblmaier, H. (2022). The Interplay between the Internet of Things and Agriculture: A Bibliometric Analysis and Research Agenda. In: *Internet of Things* (Vol. 19, p. 100580). Elsevier BV. https://doi.org/10.1016/j.iot.2022.100580

Rekha, P., Sumathi, K., Samyuktha, S., Saranya, A., Tharunya, G., & Prabha, R. (2020). Sensor Based Waste Water Monitoring for Agriculture Using IoT. Presented at the 2020 6th International Conference on Advanced Computing and Communication Systems (ICACCS). https://doi.org/10.1109/icaccs48705.2020.9074292

Rettore de Araujo Zanella, A., da Silva, E., & Pessoa Albini, L. C. (2020). Security Challenges to Smart Agriculture: Current State, Key Issues, and Future Directions. In: *Array* (Vol. 8, p. 100048). Elsevier BV. https://doi.org/10.1016/j.array.2020.100048

Rosa, R. L., Dehollain, C., Costanza, M., Speciale, A., Viola, F., & Livreri, P. (2022). A Battery-Free Wireless Smart Sensor Platform with Bluetooth Low Energy Connectivity for Smart Agriculture. Presented at the 2022 IEEE 21st Mediterranean Electrotechnical Conference (MELECON). https://doi.org/10.1109/melecon53508.2022.9842920

Saad, A., Benyamina, A. E. H., & Gamatie, A. (2020). *Water Management in Agriculture: A Survey on Current Challenges and Technological Solutions.* Institute of Electrical and Electronics Engineers (IEEE). https://doi.org/10.1109/access.2020.2974977

Sekaran, K., Meqdad, M. N., Kumar, P., Rajan, S., & Kadry, S. (2020). Smart Agriculture Management System Using Internet of Things. In: *Telkomnika (Telecommunication Computing Electronics and Control)* (Vol. 18, Issue 3, p. 1275). Universitas Ahmad Dahlan. https://doi.org/10.12928/telkomnika.v18i3.14029

Shaikh, F. K., Karim, S., Zeadally, S., & Nebhen, J. (2022). *Recent Trends in Internet-of-Things-Enabled Sensor Technologies for Smart Agriculture.* Institute of Electrical and Electronics Engineers (IEEE). https://doi.org/10.1109/jiot.2022.3210154

Tripathy, P. K., Tripathy, A. K., Agarwal, A., & Mohanty, S. P. (2021). *MyGreen: An IoT-Enabled Smart Greenhouse for Sustainable Agriculture.* Institute of Electrical and Electronics Engineers (IEEE). https://doi.org/10.1109/mce.2021.3055930

Tocco Di, J., Lo Presti, D., Massaroni, C., Cinti, S., Cimini, S., De Gara, L., & Schena, E. (2023). *Plant-Wear: A Multi-sensor Plant Wearable Platform for Growth and Microclimate Monitoring.* MDPI AG. https://doi.org/10.3390/s23010549

Ullo, S. L., & Sinha, G. R. (2021). *Advances in IoT and Smart Sensors for Remote Sensing and Agriculture Applications.* MDPI AG. https://doi.org/10.3390/rs13132585

Vunnava, S. L., Yendluri, S. C., & Dhuli, S. (2021). IoT Based Novel Hydration System for Smart Agriculture Applications. Presented at the 2021 10th IEEE International Conference on Communication Systems and Network Technologies (CSNT). https://doi.org/10.1109/csnt51715.2021.9509597

Wang, L., Xu, L., Zheng, Z., Liu, S., Li, X., Cao, L., & Sun, C. (2021). *Smart Contract-Based Agricultural Food Supply Chain Traceability.* Institute of Electrical and Electronics Engineers (IEEE). https://doi.org/10.1109/access.2021.3050112

Wójcicki, K., Biegańska, M., Paliwoda, B., & Górna, J. (2022). *Internet of Things in Industry: Research Profiling, Application, Challenges and Opportunities—A Review.* MDPI AG. https://doi.org/10.3390/en15051806

Yang, X., Shu, L., Chen, J., Ferrag, M. A., Wu, J., Nurellari, E., & Huang, K. (2021). *A Survey on Smart Agriculture: Development Modes, Technologies, and Security and Privacy Challenges.* Institute of Electrical and Electronics Engineers (IEEE). https://doi.org/10.1109/jas.2020.1003536

Zhang, M., Yan, X., Lu, Z., Bai, Q., Zhang, Y., Wang, D., & Yin, Y. (2021). *Effects of Micropore Group Spacing and Irrigation Amount on Soil Respiration and Yield of Tomato with Microsprinkler Irrigation under Plastic Film in Greenhouse* (L. Xu, Ed.). Hindawi Limited. https://doi.org/10.1155/2021/6658059

20 Federated Learning (FL) with Internet of Things (IoT)-Powered Cooperative Communication for Smart System

Chellaswamy C., Sriram A., Geetha T. S., and Arul S.

INTRODUCTION

The integration of federated learning (FL) with internet of things (IoT)-powered cooperative communication and security systems in smart energy systems has acquired a substantial amount of attention during the last several years. This chapter aims to explore the significance and potential of this integration by examining various aspects related to efficient data analysis, communication mechanisms, and security considerations.

Rajagopal et al. (2023) propose a smart decision-making module called FedSDM that utilizes FL to make accurate decisions based on ECG data in the internet of things systems that incorporate Edge, Fog, and Cloud computing. The module ensures data privacy and security while training machine learning models on decentralized ECG data collected from IoT devices.

Yaacoub et al. (2023) examine the security concerns related to FL in IoT systems. They discuss the challenges, limitations, and potential solutions to improve the safety of FL models in IoT environments. Their focus is on safeguarding the privacy of sensitive data during the training process and ensuring the integrity and authenticity of the FL system.

Application of machine learning algorithms in addressing problems and difficulties relating to security in IoT networks has been analyzed by Anuradha and Rajini (2019). They explore the vulnerabilities and threats faced by IoT networks and discuss the potential use of machine learning algorithms to enhance IoT security. The

 DOI: 10.1201/9781003434269-20

paper emphasizes the significance of leveraging machine learning techniques for detecting and mitigating security threats in IoT systems.

Rahman et al. (2023) investigate the integration of information-centric networking (ICN) and IoT combined with FL. They discuss the concepts, security-privacy concerns, applications, and future prospects of combining ICN-IoT with FL. The authors explore the potential benefits of merging ICN and FL to improve data management, privacy, and communication efficiency in IoT environments.

Al-Huthaifi et al. (2023) conduct a survey focusing on the privacy and security aspects of FL in smart cities. They examine the challenges, threats, and privacy-preserving techniques in FL systems deployed in smart city environments. The survey covers various aspects, including data privacy, authentication, access control, and secure aggregation, with an emphasis on applications in smart city contexts.

Da Silva et al. (2023) present a systematic review that explores different approaches and techniques for optimizing FL in IoT applications. Their review focuses on resource optimization and discusses various methods to improve the performance and scalability of FL in the context of IoT. Zhang et al. (2023) propose CCM-FL, a covert communication mechanism designed for FL in crowd-sensing IoT. The authors investigate techniques that enable covert communication between IoT devices while maintaining data privacy and security. The paper explores methods such as steganography and encryption to facilitate secure communication in FL systems.

Singh et al. (2022) propose a framework that ensures privacy preservation of IoT healthcare data using blockchain and FL technology. Their framework guarantees the privacy of sensitive healthcare data during the training process. By combining FL with blockchain, the authors enhance data privacy, transparency, and accountability in healthcare systems.

Abdel-Basset et al. (2022) propose a privacy-preserved FL approach for learning from non-iid data in fog-assisted IoT. Their approach enhances privacy and data-sharing capabilities in fog-enabled IoT environments. The authors focus on addressing the challenges of heterogeneity and privacy concerns in FL scenarios.

Abdellatif et al. (2022) present a communication-efficient hierarchical FL technique for dealing with unbalanced data in heterogeneous IoT systems. Their approach improves the efficiency of FL in heterogeneous IoT environments. The hierarchical structure allows for efficient model updates and aggregation, addressing the challenges posed by imbalanced and diverse IoT data sources.

Grasso et al. (2022) propose H-HOME, a learning framework based on FL Ad-Hoc Networks (FANETs) that provide edge computing for delay-constrained IoT systems. Their framework facilitates efficient computation offloading and resource management in FANETs, enhancing the performance and responsiveness of IoT applications with strict latency requirements.

Mohammed et al. (2023) propose an energy-efficient distributed FL offloading and scheduling healthcare system in blockchain-based networks (Khang et al., 2022c). Their approach optimizes computation offloading and scheduling to minimize energy consumption in healthcare systems. The paper addresses the energy efficiency challenges in FL systems deployed in healthcare environments.

Shin and Lim (2023) propose an FL-based load balancing and energy-efficient approach for unmanned aerial vehicle (UAV) enabled mobile edge computing (MEC) systems in vehicular networks. Their approach improves resource utilization and energy efficiency in vehicular IoT scenarios. The authors focus on optimized load balancing and task allocation in UAV-enabled MEC systems for improved performance.

Djenouri et al. (2023) investigate the potential of deep FL for smart city edge-based applications. They explore the advantages and challenges of applying federated deep learning in enabling edge-based smart city applications. The paper discusses how federated deep learning can enhance the intelligence and efficiency of smart city systems (Khang et al., 2022a).

Chen and Liu (2022) propose a FL deep reinforcement-based approach for resource allocation and task offloading in edge-based smart city networks which enhance the performance of the system. By utilizing deep reinforcement learning, the authors adapt the approach to dynamic environments and improve resource utilization in smart city scenarios (Figure 20.1).

Putra et al. (2021) presented an edge computing framework based on federated compressed learning that ensures data privacy, this framework enables secure data sharing and prediction while preserving privacy. Moreover, author has focused on efficiently compressing and aggregating data from distributed IoT devices to improve prediction accuracy while protecting sensitive information.

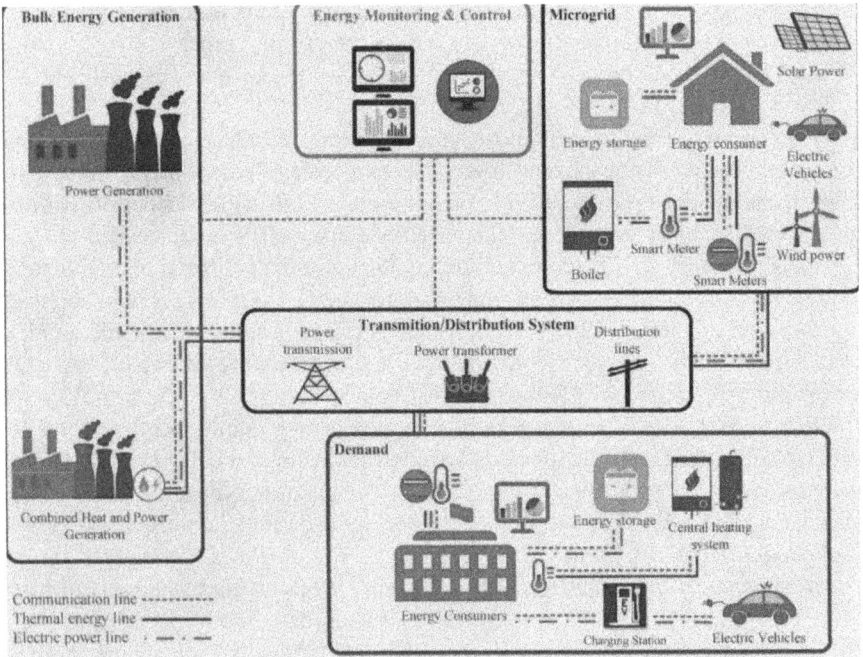

FIGURE 20.1 IoT devices used throughout the world for different smart systems

Yu et al. (2022) propose a blockchain-based secure FL system architecture that enhances security and privacy in FL scenarios. The authors discuss the applications and benefits of blockchain in securing FL. The paper explores the use of blockchain for secure model aggregation, data privacy protection, and decentralized control in FL systems.

Bi et al. (2022) offer a decentralized trust management system for IoT applications that makes use of FL, which is enabled by blockchain technology. Their system enhances trust, privacy, and security in IoT environments. By leveraging blockchain technology, the system enables secure and transparent interactions among IoT devices, ensuring trust and data integrity in FL settings.

Lu et al. (2020) investigate the use of blockchain technology to enable asynchronous FL in order to facilitate the safe exchange of data in the internet of vehicles (IoV). Their approach ensures data privacy and security while enabling efficient data sharing in IoV environments. The authors focus on utilizing blockchain to establish a decentralized and secure data-sharing framework in vehicular IoT scenarios.

According to Rehman et al.'s (2022) research, a secure healthcare 5.0 system might be built using blockchain technology in conjunction with FL. Their system enhances data security, privacy, and collaboration in healthcare applications. The authors discuss the integration of blockchain and FL to improve data management, access control, and privacy protection in healthcare systems (Khanh and Khang, 2021).

Radio fingerprinting is investigated by Halder and Newe (2022) for the purpose of anomaly detection utilizing FL in the LoRa-enabled industrial internet of things (IIoT). Their approach enables anomaly detection in IIoT environments while preserving data privacy. The authors focus on utilizing FL techniques to detect abnormal behavior and security threats in IIoT systems without compromising data privacy (Figure 20.2).

FL-based misbehavior detection system is introduced by Jai and Vetriselvi (2023) in 6G-enabled internet of vehicles (IoV) including emergency message dissemination. Their system enhances the security and reliability of emergency message dissemination in IoV scenarios. The authors explore the use of FL to detect and prevent malicious activities in the context of IoV, thereby improving the trustworthiness of emergency communication.

Dong et al. (2023) propose an affordable federated edge learning framework that efficiently allocates computational resources in FL systems with edge devices. Their framework aims to minimize resource consumption and improve the affordability of FL by efficiently estimating the contribution of edge devices using Shapley values.

In summary, these papers contribute to the advancement of FL in IoT environments by addressing various aspects such as security, privacy, resource optimization, communication efficiency, and application-specific challenges. The proposed solutions and frameworks enhance the performance, privacy, and security of FL systems in IoT contexts, enabling the effective utilization of distributed IoT data while preserving data integrity and user privacy. Various advantages of FL motivate the author to propose an FL with IoT (FL-IoT) powered

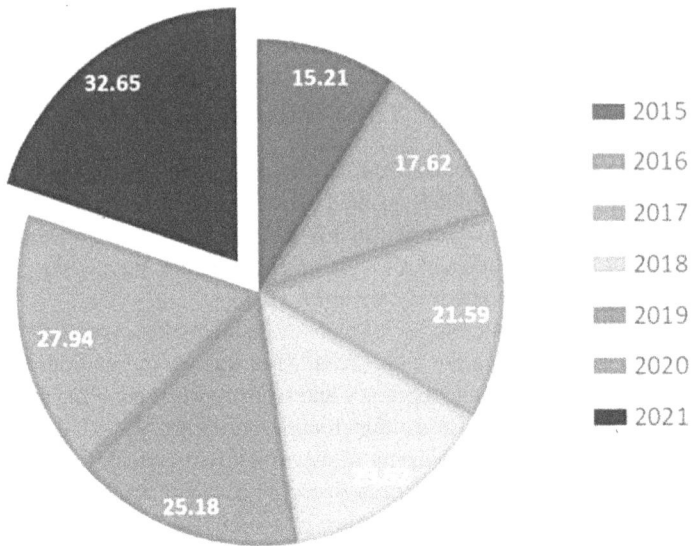

FIGURE 20.2   Number of IoT devices used in smart systems (year-wise in billions)

cooperative communication for smart systems. The major contribution of this chapter includes the following:

- Developing FL-IoT enables distributed processing of data across devices, which reduces the burden on centralized servers and network bandwidth.
- The collaborative approach in FL leads to improved performance of the model due to the use of a more diverse set of data and perspectives in model training.
- Two different datasets have been used to study the performance of the suggested method.

The following organizational conventions have been applied to the remaining portions of the paper: The section "Materials and Methods" describes the basics of FL and IoT systems, the results of the proposed FL-IoT-powered cooperative communication for smart systems have been discussed in the section "Results and Discussion." Finally, the conclusion is provided in the last section "Conclusion."

MATERIALS AND METHODS

The integration of FL with IoT-powered systems has gained significant attention in various domains, including smart energy systems, healthcare, smart cities, and internet of vehicles (IoV). This section aims to provide a detailed exploration of the potential of FL in conjunction with IoT technologies, security and privacy problems, and open issues (Figure 20.3).

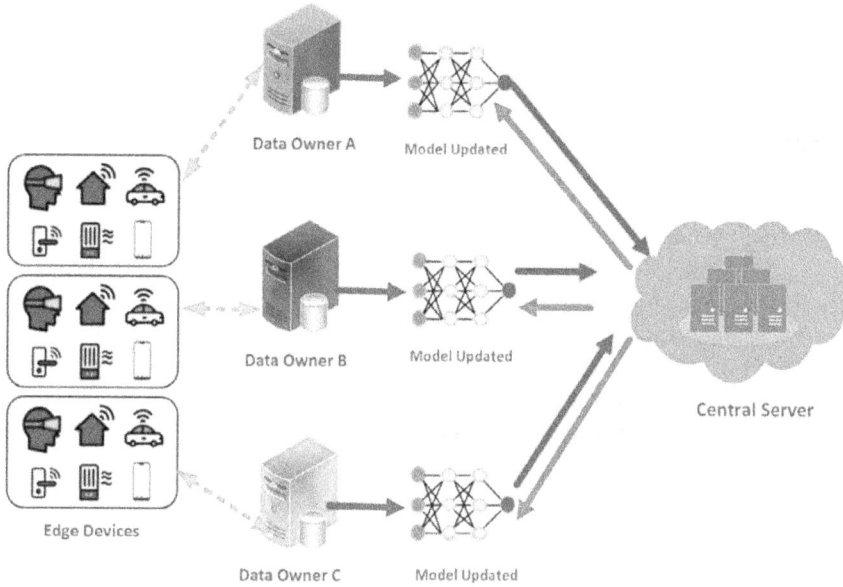

FIGURE 20.3   Integration of IoT-powered smart devices with FL

FL-BASED IoT

FL is a decentralized machine learning approach that allows IoT devices to collaboratively train models without sharing raw data. In this context, IoT devices, equipped with computational power and data storage capabilities, participate in model training while preserving data privacy. The distributed nature of IoT devices enables local model updates using their respective data, and the aggregated updates are shared to create a global model. Federated learning (FL) has great potential in IoT, as it presents a powerful tool for training machine learning models using data from multiple IoT devices distributed in different locations. Limited connectivity can lead to challenges in federated learning with IoT and reduce the effectiveness of the approach. However, there are methods to mitigate these challenges, for instance by improving the network infrastructure, by designing better communication protocols for the IoT devices to provision robust communication even with limited network connectivity, and by optimizing algorithms to work efficiently with limited network capabilities. Figure 20.3 shows various IoT devices integrated with FL.

Edge computing can be a good solution for FL-IoT, particularly when network connectivity is limited. Edge computing involves processing data at the edge of the network, closer to the source of the data, rather than sending all the data to a centralized server for processing. In the case of FL-IoT, edge computing can be used to consolidate and process data from multiple IoT devices before sending it to a central server. This can greatly reduce communication costs and latency since data is processed locally before being sent to the central server.

Additionally, edge computing can be used to train the local models of IoT devices. This approach can leverage the computational power and storage capabilities of edge devices, which are often more powerful than the IoT devices themselves. By using edge devices to train local models, communication costs are reduced, and the latency associated with sending data to a centralized server is eliminated. Moreover, edge devices can coordinate with each other to allow distributed federated learning while minimizing the burden on the central server. This decentralization can also help address issues such as data privacy, as data local to a particular device can be trained on that device.

Security in federated IoT with Edge computing is of utmost importance since it involves processing sensitive data in distributed environments. Security can be ensured in federated IoT with edge computing by authentication, secure communication, confidentiality, secure storage, continuous monitoring, and intellectual property protection mechanisms. These techniques should be carefully implemented and integrated to ensure the security of data and model training in federated IoT with edge computing environments.

SECURITY AND PRIVACY OF FL-IoT NETWORK

The privacy and security of FL-IoT networks are critical considerations. Ensuring data privacy and protection against malicious attacks is of utmost importance when leveraging IoT devices for FL. Techniques such as secure aggregation protocols, differential privacy, and encryption algorithms are employed to enhance the privacy and security of FL-IoT networks. Additionally, robust authentication mechanisms and access control policies are implemented to prevent unauthorized access to sensitive data (Khang et al., 2022b).

By exploring the integration of FL-IoT-powered systems, this research aims to shed light on the potential applications and benefits of this combination. The investigation of different IoT layers utilized in the electric grid highlights the diverse range of devices involved in data collection and monitoring. Understanding the principles of FL in conjunction with IoT provides insights into collaborative model training while preserving data privacy.

Furthermore, the exploration of FL-IoT applications and services showcases the impact of this integration in domains such as smart energy systems, healthcare, smart cities, and IoV. Additionally, considering the security and privacy aspects ensures that the FL-IoT framework is implemented in a secure and trustworthy manner. Maintaining security and privacy in FL-IoT systems involves several key practices, including the following:

- **Secure Communication Mechanisms**: Provide secure communication channels between IoT devices, edge nodes, and servers to ensure confidentiality, authentication, and data integrity. Implementation of Secure Sockets Layer (SSL) or Transport Layer Security (TLS) protocols can help guarantee secure communications.
- **Data Privacy and Anonymization**: Anonymized data can be used wherever large amounts of data of a sensitive nature are generated by IoT

networks. Additionally, privacy-preserving techniques like differential privacy, noisy labeling, and secure multi-party computation can help protect sensitive data better.

- **Model Security**: Protect models from unauthorized access, forgery, and tampering. Techniques such as digital signatures, digital watermarking, and model obfuscation could provide model security.
- **Access Control**: Provides restrictive access control to IoT devices and other resources. The authentication mechanisms ensure that only the authorized parties can access the resources.
- **Federated Learning Algorithms Protection**: Implement different safeguards to prevent model inversion attack or membership inference attack.
- **Threat Detection and Response**: Maintain continuous monitoring of the federated learning system to detect possible security breaches, malware, and abnormal activities. This helps in providing proactive measures to ensure data privacy and security.

CHALLENGES OF FL-IoT

FL is an exciting new approach for IoT systems since it has the potential to provide a number of advantages, some of which have already been mentioned (protection of user privacy, scalability, and decreased data transmission). However, as shown in Figure 20.4, FL-IoT systems face a number of challenges. In order to fully fulfill the promise of FL, many issues need to be overcome, including the following: Constraints imposed by IoT devices: IoT devices that have low computing and resource capabilities may not be able to participate in the local training model since FL may be computationally costly.

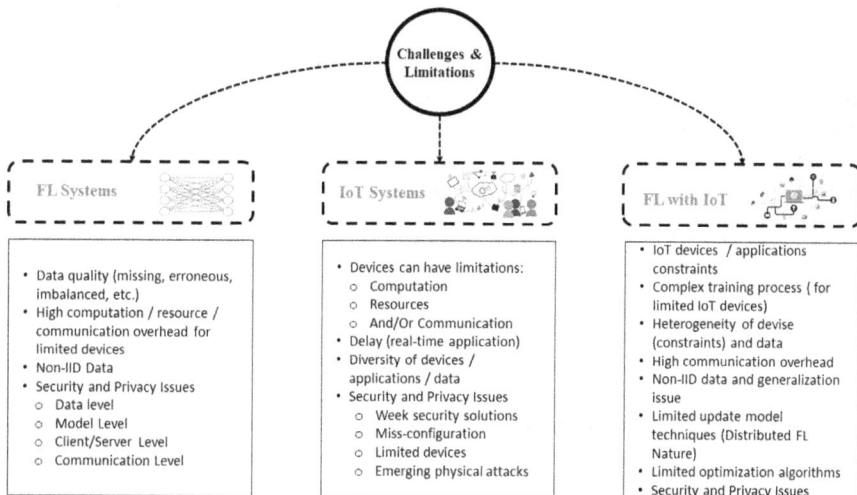

FIGURE 20.4   Challenges of IoT and/or FL

The IoT devices that have low computational capabilities might nonetheless potentially contribute to the training process. Various Algorithms for Optimization: Optimization algorithms are essential to the functioning of trained local models at IoT/edge devices, which are decentralized. On the other hand, conventional optimization methods may not be appropriate for the internet of things because of the limitations imposed by IoT devices in terms of computing and resources, in addition to the fact that IoT data is spread. Creating novel optimization algorithms that are able to successfully train models on restricted IoT devices and maintaining good convergence rate, high precision, and accuracy is a goal that has been set.

In general, FL is used to train models based on a particular group of devices, which cannot be regarded as representative of the population as a whole. The generalization of an FL-IoT system is the ability to apply the system to other IoT use cases and scenarios. To generalize an FL-IoT system, certain considerations need to be taken into account: Scalability: Designing an FL-IoT system such that it can scale to accommodate more IoT devices as the need arises. Heterogeneity: Designing the system to handle the diversity of devices, data types, and data formats used across different domains. Communication: Developing communication frameworks that are optimized to handle effective communication.

Explainability: It refers to the ability to understand and interpret the decisions and actions taken by an artificial intelligence (AI) system (Khang et al., 2023a). In the context of FL-IoT systems, Explainability is crucial for ensuring the accuracy, transparency, and trustworthiness of the AI models that are used for decision-making. In a typical FL-IoT system, data is distributed across multiple devices and is used to train machine learning models locally. These models are then sent to a central server, where they are aggregated to create a global model that can be used for inference. However, since the local models are trained on different data sets, they may have different decision-making criteria, making it difficult to interpret the overall decision of the global model.

To address this issue, explanations can be added to the FL-IoT system, which helps to provide insight into the decision-making process. This can be done by using techniques such as model interpretability, explanation generation, and visualization techniques. Model interpretability allows users to understand how the model arrived at a specific decision by exposing its internal workings.

Explanation generation generates natural language explanations of the decision-making process, while visualization techniques create diagrams and graphs that help users understand the model's behavior. By adding Explainability to FL-IoT systems, we can improve transparency and accountability and ensure that the decisions made by AI systems are not only accurate but also understandable to all stakeholders, including regulators and end-users. This helps to build trust in AI systems and increase their adoption in various applications (Khang et al., 2023b).

RESULTS AND DISCUSSION

CYBERSECURITY IN IoT-BASED SMART SYSTEM

The smart system is becoming an increasingly crucial component of it as technology continues to improve and the internet of things grows more widespread in all facets (Unsal et al., 2021). On the other hand, using a single communication topology for the whole of the network in addition to other horrible acts of cybercrime, all of which are able to cause network systems to diverge, are all examples of problematic behavior. If the nation's power infrastructure is destroyed or rendered inoperable, the country may experience a catastrophic economic downturn. Therefore, both cyber and physical security are required for a smart grid.

A national economic disaster may befall the nation in the event that its energy infrastructure is destroyed or rendered inoperable. As a result, it is essential to have both cyber and physical security in smart systems. Security and privacy are of utmost importance for smart systems since they deal with sensitive data and often interact with other devices or networks. The realistic strategy that will be suitable for implementation is The Grid Connection (NIPS), which limits access to networks and can defend against a wide variety of cyberattacks, is a component of the Network Intrusion Prevention System (NIPS). It is possible for the safety of intelligent systems to have a substantial effect on the infrastructure of a country.

Infrastructure of ritical industries such as energy, transportation, healthcare, and public services are rapidly adopting the use of intelligent systems. These systems are linked and depend on devices that are networked, sensors, and the exchange of data in order to maximize the effectiveness of operations, increase efficiency, and expand service offerings. However, in the event that these smart technologies are hacked or otherwise compromised, they have the potential to pose significant dangers to the nation's infrastructure. The NIPS is a realistic method that will be ideal for deployment in smart systems since it regulates network access and can protect against many distinct types of cyberattacks.

Cyber intrusion detection that is based on FL-IoT may increase the safety of FL-IoT devices by offering real-time monitoring, centralized administration, and enhanced protection. Nevertheless, implementation is difficult and costly. It is simple and inexpensive to create a cyber intrusion detection system that is not FL-IoT based; however, this kind of system does not provide real-time monitoring or centralized administration. The progress of FL-IoT-based cyber intrusion detection is shown using a variety of different features. The data that was utilized for the comparison came from a variety of different sources, and it was compared using something called a confusion matrix.

In the field of machine learning, one important tool for determining how well a classification model is doing is called a confusion matrix. This is a table that provides a summary of the categorization findings, including the frequencies of true positives, true negatives, false positives, and false negatives. The confusion matrix consists of four distinct values, which are as follows: True Positive (TP): The predicted positive class matches the actual positive class. False Positive (FP): The predicted positive class does not match the actual negative class. False Negative (FN): The predicted negative class does not match the actual positive class. True Negative (TN): The predicted negative class matches the actual negative class. Using these values, we can calculate several metrics that provide more information about the performance of the model, such as accuracy, precision, recall, and F1 score. Table 20.1 provides a comprehensive analysis that is numerically based and shows the state of the art. If

TABLE 20.1

Cyber Intrusion Detection for the FL-IoT, IoT-based, and Non-IoT-based Systems

Parameters	Formula	Non-IoT Based	IoT Based	FL-IoT
Sensitivity	TP/(TP+FN)	0.68	0.86	0.91
Precision	TP/(TP+FP)	0.69	0.89	0.95
Accuracy	2TP/(2TP+FN+FP)	0.70	0.90	0.96
F1-score	(TP+TN)/(TP+TN+FP+FN)	0.69	0.91	0.94

an intrusion occurs, it is likely to be a false negative (FN) because the model failed to detect the abnormal activity.

To detect an intrusion, the model's precision and recall should be high. A high precision value means that when the model predicts an abnormal activity, it is usually correct. A high recall value means that the model is good at identifying abnormal activities. If the model's precision and recall values are high and it detects an abnormal activity, it is an intrusion. The security team can receive an alert from the intrusion detection system and investigate the attack further.

DENIAL OF SERVICE

Denial of service (DOS) attacks pose a significant cyber security threat in IoT-powered smart energy systems. DOS attacks aim to disrupt the availability and functionality of critical services by overwhelming the targeted system with a flood of illegitimate requests or by exploiting vulnerabilities in the system's infrastructure. In the context of smart energy systems, a successful DOS attack can have severe consequences, such as disruption of energy generation, transmission, or distribution processes. This chapter focuses on investigating the vulnerabilities that make smart energy systems susceptible to DOS attacks and proposes mitigation strategies to enhance the system's resilience (Tuor et al., 2017). By analyzing the potential attack vectors and the impact of DOS attacks on the system, this research aims to develop robust defense mechanisms to safeguard against such threats.

To achieve this objective, the research will first analyze the characteristics of DOS attacks in IoT-based smart energy systems. It will explore various types of DOS attacks, including network-based attacks, application-layer attacks, and resource depletion attacks, which can target different components of the system. Next, the research will examine the vulnerabilities in IoT-powered cooperative communication systems that could be exploited to launch DOS attacks. This includes analyzing the communication protocols, network infrastructure, and the coordination mechanisms employed by IoT devices in the smart energy system. Furthermore, the research will investigate the impact of DOS attacks on the system's performance and reliability. The DOS attack statistics is highlighted in Figure 20.5.

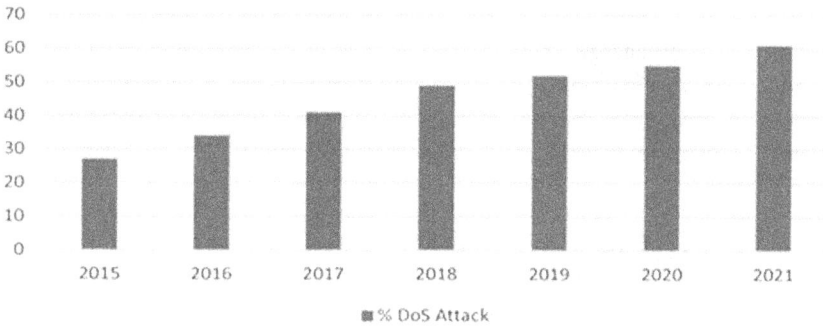

FIGURE 20.5   Statistics of DOS attack on smart systems

To mitigate the risks associated with DOS attacks, the research will propose countermeasures and defense mechanisms. This includes the development of intrusion detection and prevention systems tailored specifically for IoT-based smart energy systems. These systems will leverage machine learning techniques to detect anomalous traffic patterns, identify potential attackers, and trigger appropriate response mechanisms. Additionally, the research will explore the utilization of traffic filtering and rate limiting techniques to mitigate the impact of DOS attacks. By employing adaptive algorithms and dynamic policies, the system can intelligently manage incoming traffic and identify and block malicious requests.

Lastly, the research will evaluate the effectiveness of the proposed defense mechanisms through extensive simulations and experiments. It will measure the system's resilience against various DOS attack scenarios, considering different levels of attack intensity and sophistication. By conducting this research, we aim to contribute to the development of secure and resilient smart energy systems by addressing the specific challenges posed by DOS attacks. The findings of this research will provide valuable insights to system designers, operators, and policymakers in the field of smart grid cybersecurity, facilitating the development of effective strategies to protect against DOS attacks and ensuring the reliable and uninterrupted operation of IoT-powered cooperative communication systems in smart energy systems.

FALSE DATA INJECTION

False data injection (FDI) is a critical cybersecurity threat in IoT-powered smart energy systems, as it can compromise the integrity and reliability of data used for decision-making and control purposes. Malicious actors can exploit vulnerabilities in the IoT network infrastructure to inject fabricated or manipulated data, leading to inaccurate system state estimation, suboptimal control actions, and potentially causing disruptions in the grid operation.

This chapter focuses on investigating the potential vulnerabilities that can be exploited for false data injection attacks in smart energy systems and proposes mechanisms to detect and mitigate these attacks effectively. To begin, the research will analyze the various attack vectors and techniques used for false data injection in IoT-powered smart energy systems. It will explore how attackers can exploit vulnerabilities in communication protocols, compromised devices, or unauthorized access to the system to inject falsified data. Next, the research will examine the potential impact of false data injection attacks on the operation of smart energy systems (Figure 20.6).

By quantifying the consequences of such attacks, the research aims to highlight the urgency and importance of addressing this security challenge. To detect and mitigate false data injection attacks, the research will propose novel anomaly detection algorithms and data validation techniques. These techniques will leverage machine learning and statistical analysis methods to identify deviations from expected data patterns and detect anomalies that may indicate the presence of false data.

Furthermore, the research will explore the utilization of cryptographic techniques such as digital signatures and secure data exchange protocols to enhance the authenticity and integrity of data transmitted within the smart energy system. By employing these techniques, the system can ensure the origin and integrity of data, making it more resilient to false data injection attacks. The proposed detection and mitigation mechanisms will be evaluated through extensive simulations and experiments. The research will assess the effectiveness of the techniques in detecting and mitigating false data injection attacks under different scenarios, considering various levels of attack sophistication and system complexities.

CONCLUSION

Research on FL with IoT-powered cooperative communication and security systems in smart energy systems has provided valuable insights into the integration of these technologies and their implications for the grid's performance, security, and privacy. The research journey covered various aspects, including the types of IoT layers utilized in the electric grid, the relationship between FL-IoT, the applications and services enabled by FL-IoT, the potential of FL in 6G communication systems, and

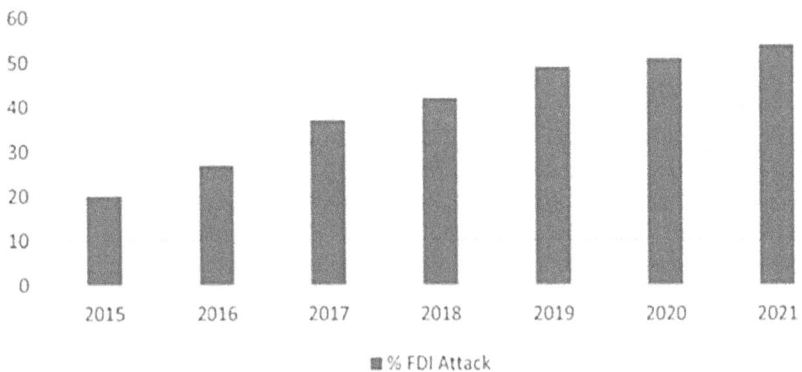

FIGURE 20.6   Statistics of FDI attack on the IoT-based smart systems

the critical considerations of security and privacy in FL-IoT networks. The investigation into the types of IoT layers used in the electric grid revealed the diverse range of sensors, actuators, and other IoT devices employed to enable data collection, monitoring, and control within the grid infrastructure. This understanding is crucial for comprehending the data ecosystem of smart energy systems and how FL can be leveraged effectively.

The exploration of the integration of FL-IoT sheds light on their synergistic relationship. FL harnesses the distributed nature of IoT devices to enable collaborative model training without centralizing data. This approach ensures data privacy while enabling collective learning and improved decision-making in smart energy systems. The research also highlighted the wide array of FL-IoT applications and services. From predictive maintenance to load forecasting, and energy optimization to anomaly detection, FL-IoT offers significant potential for enhancing the efficiency and performance of smart energy systems.

Additionally, the study investigated the role of FL in the development of 6G communication systems. It revealed the potential of FL to improve the speed, reliability, and efficiency of communication in future-generation networks, enabling real-time decision-making and further enhancing the capabilities of smart energy systems. Furthermore, the research delved into the critical aspects of security and privacy in FL-IoT networks. Recognizing the vulnerabilities and challenges related to data privacy, secure communication, and authentication mechanisms, the research emphasized the need to develop robust security measures to safeguard FL-IoT systems from potential threats and ensure the integrity and privacy of data (Khang et al., 2024).

FUTURE SCOPE OF WORK

In terms of results, the research uncovered the significance of addressing cybersecurity in FL-IoT-based smart grids. Specifically, it focused on the threats of denial of service (DOS) attacks and false data injection. DOS attacks can disrupt the availability and functionality of critical services, while false data injection compromises the integrity and reliability of data used for decision-making and control. The research provided insights into the vulnerabilities, potential impacts, and proposed mitigation strategies for both DOS attacks and false data injection.

Through the development of intrusion detection and prevention systems, traffic filtering techniques, anomaly detection algorithms, and data validation mechanisms, the research aimed to enhance the resilience and security of smart energy systems against these threats. In conclusion, this research has contributed to the understanding of FL with FL-IoT-powered cooperative communication and security systems in smart energy systems. The findings have shed light on the integration of these technologies, their applications, and the challenges they face. By addressing the crucial aspects of security and privacy and providing insights into the potential threats of DOS attacks and false data injection, this research aims to assist researchers, practitioners, and policymakers in developing robust and secure smart energy systems for the future (Khang et al., 2023c).

REFERENCES

Abdel-Basset, M., Hawash, H., Moustafa, N., Razzak, I., & Abd Elfattah, M. (2022, December). Privacy preserved learning from non-iid data in fog-assisted IoT: A FL approach. *Digital Communications and Networks.* https://doi.org/10.1016/j.dcan.2022 .12.013

Abdellatif, A. A., Mhaisen, N., Mohamed, A., Erbad, A., Guizani, M., Dawy, Z., & Nasreddine, W. (2022, March). Communication-efficient hierarchical FL for IoT heterogeneous systems with imbalanced data. *Future Generation Computer Systems*, 128, 406–419. https://doi.org/10.1016/j.future.2021.10.016

Al-Huthaifi, R., Li, T., Huang, W., Gu, J., & Li, C. (2023, June). FL in smart cities: Privacy and security survey. *Information Sciences*, 632, 833–857. https://doi.org/10.1016/j.ins .2023.03.033

Anuradha, K., & Rajini, S. N. S. (2019, September 25). Analysis of machine learning algorithm in IOT security issues and challenges. *Journal of Advanced Research in Dynamical and Control Systems*, 11(0009-Special Issue), 1030–1034. https://doi.org/10 .5373/jardcs/v11/20192668

Bi, L., Muazu, T., & Samuel, O. (2022, December 26). IoT: A decentralized trust management system using blockchain-empowered FL. *Sustainability*, 15(1), 374. https://doi.org/10 .3390/su15010374

Chen, X., & Liu, G. (2022, June 23). Federated deep reinforcement learning-based task offloading and resource allocation for smart cities in a mobile edge network. *Sensors*, 22(13), 4738. https://doi.org/10.3390/s22134738

Da Silva, L. G. F., Sadok, D. F., & Endo, P. T. (2023, May). Resource optimizing FL for use with IoT: A systematic review. *Journal of Parallel and Distributed Computing*, 175, 92–108. https://doi.org/10.1016/j.jpdc.2023.01.006

Djenouri, Y., Michalak, T. P., & Lin, J. C. W. (2023, May). Federated deep learning for smart city edge-based applications. *Future Generation Computer Systems.* https://doi.org/10 .1016/j.future.2023.04.034

Dong, L., Liu, Z., Zhang, K., Yassine, A., & Hossain, M. S. (2023, May). Affordable federated edge learning framework via efficient Shapley value estimation. *Future Generation Computer Systems.* https://doi.org/10.1016/j.future.2023.05.007

Grasso, C., Raftopoulos, R., Schembra, G., & Serrano, S. (2022, December). H-HOME: A learning framework of federated FANETs to provide edge computing to future delay-constrained IoT systems. *Computer Networks*, 219, 109449. https://doi.org/10.1016/j .comnet.2022.109449

Halder, S., & Newe, T. (2022). Radio fingerprinting for anomaly detection using FL in lora-enabled industrial Internet of things. *SSRN Electronic Journal.* https://doi.org/10.2139 /ssrn.4229633

Jai Vinita, L., & Vetriselvi, V. (2023, May). FL-based Misbehaviour detection on an emergency message dissemination scenario for the 6G-enabled Internet of Vehicles. *Ad Hoc Networks*, 144, 103153. https://doi.org/10.1016/j.adhoc.2023.103153

Khang, A., Rani, S., & Sivaraman, A. K. (2022a). *AI-Centric Smart City Ecosystems: Technologies, Design and Implementation* (1st Ed.). CRC Press. https://doi.org/10.1201 /9781003252542

Khang, A., Hahanov, V., Abbas, G. L., & Hajimahmud, V. A. (2022b). Cyber-physical-social system and incident management. *AI-Centric Smart City Ecosystems: Technologies, Design and Implementation* (1st Ed.), 2(15). CRC Press. https://doi.org/10.1201 /9781003252542-2

Khang, A., Chowdhury, S., & Sharma, S. (2022c). *The Data-Driven Blockchain Ecosystem: Fundamentals, Applications, and Emerging Technologies* (1st Ed.). CRC Press. https:// doi.org/10.1201/9781003269281

Khang, A., Gupta, S. K., Hajimahmud, V. A., Babasaheb, J., & Morris, G. (2023a). *AI-Centric Modelling and Analytics: Concepts, Designs, Technologies, and Applications* (1st Ed.). CRC Press. https://doi.org/10.1201/9781003400110

Khang, A., Gupta, S. K., Shah, V., & Misra, A. (Eds.). (2023b). *AI-Aided IoT Technologies and Applications in the Smart Business and Production.* CRC Press. https://doi.org/10.1201/9781003392224

Khang, A., Vrushank, S., & Rani, S. (2023c). *AI-Based Technologies and Applications in the Era of the Metaverse* (1st Ed.). IGI Global Press. https://doi.org/10.4018/978-1-6684-8851-5

Khang, A., Gujrati, R., Uygun, H., Tailor, R. K., & Gaur, S. S. (2024). *Data-Driven Modelling and Predictive Analytics in Business and Finance.* CRC Press. https://doi.org/10.1201/9781032600628

Khanh, H. H., & Khang, A. (2021). The role of artificial intelligence in blockchain applications. *Reinventing Manufacturing and Business Processes through Artificial Intelligence*, 2, 20–40. CRC Press. https://doi.org/10.1201/9781003145011-2

Lu, Y., Huang, X., Zhang, K., Maharjan, S., & Zhang, Y. (2020, April). Blockchain empowered asynchronous FL for secure data sharing in Internet of vehicles. *IEEE Transactions on Vehicular Technology*, 69(4), 4298–4311. https://doi.org/10.1109/tvt.2020.2973651

Mohammed, M. A., Lakhan, A., Abdulkareem, K. H., Zebari, D. A., Nedoma, J., Martinek, R., Kadry, S., & Garcia-Zapirain, B. (2023, May). Energy-efficient distributed FL offloading and scheduling healthcare system in blockchain based networks. *Internet of Things*, 100815. https://doi.org/10.1016/j.iot.2023.100815

Putra, K. T., Chen, H. C., Prayitno, Ogiela, M. R., Chou, C. L., Weng, C. E., & Shae, Z. Y. (2021, July 4). Federated Compressed Learning Edge Computing Framework with Ensuring Data Privacy for PM2.5 Prediction in Smart City Sensing Applications. *Sensors*, 21, 13, 4586, 21. https://doi.org/10.3390/s21134586

Rajagopal, S. M., Supriya, M., & Buyya, R. (2023, July). FedSDM: FL based smart decision making module for ECG data in IoT integrated Edge–Fog–Cloud computing environments. *Internet of Things*, 22, 100784. https://doi.org/10.1016/j.iot.2023.100784

Rahman, A., Hasan, K., Kundu, D., Islam, M. J., Debnath, T., Band, S. S., & Kumar, N. (2023, January). On the ICN-IoT with FL integration of communication: Concepts, security-privacy issues, applications, and future perspectives. *Future Generation Computer Systems*, 138, 61–88. https://doi.org/10.1016/j.future.2022.08.004

Rehman, A., Abbas, S., Khan, M., Ghazal, T. M., Adnan, K. M., & Mosavi, A. (2022, November). A secure healthcare 5.0 system based on blockchain technology entangled with FL technique. *Computers in Biology and Medicine*, 150, 106019. https://doi.org/10.1016/j.compbiomed.2022.106019

Shin, A., & Lim, Y. (2023, March 6). Federated-Learning-Based Energy-Efficient Load Balancing for UAV-Enabled MEC System in Vehicular Networks. *Energies*, 16(5), 2486. https://doi.org/10.3390/en16052486

Singh, S., Rathore, S., Alfarraj, O., Tolba, A., & Yoon, B. (2022, April). A framework for privacy preservation of IoT healthcare data using FL and blockchain technology. *Future Generation Computer Systems*, 129, 380–388. https://doi.org/10.1016/j.future.2021.11.028

Tuor, A., Kaplan, S., Hutchinson, B., Nichols, N., & Robinson, S. (2017). Deep learning for unsupervised insider threat detection in structured cybersecurity data streams. In *Workshops at the Thirty-First AAAI Conference on Artificial Intelligence*.

Unsal, D. B., Ustun, T. S., Hussain, S., & Onen, A. (2021). Enhancing cybersecurity in smart grids: False data injection and its mitigation, Energies, 14(9), 2657. https://www.mdpi.com/1996-1073/14/9/2657

Yaacoub, J. P. A., Noura, H. N., & Salman, O. (2023). Security of FL with IoT systems: Issues, limitations, challenges, and solutions. *Internet of Things and Cyber-Physical Systems*, 3, 155–179. https://doi.org/10.1016/j.iotcps.2023.04.001

Yu, F., Lin, H., Wang, X., Yassine, A., & Hossain, M. S. (2022, December). Blockchain-empowered secure FL system: Architecture and applications. *Computer Communications*, 196, 55–65. https://doi.org/10.1016/j.comcom.2022.09.008

Zhang, H., Zou, Y., Yin, H., Yu, D., & Cheng, X. (2023, March). CCM-FL: Covert communication mechanisms for FL in crowd sensing IoT. *Digital Communications and Networks*. https://doi.org/10.1016/j.dcan.2023.02.013

Index

.

For Product Safety Concerns and Information please contact our EU
representative GPSR@taylorandfrancis.com
Taylor & Francis Verlag GmbH, Kaufingerstraße 24, 80331 München, Germany

www.ingramcontent.com/pod-product-compliance
Lightning Source LLC
Chambersburg PA
CBHW060750220326
41598CB00022B/2389